CLIMATE CHANGE

What You Can Do Now

April 25, 2012

To Sam,

Jon thanks for all the wonderful work you do on climate change!

Rachael

LMI

Complex Problems. Practical Solutions.

CLIMATE CHANGE

What You Can Do Now

Rachael Jonassen
Michael Canes
Matt Daigle
Jeremey Alcorn
Julian Bentley
Virginia Bostock
Frank Reilly
Rich Skulte
Taylor Wilkerson
John Yasalonis

Complex Problems. Practical Solutions.

Published by LMI, which is a 501(c)(3) not-for-profit corporation. This publication advances LMI's not-for-profit purpose and mission.

ISBN 978-0-9661916-7-7

Library of Congress Control Number: 2012935162

Printed in the United States of America.

Contents

Acknowledgments

Acknowledging the many people who made this book possible is a pleasure.

Research support came from a group of talented LMI analysts and interns, including Kevin Ennis, Mark Gross, Karen Hagberg, Keith Herrmann, Julia Kalloz, Aman Luthra, Lia Marroquin, Kristina Olanders-Ipiotis, Snapper Poche, Libby Backman (University of Mary Washington), and Kelli Canada (American University). Clark Spencer and David Shepherd advised on the latest developments in computer technology for Chapter 3. Greg Wilson, of Phase2 Technology, also provided welcome input to Chapter 3. Snapper Poche substantively reviewed and scientifically edited Chapter 4. Lon Cross furnished guidance on national security issues for Chapter 8.

Kurt Kraus drafted all figures. As we refined our graphical needs, he endured much iteration and travelled with us down several blind alleys with patience. Diane Donohoe designed the cover, Barbara Taff designed the overall text, and Sharon Hall advised on layout. Kathy Myers oversaw the graphics efforts.

Rebecca Stigall of *ForeWORD* Communications served as a capable copyeditor for all chapters, sometimes making multiple passes. We greatly appreciate her help ensuring consistency between chapters. Courtney Boyce supervised the copyediting process.

Rosalie Beckham laid out the text and graphics, made corrections in response to proofreading, prepared a final proof, and rendered innumerable other support services. Marty Lyon edited the final version of the text. Bettina Worthington proofread the book front to back. Warren Murray supervised the printing process, including multiple print stages.

Special thanks to Judy Abbett, who advised on and supervised all communication, outreach, editing, and graphics support for the project from beginning to end. Judy's enthusiasm for and encouragement of this effort was essential to its success. She also provided review recommendations throughout the process.

In the LMI Library, Laura Tyler arranged for all figure permissions from the original sources, and Sara Faulk arranged publication catalog information. Laura and Jim Foti (now with the National Institute of Standards and Technology) also performed essential research. Tamara Jack advised on legal issues related to figure and table permissions, copyrights, and export control and participated in the review process.

At various stages of preparation, the chapters were subject to extensive review, internal and external. The Honorable Nelson Ford, Dr. Bill Moore, Stu Funk, Dr. Dave Gallay, Joe Zurlo, Bob Schmitt, Chris Alligood, Sue Nicholas, the Honorable George Moose, Roger Ervin, Ray Schaible, Shahab Hasan, Nora Ryan, Jeff Bennett, and Donna Bennett, all of LMI, reviewed all or parts of the prepublication draft.

Special thanks to the management group at LMI, especially Mr. Ford and Dr. Moore, for their vision and support throughout this project. The LMI Research Institute funded the book preparation.

We strove to make this book a practical guide for the many managers who will face the enormous challenge of climate change in the coming years. In doing so, we relied on many reviewers in the private, federal, and academic sectors to help concentrate our efforts and avoid errors. We list these individuals and their affiliations at the end of the book, and we extend our heartfelt thanks for their toil. Any remaining errors are the responsibility of the authors.

The Authors

Preface

John Selman

This book is based on a simple premise—that demand exists for practical solutions to complex problems. This is especially true when it comes to the issues surrounding climate change.

Environmental professionals and scientists are not the only ones who care, who want to understand, and who want to act. Each of us, in our own way, large or small—if we had the opportunity—would do something. I've talked to many people outside the environmental profession who've repeated this theme time after time. I respond with the question, "What's preventing you from doing something right now?" The response is almost always the same: "Me? But what would I do, and where would I start?"

That's where this book comes into play. All professionals have some interaction with—and, therefore, influence on—the climate. Simply put, we all need a stable, healthy, and functioning climate to do the things we've always done and want to keep on doing; we all have an enduring interest in our climate. However, in the absence of legislation mandating that we do something, it's not clear how we should protect that interest or that climate change is even a serious problem worthy of our attention.

This book does not attempt to sway nonbelievers. It simply presents facts based on the universally accepted science of the day in the context of specific functional areas. If your professional energies focus on health, information, land, structure, vehicles, supply, or security, you now have a resource that allows you to engage in the discussion and consider what you might do right now—without waiting for legislation or regulatory guidance that may never come. You can take practical steps and have a big influence.

We chose to write about these seven specific areas for two reasons. First, taken together, they comprise almost every major source of greenhouse gas emissions. We don't claim to cover every source, but in the pages of this book, we nail the ones that count.

Second, as LMI celebrates its 50th anniversary, we like to talk about the crosscutting nature of complex issues and how we're well suited to tackle them. Climate change is one such issue, and I knew we had the authors and thought leadership needed to pull this off. When I asked our subject matter experts if they'd be interested in participating in writing this book, they all agreed without hesitation. And, to a person, conscientious professionals that they are, each demanded appropriate support to ensure their ideas and recommendations stayed grounded in the best science of the day. This mix of LMI's knowledge of climate science and subject matter expertise in other functional areas makes this undertaking special and unique.

And so we went about writing a book composed of seven functional chapters, ensuring that each is grounded in realistic, current scientific understanding.

Each of the functional authors deserves credit for tackling the subject matter with great energy and passion. They worked hard to study and incorporate existing literature, best practices, and climate change science into their respective chapters and to arrive at practical recommendations so that people, if they so choose, can do something now. These experts include Jeremey Alcorn, Julian Bentley, Francis J. Reilly, Jr., Rich Skulte, Taylor Wilkerson, and John Yasalonis. Biographical sketches at the end of this book briefly summarize their relevant background and specific contributions.

These functional authors are the first piece of the puzzle that is this book. The contributions of Dr. Rachael Jonassen and Dr. Mike Canes are the second piece. This book simply would not have happened without Rachael's ability to fully understand the way that climate change impacts each of the functional areas written about in this book. All of us at LMI admire her mastery of the science behind climate change issues. Mike complemented Rachael's contribution by ensuring each chapter had the precision, accuracy, and consistency our clients have come to expect. His substantial career in energy and environmental economics proved of immense value as he continually suggested improvements in the chapters. Rachael and Mike, while not the lead for any of the functional chapters, made considerable contributions to every chapter in this book, and their wisdom is imparted on each and every page. Assisting Rachael and Mike was Virginia Bostock, who coordinated many of the moving parts to get this book published. In addition to overall project support, Virginia scoured the Internet looking for relevant articles and images, provided reviews, continually updated content and layout, and worked with authors, editors, graphics, and support staff. Their combined effort collaborating with and educating the authors is an accomplishment in itself.

The third piece of the puzzle comes in the form of contributions from a talented LMI writer, Matt Daigle. I engaged Matt in this effort because I needed someone who could take our technically oriented writing and turn it into prose understandable to any reader, especially those not educated in the intricacies of climate change science. He did that and so much more. During the process, he helped the authors achieve a consistent style, added his own wisdom to the discussion in every chapter, and gave the book accessibility and style that would have been absent had he not contributed.

Last, I thank each of the external reviewers who examined early drafts of individual chapters. They are the fourth and final puzzle piece, and their comments were a major catalyst in improving the chapters. I was amazed at the time people were willing to invest to help us make the book better and more meaningful. When people freely share such extraordinary amounts of time on something like this, it shows how deeply committed they are and how much they really care. My hope is that each external reviewer sees his or her contribution in the pages ahead. All external reviewers are listed in Appendix B.

Our book is the combination of these four puzzle pieces. After you've read it, even if only a chapter is of interest to you, I hope you'll feel empowered and armed with new knowledge of a role you can take to address climate-related issues. I encourage you not to wait for political or governmental action but to do what you can now.

John Selman
Program Director
Energy and Environment Group
LMI

Action

If we could first know where we are, and whither we are tending, we could then better judge what to do, and how to do it.

—Abraham Lincoln

Skate to where the puck is going to be, not to where it has been.

—Walter Gretzky, to son Wayne

In mid-2011, a study by researchers at Yale and George Mason Universities revealed that most Americans want scientists, not politicians, to lead the climate change debate. By the time that study began, we'd already begun working on this book, and though the study has now been completed, there is still a piece missing.

This is not a political book. There is no agenda in the pages you're about to read. We're not debating the cause of what a vast majority of scientists have come to recognize is happening to our planet's climate; we're not casting blame, nor do we present theories as to why it's happening. It just is, and it's time to move forward.

It's also time to bring actual decision makers to the table—the functional managers, the people charged day to day with facing the impacts of climate change head-on.

We stand at a moment of opportunity. Now is the time to determine the outcomes of these changes and how we will respond to them. This book discusses—and presents tangible recommendations for—plotting the course of action needed in the decades to come.

This book is a way for the discourse on managing climate-related effects to change hands, encouraging scientists and politicians to pass the baton to the *climate users*.

Who are climate users? Where do we find them? Frankly, we hope you are one, and you probably are. If you have a feeling that climate change is going to impact the decisions you'll be making over the next 10 years, then you definitely fit the bill.

The term "climate user" refers to you, your colleagues, your suppliers and partners—anyone who doesn't work on climate change day to day but who will soon begin experiencing the impacts of these changes.

You may not study the science of climate change or set related policy, but you certainly have a reason to pay attention: it's going to affect you every day when you arrive at your job.

Unfortunately, until now, you've been left out of the discussion.

LMI developed the climate users concept as a way of broadening the climate change dialogue. Our aim is to reach functional decision makers in the private and public sectors, the military, and business partners and suppliers—the people who must ensure that climate change doesn't ruin all the good things society has achieved.

Climate users are members of every field and discipline, but for the purpose of this book, we've identified stakeholders in seven critical groups: health, information, land, structure, vehicles, supply, and security. In these areas, the public and private sectors will see a significant opportunity to mitigate and adapt to our planet's changing climate.

Climate users in these areas are significant, not just because they've been overlooked in the climate change discussion, but also because they wield great power to enact significant changes for their organizations. These functional managers are stakeholders in the sense that they have a vested interest in how their organizations respond to a changing climate. The success of their jobs depends on their ability to understand what is happening and how they can act on behalf of their organization. These seven areas are important because they represent the bulk of an organization's critical operations and activities. Strong organizations pay close attention to each of these areas and assign talented individuals to oversee them.

That's why this book takes a very careful approach to the science involved. It's simplified without being dumbed down. The topics discussed in each chapter constitute the bulk of considerations for an organization. Organizations make daily decisions on the basis of their people, assets, and processes; their functional managers are asked to facilitate those decisions, ensuring they are correctly executed.

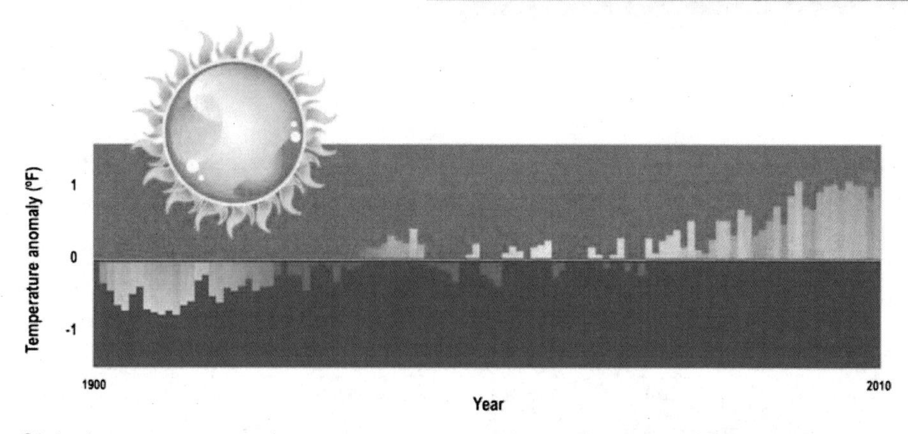

Global average annual temperatures have been above the 20th century norm for the past 34 years.

Therein lies this book's balancing act. Some of the functional areas we're addressing appeal to a specific audience. For example, if you ask someone on the street about international instability and how it will affect humanitarian aid, you might be answered with a blank stare. But that's one element of the discussion in the "Security" chapter. Likewise, asking a federal employee how zoning changes in Kennewick, WA, will impact climate change may earn you a look of bewilderment. But, how zoning can be a tool for adaptation is part of our "Land" chapter.

The goal, then, is to not discount the various levels on which these discussions take place, but to encourage an understanding of how it all fits together. We're writing to stakeholders, not just in the top levels of government, but anyone who has a voice in the matter. When we see how all of our actions fit together—regardless of disciplinary area—we can holistically respond to climate change.

The world we envision over the next several decades is not a post-apocalyptic wasteland, but one impacted by real and potential changes in our climate. With this in mind, we want to support stakeholders as they search for practical, efficient, and creative solutions to climate change problems.

But first, let's take a look at where we're coming from.

Our Background

This book is a companion to LMI's *A Federal Leader's Guide to Climate Change*. That book is directed toward federal policy leaders new to the issue of climate change. It is intended to bring federal leaders up to date on the latest science and implications of climate change for humans and to present actions that can be taken.

What's different about this book is the broadening of the audience to include functional managers.

We focus on helping decision makers who must deal with climate change and its ramifications make tough, well-informed decisions—

whether they're managers, executives, or stakeholders. It's a broad group, and their individual ability to affect their organization will vary. Their place in the hierarchy also varies, but all levels of government and business have a role to play in the upcoming discussion. Our goal is to see these issues considered and drawn into the everyday conversation of organizational goals.

This effort is unique in two ways. First, the stakeholders we have in mind work in functional areas, not just in environmental shops. A stakeholder may work for a particular organization as a functional manager or sit on a civic body that oversees a particular set of activities. Another may have an advisory role in how an organization develops its strategy in a particular area. Any individual interested in the results of an action is a stakeholder.

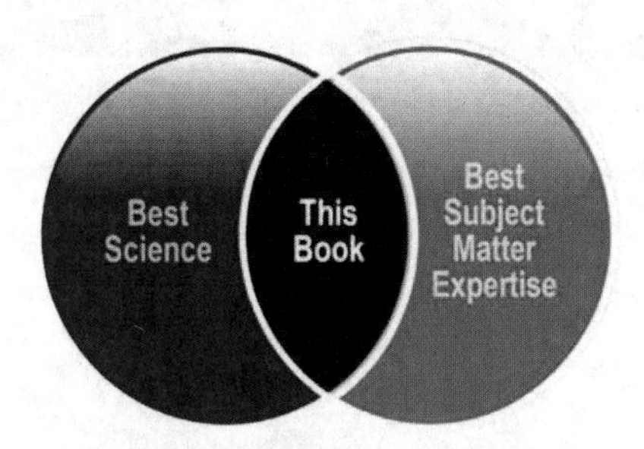

This book represents a combination of the best science on climate change and the best subject matter expertise in seven fields of management.

Stakeholders include managers working in health, information, land, infrastructure, vehicles, supply, and security. In the past, LMI has proven its ability to innovate and support the public sector in these areas, which face increasing challenges because of the climate change problem. Our experience has allowed us to draw from our nationally regarded experts *in those fields* to look at climate change as it relates to each expert's area of authority. LMI's own climate change experts bolstered their efforts, working side by side with them to develop recommendations in the context of each area of knowledge.

This combination of science and expertise gives us a unique understanding of climate change mitigation and adaptation. This new approach tackles climate change at a practical level where it impacts the operations of organizations.

The second way this book differs is that it doesn't ignore the role and contribution of the private sector. Thousands of firms supply the federal government with goods and services, many of which innovate and operate their businesses in a way that considers their role in the overall environment. We especially want to speak to federal suppliers because the government will need to form partnerships to reduce its greenhouse gas (GHG) emissions and adapt to the climate changes now under way.

Our Baseline

Hot, hot, hot. Or at least hot*ter*. That was the story of 2011.

The number to know is 2,755. It's the number of daily records for warmest minimum temperature broken or tied in a single month (July 2011) at US weather stations. In fact, in 2011 more than 300 weather stations across the country reached their warmest minimum temperature ever recorded on any date.[1]

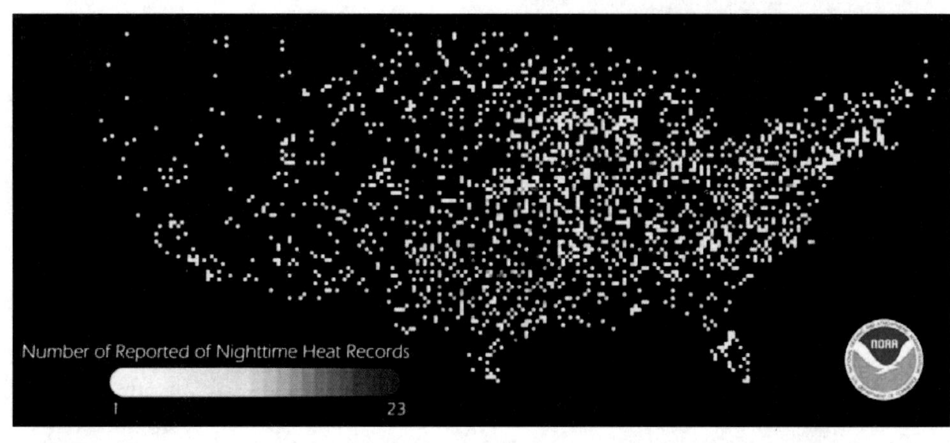

Number of weather stations in the continental United States where daily records for warmest minimum temperature were either broken or tied in the month of July 2011.

January through August 2011 marked the 34th consecutive January–August period recorded as warmer than the long-term average. Global land temperatures in August 2011 were the second warmest on record. All of the top 15 years of high global average temperatures have occurred since 1995.[2]

As the climate gets warmer globally, it spurs changes locally.

Across the United States, the impacts of climate change are already in evidence, and, in response, 42 states have passed legislation related to climate change adaptation or mitigation, meaning we can't profile this effort on the basis of red state or blue state leanings.

Some cities and municipalities have followed suit on legislating actions. More policy is expected to follow. Through all of this, climate change increasingly crosses different sectors.

Our Format

Scattered action has been taken, but a comprehensive, practical strategy is lacking. And, talk has been sparse on how to apply a strategy to specific areas.

Footnotes and Endnotes. This book uses footnotes (superscript letters) to expand on concepts through background information found at the bottom of the page. It uses endnotes (superscript numbers) to refer to source material found at the end of each chapter.

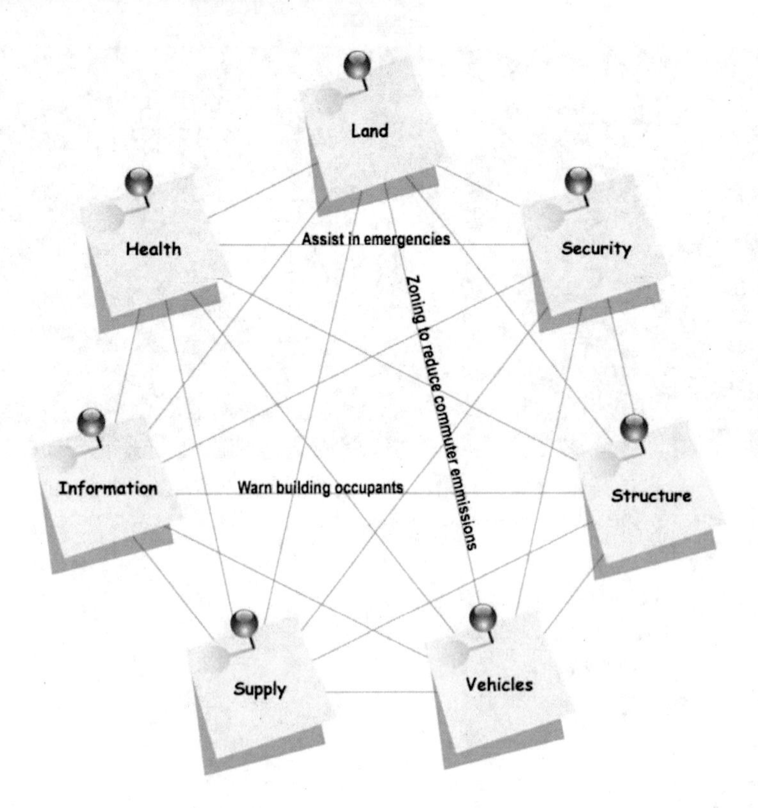

For example, a vehicle fleet manager may find that land use or public health may affect his or her ability to adapt to climate change. Climate change impacts on structures may constrain the ability to help mitigate emissions. The ability to respond to all of these challenges will greatly impact the efficiency and operation of the organizational supply chain.

That's why this book doesn't shy away from the fact that our seven unique functional areas overlap. Issues that pop up in the "Land" chapter, for example, are seen in other chapters. In fact, the "Security" chapter is the culmination of all the issues discussed in the pages leading up to it.

Our intent is to make this book as cohesive as possible, to impress upon you how each area is a part of a larger picture. For this reason, among others, we encourage readers to consider all chapters.

Each chapter focuses on a distinctive area of operation, but we use a common format to help paint a cohesive picture. Each chapter opens with a discussion on how climate change relates to a particular field. This background may be old news to seasoned professionals, but keeps the book broad enough that everyone can grasp the issues and concepts. We don't assume our readers are expert in a particular area.

Each chapter has two parts: mitigation and adaptation. In each part, readers will find five recommendations for addressing the challenges presented by climate change. These recommendations are practical yet innovative. They're rooted in best practices applied in other areas. And again, they're rooted in the convergence of best science and best subject matter expertise.

Why five recommendations? That number offers a sampling of real, practical, affordable, and proven steps that anyone can take to mitigate the effects of climate change. Examining every available option is impossible, so we looked at what made the most sense for each subject area.

Different recommendations speak to different audiences. Some are better suited for a federal level; others can only be accomplished locally. These differences reflect the need for a comprehensive strategy that engages *all stakeholders*—not just federal managers, not just the private sector, and certainly not just local authorities.

Each recommendation has three parts. We begin by laying the groundwork, documenting a particular set of challenges brought on by climate change. We then introduce solutions—tangible, realistic ideas—for overcoming the projected challenges.

Finally, we present examples of where the concepts and ideas offered as solutions are actually being used with success. Our examples show that no one is alone in addressing climate change–related issues and that organizations have lessons to share from working knowledge.

The concluding chapter combines all of our recommendations with the themes that pervade the book, advocating a coherent strategy for mitigation and adaptation across functional areas that considers the need for innovation and prudent use of available resources.

Our Time Frame

Climate change is with us *now*, and now is the time to act. With that in mind, we set this book in a time frame from now through 2025. Managers and stakeholders will find that the period leading to the quarter-century mark is where the critical decisions should be made and that those decisions will carry their organizations well past that date.

Some actions, such as a plan for the coming year, require consideration of shorter time spans, but any overarching strategy should focus on a decade of actions. It's not so long as to be unrealistic, nor too short to respond to serious challenges.

In our effort to use the best available science, our climate change work is informed by the Intergovernmental Panel on Climate Change (IPCC). The IPCC—the United Nations' scientific arm for all climate change issues—has established potential emission scenarios (*not* predictions, but plausible future conditions developed under certain assumptions).[a]

When should you begin to mitigate and adapt?

[a] The IPCC takes projections of population, demographics, economic growth, energy supply and demand, land use, and technological developments and inputs them into climate models that estimate emissions of green-

To represent a range of driving forces and resultant emissions the IPCC considers four "qualitative storylines" called "families": A1, A2, B1, and B2. From these four families, alternative scenarios are developed in six scenario groups (A1B, A1FI, A1T, A2, B1, and B2). You can see in Figure 1-1, how three of these scenarios will look after 2025 is uncertain.

Figure 1-1. Future Increases in Temperature Depend on Choices Made Today

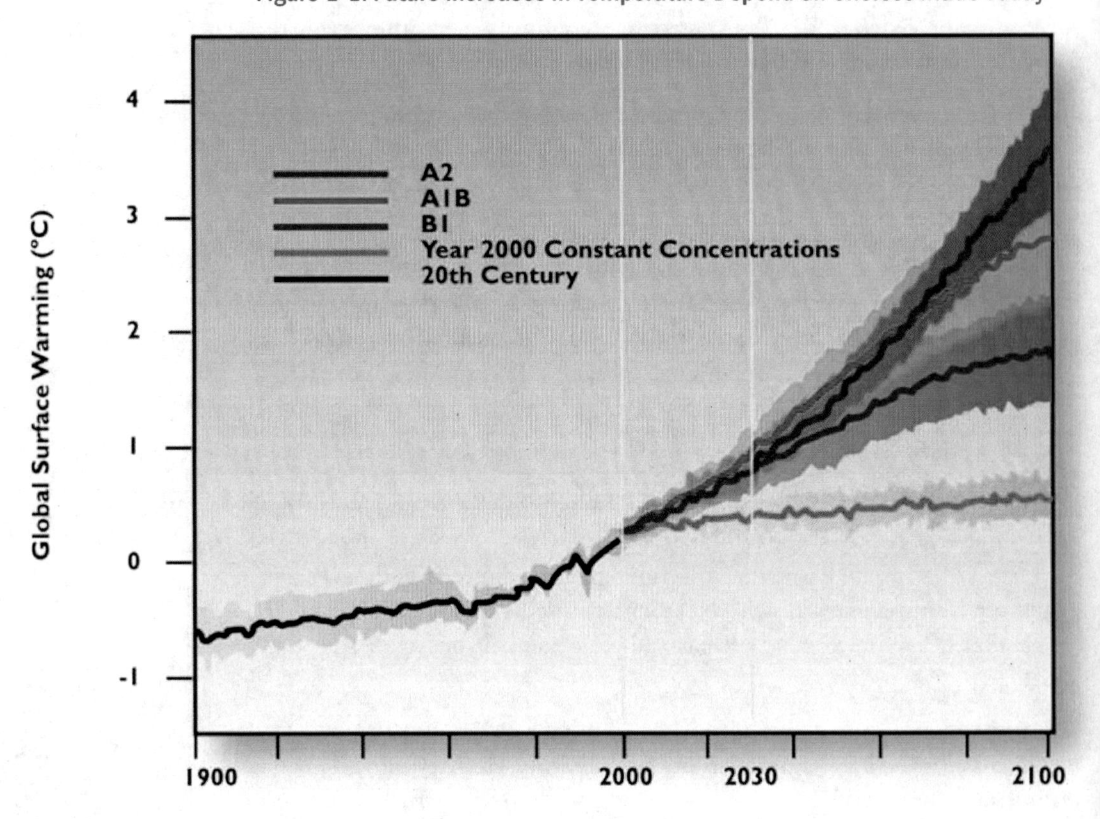

How plans evolve depends on how much mitigation is completed in the near future (more on *mitigation*, and its counterpart *adaptation*, in a second). Therefore, focusing on the period to 2025 simplifies the effort to select needed actions and supplies a foundation for considering things far past that date.

---

house and other significant gases—mostly those resulting from human activities in a number of sectors.

Our Science

Our efforts to present a meaningful discussion on climate change challenges were best served by using only well-founded science. Some of our most prominent sources include the *IPCC Fourth Assessment Report* (known as the AR4; see the CliCKE database for quick access)[3] and the US Global Change Research Program (USGCRP).[4] We formed this book around key pieces from these resources:

- Warming of the climate system is clear and unmistakable.
- Human activity adds GHGs to the atmosphere and very likely is responsible for a good part of observed changes in the climate.
- In the United States, climate changes are under way and are projected to grow.
- Climate change will stress water resources, a challenge to crop and livestock production.
- Coastal areas are increasingly at risk from sea level rise and storm surge.
- Threats to human health will increase.
- Certain climate and ecosystems thresholds will be crossed, leading to large, worldwide changes.
- Polar regions are warming more rapidly than the rest of the earth, melting glaciers and ice sheets.

Melting sea ice in the Arctic may be a harbinger of climate change impacts.

The work of the National Academy of Sciences informed our adaptation discussion. In summary, most scientists who have studied the interrelation of GHG and climate expect that the continuing rise in emissions will profoundly affect the earth's climate; many of the effects will emerge slowly in the coming years, but the changes will be steady.

Our Choices: Mitigation or Adaptation?

Mitigation and adaptation are the two areas most often discussed in coping with climate change impacts. These two concepts, however, differ greatly. *Mitigation* is an effort to stop or slow climate change, usually by reducing the GHG emissions driving the problem. *Adaptation*, on the other hand, is a response—actions necessitated by the actual or anticipated impacts of climate change.

They can be used in concert: mitigation reduces impacts, making it easier to adapt.

For the immediate future, the mitigation challenge is greater than that of adaptation (Figure 1-2). The mitigation challenge will grow with time, though in a predictable way, which means that delaying efforts to combat the problem will require greater effort down the line;[b] we will need to make up the difference, and the final target will become harder to reach as time passes.

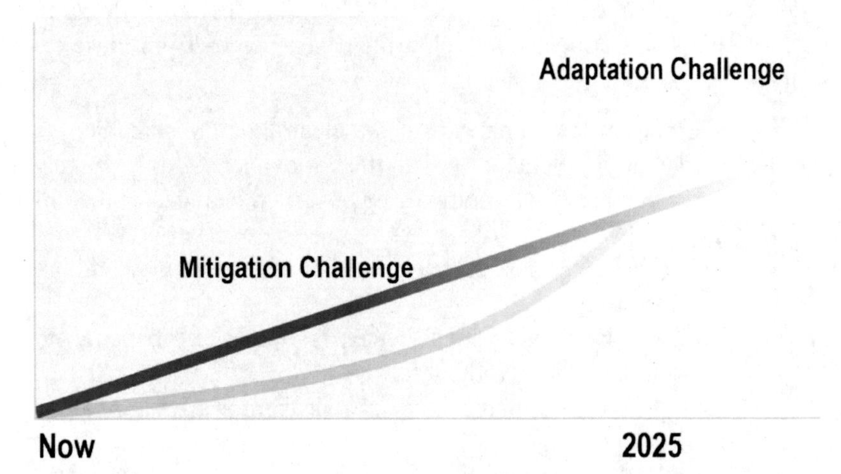

Figure 1-2. The Mitigation and Adaptation Challenges Will Grow Over Time

The need for adaptation, though, is like a train leaving the station. Today, it moves more slowly than the need for mitigation, but as it gains steam, it will eventually become the more urgent of the two. Adaptation will require a greater overall effort and have a larger price tag. The longer we postpone our efforts to mitigate, the larger the adaptation challenge will grow and the sooner it will arrive.

Each of these challenges will require a significant part of your attention by the dawn of 2025, so stakeholders must be prepared to focus on both. Throughout this book, we focus on mitigation and adaptation equally.

Mitigation

What are we mitigating? Mitigation currently targets six classes of GHG (Table 1-1). Each class varies in the degree to which it affects

[b] For example, President Obama's Executive Order (EO) 13514, "Federal Leadership in Environmental, Energy, and Economic Performance," requires federal agencies to reduce GHG emissions by 28 percent before 2020. By 2050, the reduction goal is 80 percent.

the climate, its *global warming potential*, or GWP.[c] The atmospheric lifetimes of each gas varies as well.[d]

Table 1-1. Characteristics of the Principal Kyoto GHGs

GHG Name	Atmospheric Lifetime (years)	100-Year GWP
Carbon dioxide (CO_2)	50 – 200+	1
Methane	12	25
Nitrous oxide	114	298
Sulfur hexaflouride	3,200	22,800
Perfluorocarbons (PFC), Hydrofluorocarbons (HFC), and other synthesized gases	5 – 100	146 – ~10,900

These gases come from three categories of sources, called "scopes," defined as follows:

- *Scope 1.* Direct GHG emissions from sources owned or controlled by an organization. For example, emissions from combustion in owned or controlled boilers, furnaces, and vehicles.
- *Scope 2.* GHG emissions from the generation of purchased electricity by an organization.
- *Scope 3.* All other indirect emissions. These emissions come from an organization's activities, but from sources the organization either does not own or control, such as third-party deliveries, business travel activities, and use of sold products and services. Suppliers to the federal government have their emissions counted among the government's Scope 3 emissions.[e]

---

[c] GWP is a measure of how much heat a GHG traps in the atmosphere relative to CO_2. Higher numbers are associated with more potential to trap heat in the atmosphere.

[d] This list does not include GHGs already regulated by 1989's Montreal Protocol on Substances That Deplete the Ozone Layer; the list could also grow in the future.

[e] The general definition of the scopes of GHG emissions are set by the Greenhouse Gas Protocol (GHG Protocol), an international accounting tool for stakeholders to understand, quantify, and manage GHG emissions.

GHG Mitigation Programs

Compared with other nations, the United States has only a fledging national GHG emissions-reduction program, overseen by the Environmental Protection Agency (EPA). Recent federal action has focused on reducing the GHG emissions of government agencies.

In October 2009, President Barack Obama issued EO 13514 to "make reduction of GHG emissions a priority for Federal agencies." EO 13514 was followed by guidance from the Council on Environmental Quality (CEQ), which requires executive branch agencies to set goals to reduce their Scope 1 and Scope 2 GHG emissions.

In this case, agencies must set a goal for fiscal year (FY) 2020 with their 2008 levels as the baseline. Agencies have set specific goals (Table 1-2) that collectively promise a GHG reduction of 28 percent. These agencies have developed plans to accomplish their goals and are beginning to execute them.

The council's guidance for EO 13514 also requires agencies to set a goal of reducing certain classes of Scope 3 emissions by FY 2020 relative to 2008 levels. The collective goal (Table 1-2) is a 13 percent reduction for Scope 3 emissions.

Federal agencies also are called upon to look at which vendors and contractors can help them in this effort, by recommending a strategy for working with their community of suppliers to track and reduce Scope 3 GHG emissions related to the products and services supplied to the federal government. At some point, federal contracts will include conditions for helping reach emission-reduction goals (already under way through the *Federal Acquisition Regulation*).

The federal sector is certainly not the only place where mitigation action is under way. Several regions have targeted reducing GHG in their power sectors through economic activity. One significant effort is the Regional Greenhouse Gas Initiative, a cap-and-trade system on power plant emissions, in which nine states participate—Connecticut, Delaware, Maine, Maryland, Massachusetts, New Hampshire, New York, Rhode Island, and Vermont—and several other states and Canadian provinces observe (Pennsylvania, Québec, New Brunswick, and Ontario).

Table 1-2. Major Agency GHG Mitigation Goals (Percentage Reduction) for 2020 under EO 13514

Agency	Scopes 1 and 2	Scope 3
Commerce	21	6
Defense	34	13.5
Energy	28	13
Education	29	3
Interior	20	9
Justice	16.4	3.8
Labor	27.8	23.4
Transportation	12.3	11
EPA	25	8
GSA	28	44
HHS	10.4	3.3
HUD	47.4	16.2
State	20	2
Treasury	33	11
USDA	21	7
Postal Service	20	20
Veterans Affairs	29.6	10

Meanwhile, California recently enacted AB 32, the Global Warming Solutions Act of 2006, which sets a target of returning state emissions to their 1990 level by 2020. It also seeks an 80 percent reduction of GHG by 2050. California now requires its largest industrial facilities to assess their energy efficiency to identify where they can reduce emissions.

Some private-sector organizations also are on board, voluntarily agreeing to limit their GHG emissions. Many participate in EPA's Green Power Partnership program (including Intel, Kohl's, Starbucks, and Johnson & Johnson) and in privately run programs by The

Climate Registry (such as The Hershey Company, Red Bull North America, Inc., and Walmart).

Finally, many individuals are undertaking actions to reduce their own personal GHG emissions. These include purchasing energy-efficient products such as hybrid motor vehicles, changing commuting habits, planting trees, and so on.

Adaptation

The US Global Change Research Program expects climate change to bring increases in heavy downpours.

Because we're already experiencing the impact of climate change, adaptation is necessary. Major organizations have emphasized the need to adapt, including the National Research Council, which recommends a risk-based approach to climate change adaptation. (Risk is a big part of this book; we detail risks in each chapter and provide examples.)

We must adapt to a vast array of expected impacts. Organizations such as the IPCC, USGCRP, National Academy of Sciences, EPA, and the federal Interagency Adaptation Working Group cite numerous issues to which we will have to adapt, including

- increases in heavy downpours,
- rising temperature,
- rising sea level,
- thawing permafrost,
- lengthening growing seasons,
- lengthening ice-free seasons on water bodies,
- earlier snowmelt, and
- alterations in river flows.

These impacts are projected via climate models using the various IPCC emission scenarios discussed earlier (and in the "Information" and "Security" chapters).

GHG Adaptation Programs

Because of EO 13514, federal agencies are developing adaptation plans, while CEQ is reporting on progress toward a national adaptation strategy. An interagency working group has formulated this strategy, and agencies are expected to implement it in their operational areas.

Meanwhile, the private sector is looking at adaptation and the economic benefits of minimizing potential damage to their organizations' insured assets.

For example, insurance companies are beginning to put price tags on what climate change impacts will mean to their products. The National Association of Insurance Commissioners has adopted a mandatory requirement that insurance companies disclose their financial risks from climate change. They're also asked to show the actions they are taking to respond to those risks, and all insurance companies with annual premiums of $500 million or more are completing a yearly survey on climate risk disclosure.[5]

The Securities and Exchange Commission, meanwhile, requires US-registered corporations to report their climate change–related risks, including the effects of legislation and international agreements, indirect effects of regulation, and potential physical impacts.[6]

Legal precedent also mandates that federally funded projects include climate change implications in any required environmental impact statements.

In short, the range of stakeholder responsibilities is already large—and growing.

Planning for Adaptation

What sorts of adaptation decisions might functional managers make between now and 2025? They could involve changes in activities or capital investment in construction of private or public buildings and public infrastructure such as levees, dams, airports, and bridges.

Mitigation of emissions from aircraft is already required in European countries.

Examples of this kind of adaptation planning already exist. The US Army Corps of Engineers has issued guidance requiring the consideration of climate change impacts in projects that may be affected by sea level rise.[7] It concluded that "virtually all" of its infrastructure investment will require adaptation. The Federal Emergency Management Agency is updating its risk mapping, assessment, and planning to better reflect the effects of climate change. It now includes changing rainfall and hurricane patterns and intensities in its data collection.[8] Another example is the Forest Service *National Roadmap for Responding to Climate Change*,[9] which is tied to a detailed performance scorecard.[10]

Our Recommendations

We offer recommendations because functional managers need and want practical advice.

Individual chapters introduce specific scientific and technological challenges relevant to each area in a way that non-scientific managers can appreciate, including how to keep climate change discussions themselves from getting bogged down in the nuances of this vast challenge.

This book's mitigation recommendations consider six major emission sources (Table 1-3).

Table 1-3. Principal Sources of GHG Emissions by Scope

Source	Scope	Example
Stationary	1	Power plants burning coal or natural gas
Mobile	1	Vehicles using gasoline
Fugitive	1	Leakage from refrigeration condensors
Electricity	2	On-site lighting
Travel	3	Emissions of aircraft exhaust used in business travel
Supplies	3	Energy expended to produce office chairs

Our adaptation recommendations focus on challenges the scientific community judge to be highly likely. The actions considered are those all stakeholders can embrace, whether in the executive branch, local government, or private sector.

In particular, we illustrate three analytical tools essential for decision making: cost-benefit analysis, risk assessment, and life-cycle analysis. Within the various chapters, we illustrate the best use of these three tools. The concluding chapter expands on this area.

Evaluating the Recommendations

Example of Recommendation Rating

Adaptation Potential	★★
Operational Impacts	★
Financial Impacts	★★★
Feasibility and Timing	★★

Each of our recommendations includes an evaluation of its potential, operational and financial impacts, and feasibility and timing.

We certainly expect our work to generate questions, including "Will this really help to reduce emissions?" and "How expensive will this be to implement?" We don't shy away from these questions, and we have developed a system to help answer them.

We've identified four areas of concern for most managers who will assess these recommendations:

1. Mitigation or adaptation potential
2. Operational impacts
3. Financial impacts

4. Feasibility and timing.

In each area, we apply our best judgment to rate each recommendation, using a three-star system (Table 1-4).

Table 1-4. Mitigation Recommendation Ratings

Criteria	Description	Good ★	Better ★★	Best ★★★
GHG reduction potential	GHG reduction potential of the recommendation	Recommendation will provide minimal GHG reductions over 10 years	Recommendation will provide moderate GHG reductions (up to 10% of the baseline) over 10 years	Recommendation will provide significant GHG reductions (more than 10% of the baseline) over 10 years
Operational impacts	Impact the recommendation will have on the ability to consistently execute mission-essential processes	Recommendation usually requires significant changes in business practices and additional resources or new skill sets	Recommendation can usually be implemented with minor changes to business practices that may require some temporary resources and training	Recommendation can usually be implemented with current staffing and resources and minimal process changes
Financial impacts	Financial cost or benefit of the recommendation	Recommendation typically involves a net cost to the organization	Recommendation typically has little to no net financial impact on the organization	Recommendation typically provides a positive net return to the organization
Feasibility and timing	Technical feasibility and implementation timing for the recommendation	Technology supporting the recommendation is still in research phases or the implementation timeline is greater than 5 years	Technology supporting the recommendation is new and fairly unproven or the implementation timeline is up to 5 years	Technology supporting the recommendation is mature and proven or the implementation timeline is 1 year or less

For our adaptation recommendations, only the first criterion changes: "Adaptation potential" replaces "GHG reduction potential" (Table 1-5); we otherwise retain the same system of evaluation. In regard to adaptation, we focus on whether the effect of the measure is only long term and whether it is significant.

Table 1-5. Adaptation Recommendation Ratings

Criteria	Description	Good ★	Better ★★	Best ★★★
Adaptation potential	Potential for the recommendation to secure the organization against the effects of climate change	Recommendation will have little effect over the next 10 years because the threats are not likely over that period, but it will do so after	Recommendation will somewhat reduce the organization's vulnerability to likely climate change threats over both the next 10 years and after	Recommendation will significantly reduce or eliminate the organization's vulnerability to likely climate change threats over the next 10 years and after

We offer only general solutions to adaptation. Exact answers will depend on the specific situation and how recommendations are implemented. Answers also will change over time as requirements for adaptation and mitigation grow and pertinent regulations, economic circumstances, and local climate challenges evolve.

Your Reality

This background discussion of emerging GHG emission constraints and adaptation challenges demonstrates that nearly everyone will soon be asked to contribute to future mitigation and adaptation efforts.

So why should you care?

We could say that being a good steward of the planet is "the right thing to do." But as you read this book, you'll also see that it's in your organization's operational and business interests.

Money can be saved and efficiencies found. Your organization can remain competitive and relevant, while making progress in reducing GHG emissions and preparing for potential impacts of a changing climate.

It's good business.

Between now and 2025, organizational leaders could require managers to become involved in planning and identifying tangible adaptation and mitigation strategies. They will expect a plan that can significantly change the organization's emissions profile and its ability to adapt to the expected consequences of climate change.

Your response to climate change challenges will affect your overall job performance.

As a stakeholder, you will be expected to play a part as all levels of government and the private sector act to reduce their GHG emissions and adapt to climate change. You will need to develop processes, products, and services that respond to these new demands and to move quickly to create cost-effective solutions.

Stakeholders will be evaluated in part on the quality of their response to this demand, and so it is in your best interest to act. Through this book, we hope to better prepare stakeholders to meet organizational climate change mitigation and adaptation strategy requests. Our goal is to help them better understand the mitigation and adaptation actions that hold promise in each area of responsibility.

The better informed managers are about practical, cost-effective opportunities, the greater the value they will add to an organization. The book's ideas can help ensure that every action—whether forced or elective (as in good business)—is well placed.

In the end, we hope that our approach will bring together functional communities as a means for greater openness, more profound discussion, and more significant efforts to mitigate and adapt to changes that are already under way.

---

[1] University Corporation for Atmospheric Research (UCAR), "2,755...," *UCAR Magazine: In The Air*, accessed December 5, 2011, http://www2.ucar.edu/magazine/in-the-air/2755.
[2] National Oceanic and Atmospheric Administration (NOAA), "NCDC 2010 Annual State of the Climate Report — Supplemental Figures and Information," *NOAA News*, January 12, 2011, accessed December 5, 2011, http://www.noaanews.noaa.gov/stories2011/20110112_globalstats_sup.html.
[3] http://clicke.lmi.org.
[4] Thomas R. Karl, Jerry M. Melillo, and Thomas C. Peterson, (eds.), *Global Climate Change Impacts in the United States*, Cambridge University Press, 2009.
[5] The surveys must be submitted in the state where the insurance company is located. See *Insurance Regulators Adopt Climate Change Risk Disclosure*, http://www.naic.org/Releases/2009_docs/climate_change_risk_disclosure_adopted.htm.
[6] Section 501.15. For an informative discussion of the SEC regulation, see http://www.sec.gov/rules/interp/2010/33-9106.pdf.
[7] See for example http://140.194.76.129/publications/eng-circulars/ec1110-2-6070/entire.pdf and http://140.194.76.129/publications/eng-circulars/ec1165-2-211/entire.pdf.
[8] See "Risk Mapping, Assessment, and Planning (Risk MAP): Fiscal Year 2009 Flood Mapping Production Plan," Version 1, May 2009, http://www.fema.gov/library/viewRecord.do?id=3680.
[9] US Department of Agriculture, *National Roadmap for Responding to Climate Change*, February 2001, http://www.fs.fed.us/climatechange/pdf/Roadmapfinal.pdf.
[10] The Forest Service Climate Change Performance Scorecard, 2010, http://www.fs.fed.us/climatechange/pdf/Scorecard.pdf.

Health

As a family prepares dinner and the children settle down at the kitchen table, the sounds of the evening news drone from the TV in the next room. The mother glances up as the meteorologist begins his forecast. He notes that the heat wave will continue, with another 4 days of temperatures in the triple digits. In the kitchen, the mother sighs. Tomorrow, like the past 16 days, will be too hot for the children to play outside. It's just too risky because of their asthma, but that's life in Minneapolis in the year 2025.

Where Health and Climate Collide

Health presents us with a paradox. It's personal, as local as something can possibly be. At the same time it's global, affecting communities on the grandest scale imaginable. It means a delicate balancing act for stakeholders who operate in the realm of public health, especially when they must engage in the mitigation and adaptation of climate change impacts.

A recent study in the journal *PLoS ONE* reveals that local public health officials are aware of the risk climate change represents for their locale.[1] The same study, however, found that few had made it a top priority for their department.

But it's not for a lack of health as a topic of discussion. Health is a staple of television news and social conversation: How are you feeling? Who is sick in your county or state? What is making people sick? What are the latest breakthroughs in treatments? The lag time between the emergence of a new public health threat and community awareness of its impact is usually short.

Climate change and its impacts don't work in the same way. The steady emergence of these impacts, though providing ample opportunity for stakeholders to react and respond, isn't sudden enough to "scare up" a public reaction. Without greater public awareness of the real relationship between public health and climate change, stakeholders will face an uphill battle before they're completely empowered to adjust their processes to meet the coming need.

It's akin to cultural shifts that have succeeded in identifying causes of cancer (smoking, sun exposure) and helping the public understand what is at stake. When the risks associated with smoking reached a critical mass, local and state governments had the needed support to enact laws that would limit public exposure to secondhand smoke. If the public understands how its health is at risk, it'll support action to prevent the threat.

If the local climate can negatively impact human health, the obvious extension is that when it changes, so will the threats. Some will be new; others will become more potent versions of those we currently face. The areas where local climate and affected environment cross over with human health include

- the weather,
- our atmosphere and the quality of our air,
- the water cycle and the quality of our drinking water,
- land use and food production and distribution,
- our oceans, and
- all of the above—the combined effect of all these factors on existing ecosystems.

We're not sure how global climate change will affect things on a hyperlocal level—predicting the neighborhood weather for tomorrow is easier said than done, never mind for the year 2025. However, we do have projections, and most of them suggest the earth is likely to get warmer and that we will experience a greater number of extreme weather events.

These projections tell us we will experience extremes in hot and cold temperatures,[a] which we know can lead directly to serious illness, injury, and loss of life. We know that increases in solar ultraviolet (UV) radiation compound the matter. We know natural systems are

[a] This is a key point because it illustrates why "global warming" is an unfortunate and inaccurate battle cry adopted early on by environmentalists and the media. Not only is "climate change" more accurate, it allows for a wider picture of impacts to be shown.

subject to climate-related disturbances such as changes in the range of disease *vectors* (the person, animal, or microorganism carrying and transmitting the disease) and the amount of output from our agricultural systems.

These results of climate change on human health all directly impact where and how often we observe cases of infectious diseases. They likewise directly affect instances of populations with inadequate nutrition. We can also link warmer temperatures to increases in air and water pollution and possible harm to human health.

So locale and region play a large role in the varying extent and nature of climate change impacts on human health, but we need to consider other things, including the relative vulnerability of population groups, the extent and duration of exposure, and the society's ability to adapt to the change. This isn't easy—these factors are complex, and assessing the impacts related only to climate change is difficult.

What's especially tricky in assessing these impacts is that climate change can actually yield health *benefits*, especially on a regional basis. We may see fewer deaths due to exposure to cold, for example.

In the subsections that follow, we consider the primary sources of human health effects related to climate change:

- Temperatures
- Hurricanes, floods, and droughts
- Air pollution and related effects
- Waterborne diseases
- Food-related illness and diseases
- Vector-borne and zoonotic diseases.[2]

We discuss others later in this chapter.

Stakeholders can view this as our road map for those with a vested interest in public health—an explanation of the issues and actions they are likely to face over the next 10 years and beyond.

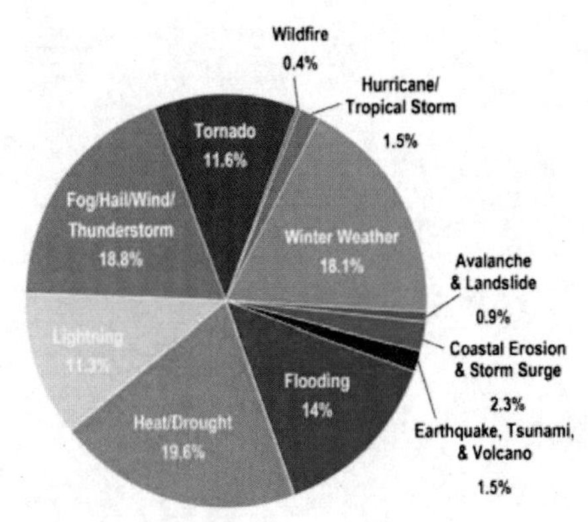

Distribution of deaths from 1970 to 2004 for 11 hazard categories.

Human Health Effects

Not many places in the United States will be immune to the human health impacts of climate change. Their variety is astounding—illnesses and deaths related to temperature extremes; regional changes in precipitation; increased solar UV radiation; flooding; severe storms; increases in diseases borne by food, water, and insects; challenges to food production and distribution; social and mental stress; and the effects of rising sea levels and flooding on drinking water and wastewater infrastructure. Somewhere, somehow, one or more of these will find its way into local communities across the country.

When this happens, people will look to their city, county, state, and federal agencies to respond, or in some cases, to have already acted. They will place the burden on the public sector and its partners to recognize, prepare for, and address the catastrophic public health outcomes of climate change through mitigation and adaptation.

Temperature Effects

Extended heat waves are a major concern. We've already seen major cities with sizable at-risk populations suffer through the effects of long, hot days. August 2003 comes to mind—that's when Europe endured a sweltering 14-day stretch of uncharacteristically high temperatures. In France, where summers are typically cooler, 7 days of temperatures at or above 104°F resulted in more than 14,800 heat-related deaths.

If we experience more frequent and intense heat waves, we are likely to see increases in heat-related illnesses such as heat cramps, heat exhaustion, and heat stroke. They are also a threat to populations susceptible to increased cardiovascular and stroke morbidity and mortality.[b] Many of these illnesses will be at their worst in urban heat islands—city areas significantly warmer than surrounding suburbs and rural areas because of the heat retained in building materials, pavements, and roads.

When we identify the populations most susceptible to heat-related illnesses, we're looking at two key groups: the large numbers of peo-

[b] In a nutshell, *morbidity* is the state of being diseased or unhealthful, and *mortality* is the state of being subject to death.

ple taking antihistamines or related drugs and the elderly (those over 65 years old).[c]

In a July 2006 *Morbidity and Mortality Weekly Report*, the Centers for Disease Control and Prevention (CDC) pointed out the great number of heat-related deaths among those over 65. The identity of these susceptible populations is well known, and the media has emerged as a strong ally to public health officials, providing regular coverage of heat and cold threats. But the sheer size of this group is worrisome because of the imbalance that baby boomers will create at the top of the age pyramid.[d]

By 2050, there will be 88.5 million Americans 65 or older, more than double the number in 2010 (Figure 2-1).[3] This group will be susceptible to the same health issues as our current elderly—but in greater numbers.

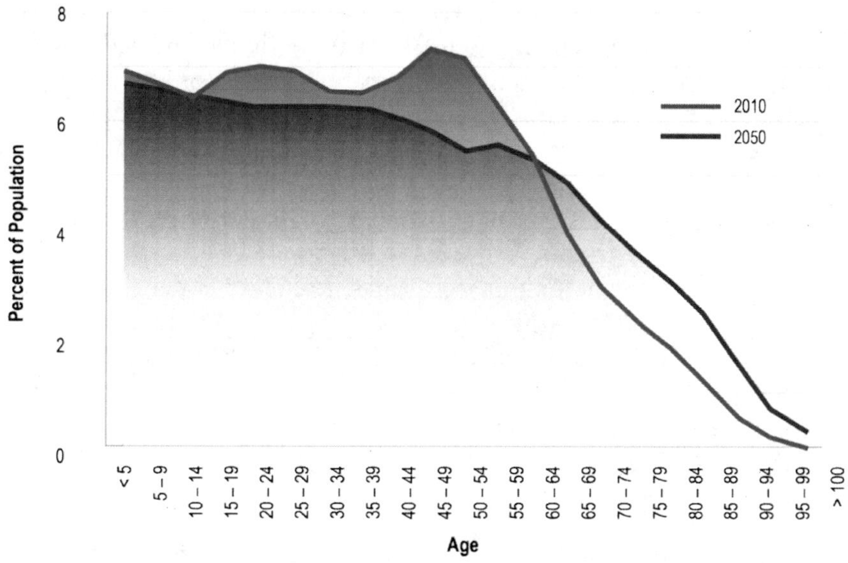

Figure 2-1. Projected Change in US Population Age Structure to 2050

If heat wave events increase over the next 10–15 years, government managers at all levels, as well as those with a role in public health, will face a formidable challenge.

[c] The tricky thing about planning? Future generations may not consider 65 to be elderly.

[d] Once, "baby boom" described any sudden increase in childbirths, but the explosion of childbirths between the end of World War II and 1957 added 77.5 million people to the American population. By the mid-1960s, the "baby boomers" were still under the age of 20, yet represented 40 percent of the American population (Landon Y. Jones, "Swinging 60s?" *Smithsonian Magazine*, January 2006).

Workers of any age who operate in certain indoor and outdoor working environments run the same risk as the elderly. These workers don't have the option of not venturing into the heat or cold. You can count on additional heat-related morbidity and mortality in a multitude of occupational settings unless managers promote work/rest cycles and provide guidance for acclimating to the heat[4]—and the workers follow their guidance.

Technology can facilitate this acclimation. The Occupational Safety and Health Administration (OSHA), for example, offers a free Heat Safety Tool mobile device app to workers and supervisors, enabling them to calculate specific work site heat indexes and the level of risk to outdoor workers. (The "Information" chapter discusses the uses of technology in climate change mitigation and adaptation).

The flipside is the cold; if temperatures increase across the board, presumably cold injuries and illness will decrease. This decrease, however, won't be enough to offset increases in heat-related injuries and illnesses from the heat wave increase. Past data on more than 6.5 million deaths between 1989 and 2000 show that severe cold increases death rates by 1.6 percent, while extreme heat was associated with a 5.7 percent increase.[5] This suggests that even total elimination of cold-related deaths from milder winters would not offset the number of extra deaths from heat waves. Furthermore, the moderated cold may decrease wintertime insect deaths, opening the door for earlier pest infestations as spring arrives. As the warm months extend deeper into what would traditionally be colder periods, the number of insect breeding cycles before a freeze would increase—with unfortunate consequences: reduced crop production from insect damage, increased disease transmission (Lyme disease, for example), and greater need for pesticide use.

Hurricanes, Floods, and Droughts

Today, we contend annually with hurricanes, floods, and droughts, and we are somewhat prepared. But when they increase in severity and frequency, they have impacts. Directly, there will be deaths; indirectly, health will be affected through lessened food production and availability. Water and waste treatment facilities will likely sustain damage from this constant barrage, contaminating drinking water supplies. In the case of drought, the reductions in water used directly by humans and in agriculture could be crippling.

We've only recently become reacquainted with the true impact of hurricanes. Because of greater warning capabilities, and a more robust set of plans for evacuating at-risk populations, annual US deaths from hurricanes are usually less than 20, despite significant coastal population growth. In 2005, Hurricane Katrina—a nightmare storm resulting in more than 1,800 deaths—was a significant exception. That storm, along with an additional 200 hurricane-related deaths that year, was a stark reminder that we must continue to improve preparation and response for huge hurricanes hitting population centers.

Other weather issues have had regional impacts. For years, the southwestern United States has experienced recurring droughts due to increased heat and changing rain patterns, and greater water demand has contributed to shortages (and possibly could lead to drought). More intense storm events can add more stress through a higher incidence of downpours, which result in immediate runoff rather than ground water replenishment. Droughts could reduce the snowmelt available for the annual replenishment of drinking supplies and hydration of crops, resulting in scarcer drinking water and lower food production.

Droughts and other extreme weather events are projected to increase in severity due to climate change.

If this happens, the little water available may become more contaminated. The land (discussed at length in the "Land" chapter) will have fewer arable acres, and when the plants and trees begin dying, they will become fuel for fires that can fill the air with pollutants.

People won't be inclined to stay put. We've already witnessed a mass migration spurred by hurricanes, in which nearly the entire population of New Orleans left the city for other areas of the country because of Hurricane Katrina's floods. Some returned to New Orleans once things stabilized, but they wouldn't return as quickly to a drought-stricken area. Regional drought has historically caused larger migrations. Generally, recovery takes longer, and the impact is wider.

The American Dust Bowl of the 1930s remains the largest migration of people in history: those who lived on millions of acres in Texas, Oklahoma, and neighboring areas of Kansas, New Mexico, and Colorado migrated in search of work so they could feed and care for themselves and their families.[6] Within a decade, an estimated 2.5 million people left the region.[7] A similar migration today could see greater numbers of people moving because the affected areas are now more populated.

Migrations big and small most frequently involve people abandoning rural areas for urban climes; the potential impact is increased crowding, amplified mental stress, and added prospects for the transmission of communicable disease.[8]

Air Pollution[9,10]

The local composition of air and the contaminants within are important factors for human populations and activities to take into account, along with the temperature, radiation, and upwind contaminant sources. Each can have a profound effect on human health. Temperature, vehicle and industrial emissions, and less air mass movement all interact to build ground-level ozone levels, causing air pollution events.[e] Too much ground-level ozone can damage the lungs (reducing lung function and inflaming tissues), resulting in more visits to the doctor.

Ground-level ozone, of course, is just one area of concern. Another is the stuff that will cause deeper physical damage to the lungs: the inhalation of airborne particulates (solid or liquid), especially those less than 2.5 microns in size (commonly referred to as particulate $matter_{2.5}$, or $PM_{2.5}$[11]).[f] Inhaling particulates this small, which become lodged deep within the lung tissue, causes respiratory health effects like decreased lung function, coughing, and chronic bronchitis.[12]

$PM_{2.5}$ comes from a variety of sources. Motor vehicles, power plants, forest fires, and other wood-burning activities (which incidentally are also major sources of CO_2) are the most common. Gases from burning fuels reacting with sunlight and water vapor can also cause their formation. The other major sources are foreign imports—such as dust storms from Asia and Africa that make their way overseas—and agriculture practices that keep soil in a disturbed state.

So Much to Discover...

The Hamilton County (Ohio) Department of Environmental Services is an excellent resource for the basics on air quality, especially particulate matter. It offers the following measure for $PM_{2.5}$: "These tiny particles are about 30 times smaller than the width of a hair on your head."

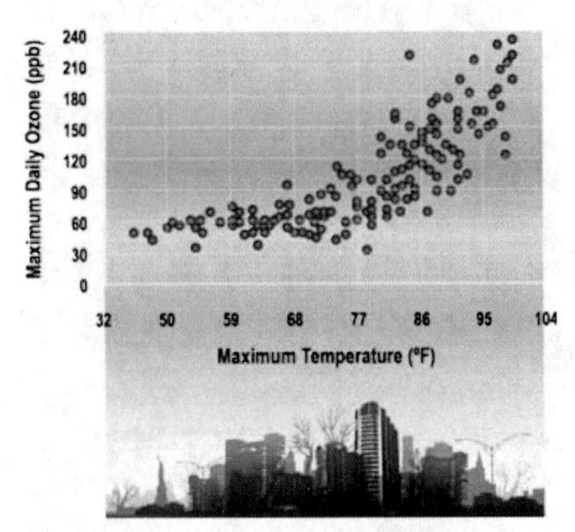

Ground-level ozone concentrations, May to October, 1988–1990 in NYC.

Non-Cancer Respiratory Illness[13]

Respiratory diseases are many—asthma, allergies, pneumonia, chronic obstructive pulmonary disease—and the higher prolonged temperatures in urban heat islands combined with air contaminants are likely to aggravate them.[14,15] Another concern is stress on the aged and very young, who are more vulnerable to cardiac illness.

Climate change could also bring about a rise in airborne allergens (aeroallergens), which each year affect a growing number of people through allergic airway symptoms. Monitoring pollen counts is now a regular event, and ragweed is a chief culprit for many people. The US

[e] Ozone is usually discussed in reference to the layer of atmosphere protecting us from harmful radiation. However, ozone at the ground level is a serious pollutant, which in high enough quantities can damage the respiratory system.

[f] A micron is a unit of measure equal to 1/1,000 of a millimeter.

Department of Agriculture (USDA) has shown that a warmer climate increases the length of growing seasons of ragweed and other plants, resulting in increased pollen production.[16]

Another concern is mold, especially if precipitation increases, raising absolute humidity. Absolute humidity combines with cooler surface temperatures to form liquid condensation on surfaces (especially when dew points are lower than the general air mass), providing mold spores with a fertile environment.

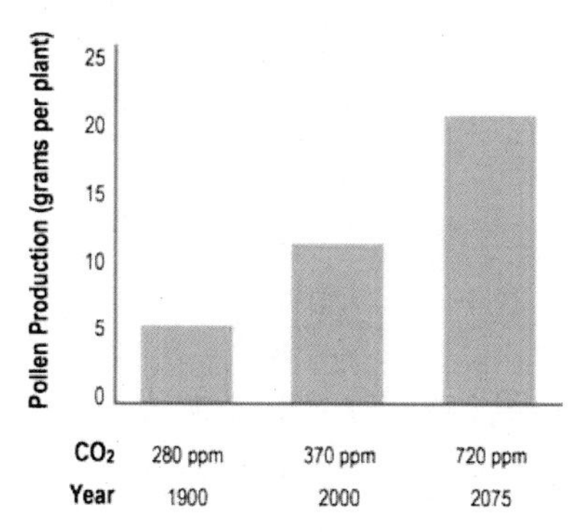

Experiments suggest that ragweed pollen concentrations will continue to increase as CO_2 concentration rises in the atmosphere.

CANCERS[17,18]

Cancer is a complex issue, certainly when considering climate change impacts. Incidences of lung cancer have been strongly linked to air pollution and fine particulates. Higher temperatures increase the evaporation of volatile and semivolatile contaminants in water and wastewater;[g] any that are potential carcinogens then reach the air we breathe. Flooding caused by heavy rainfall, snowpack melting, or rising sea levels can reach a variety of sources of potentially carcinogenic toxic chemicals or wastes; these spread when flooding results in spillage.

Finally, a cancer-related give-and-take is at work in the atmosphere. An increase of chemicals that damage the ozone layer can result in greater exposure to UV rays. The well-documented result of this is skin cancer. But as an example of the complexity of the situation, exposure to higher UV rays may also increase an individual's vitamin D levels—associated with reduced incidence of colon cancer.[19]

Waterborne Diseases

The Potomac River is traditionally a warmer waterway, peaking at around 90°F, but in the summer of 2011, following a prolonged heat wave that engulfed much of the continental United States, water temperatures in the Potomac reached 96°F. In fact, over a 3-day stretch, temperatures never fell below 91°F.[20] This difference of a few degrees illustrates how quickly our waterways can become a perfect breeding ground for bacteria—some of which could produce dangerous toxins. For example, Legionella bacteria, which cause Legionnaires' disease, grow best in 95°F waters.

---

[g] These include pesticides, herbicides, petroleum hydrocarbons, phenols, polychlorinated biphenyls, and other organic contaminants such as trichloroethane, trichloroethene, benzene, and toluene that come from agriculture, industrial, or homeowner runoff.

This example of how water can change quickly to negatively impact human health illustrates how a changing climate can quickly become a concern. Incidences of waterborne disease increase as a result of

- warmer temperatures and less safe drinking water;
- rainfall patterns (either very high or very low), which affect supply sources for safe drinking water and the systems that distribute it; and
- the combined effects of temperature and runoff on biological and chemical contamination of coastal, recreational, and surface waters.[21]

In the United States, most waterborne disease is gastrointestinal—affecting the stomach and digestive system—the result of exposure to bacteria, viruses, protozoa, or parasites in what we drink. Of course, other illnesses are related to water, including respiratory illness such as the bacteria that can cause Legionnaires' disease) and wound or skin infections (E. coli or staphylococcus bacteria and protozoan parasites). The great majority, however, are gastrointestinal. Chemical contaminants in water are also a concern.

Food-Related Illness and Diseases[22]

Higher temperatures mean that the microbes associated with food poisonings are able to multiply faster. History has shown that the longer the period of warm weather, the more cases of food poisoning reported. Food contamination also increases because the things that facilitate its spread—namely flies, rodents, and cockroaches—all see their active life cycles cover a larger part of the year as a result of higher temperatures.

Reproductive opportunities of pests like the flour beetle shown here are greater at higher temperatures.

Longer insect lives during periods of high temperatures result in more generations of insects that infest food products, especially among various flour beetles and moth larvae because of a now-shortened reproductive cycle over extended activity periods. *Vibrio* species of bacteria,[h] already common in warm marine environs like brackish waters, estuaries, and coastal bays, will have greater opportunity to spread to new areas as temperatures rise, resulting in an added risk of infections from these bacteria for people who consume undercooked shellfish.

[h] Most disease-causing strains of *Vibrio* are associated with gastroenteritis but can also infect open wounds and possibly result in septicemia.

Warmer seas and rainfall-fueled increases in fertilizer runoff will increase the growth of algal blooms (along with their associated toxins) and could result in their spread to waters that are currently colder. Similarly, runoff of pesticide and chemical residues from intensive agriculture practices can contaminate food supplies, causing direct poisoning.

Changes in rainfall and temperature patterns can damage crops and reduce the amount of available food and its quality. Quality of food is a large factor in malnutrition, especially when micronutrients are lowered by restricted dietary diversity. Reduced availability of nutritional foods can also harm human fetal development, which depends on proper nutrition during pregnancy. Furthermore, cleaning food before eating is more difficult during low water availability; a worse scenario is the widespread use of non-potable waters for this purpose, which can contaminate otherwise wholesome food supplies. We saw this scenario in the aftermath of Hurricane Katrina. The states affected by the storm had water supplies and water treatment plants contaminated with organisms and chemicals. Federal, state, and local health departments had to warn affected populations not to use the water to wash, brush teeth, or prepare food without treating it first.

Vector-Borne and Zoonotic Diseases

Vector-borne diseases result from an infectious microbe transmitted by an insect or animal—such as Lyme disease from a tick. Zoonotic diseases, on the other hand, occur when an animal directly infects a human—like a rat transmitting plague. Both disease types are in play in a changing climate.

Climate factors, such as temperature, rainfall levels, and cloud cover, profoundly affect disease-carrying insects, including mosquitoes, ticks, sand flies, and black flies. Current trends in climate change will allow the active range of these and other insects to spread, allowing for lower winter insect kills as they move to milder areas and an increase in early spring populations.

For example, ranges for ticks that carry Lyme disease and Rocky Mountain spotted fever, and for mosquitoes that carry and transmit malaria, dengue, yellow fever, and West Nile virus, have already grown.[23] A changing climate may also allow non-native species of disease-carrying insects to arrive with travelers or in cargo from overseas. Normally, climate would prevent these insects from becoming established in their new home and multiplying, but climate changes might reverse that trend, allowing these species to thrive in

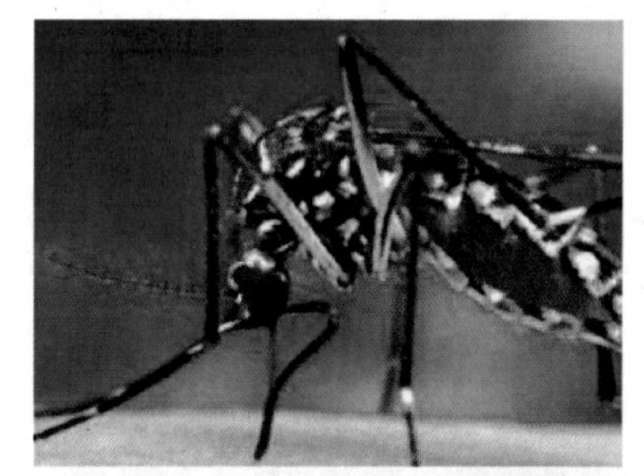

The geographic range of pests like the mosquito and tick will expand with increased temperature.

areas of the United States where they could not have previously survived.

Another example of vector-borne disease spread results from the movement of "freeze lines"[i] to higher latitudes. This movement will allow increased ranges for a particular species of snails found in the slow waters behind dams, especially when the water temperatures are right. These snails are the intermediate host to parasitic flatworms that cause schistosomiasis. Although these specific worms are not currently found in the United States, more than 200 million people worldwide are infected, and US citizens who travel to endemic areas are at risk. People who come into contact with contaminated water are exposed to this ailment, the symptoms of which include rash, fever, chills, cough, muscle ache, and inflammation of the intestines, liver, or bladder.

We are also seeing changes in the pattern and timing of bird migrations, typically movement to the north earlier and farther than before. This could help spread viral encephalitis or influenza.

Climate change can play a similar role in the spread of zoonotic disease. Rodents have long been known as transmitters of disease, blamed for the Black Death that once killed anywhere from 30 to 60 percent of Europe's population.[24] Today, we don't think of just the plague—other diseases, such as Weil's disease (leptospirosis) and hantavirus pulmonary syndrome, are emerging. Both can be deadly and their incidence can increase during heavy rain and flooding as humans and rodents become more likely to cross paths.

Public Health Mitigation and Adaptation Background

Those are a just a few of the health-related effects of climate change. We have the opportunity to mitigate and adapt—work now to reduce the things that can result in climate change and work later to reduce the negative impacts that eventually happen. Decision makers and stakeholders can join public health officials to address several areas of climate change impacts in the next decade.

---

[i] "Freeze lines" is an inexact term for the points dividing lands that experience freezes and those that don't.

Health Threat Risk Assessment: The Foundation for Action

Health threat risk assessments provide a basis for comparing and prioritizing expected risks and benefits. They allow an examination of the timing and the type of public health program interventions available. They offer perspective on the relationship of public health intervention to reductions in clinical care requirements and services, and the use of health impact assessments (HIAs).

In the immediate future, stakeholders will need to think in terms of risk. Their assessment of risk over a long period should look at gradually changing health threats and how they relate to climate change—not just mitigation and adaptation, but also the limited resources available for climate change–related interventions. They must consider the most cost-effective, evidence-based health outcomes. In simplest terms, risk assessment for public health actions considers the relationship between the probability that climate change events will cause health effects and the consequences of those health effects.

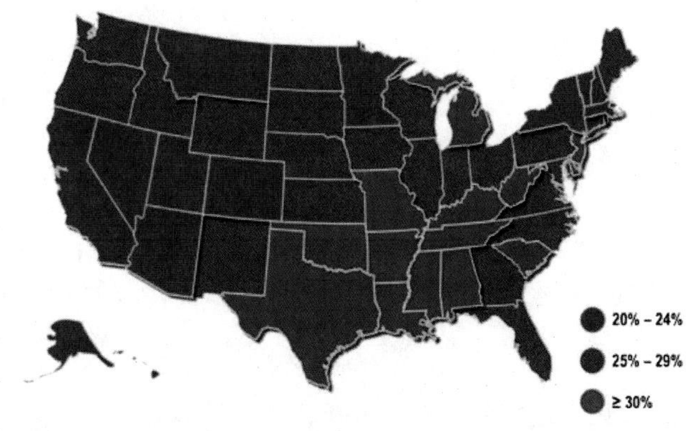

Adult obesity trends nationwide in 2010. Walking to work reduces GHG emissions and can help manage weight!

Climate change and its impact on human health, in the simplest sense, also represent a single thing: the cultural shifts that clash with the reality created by climate change. We've grown as a society to promote healthy living. We encourage people to walk or bike instead of drive so that they might increase physical activity and stave off heart disease, diabetes, and high blood pressure. We're urged to adopt healthy habits for preventing obesity, colon cancer, and low bone density. We can change our lens to view these cultural shifts as beneficial in reducing greenhouse gas (GHG) production, all the while decreasing the number of people who may require clinical care for those diseases.[25]

Consider the various groups of people in the United States, who, through preventive care, could cease to impact the health system:

- 13.5 million people with coronary heart disease
- 1.5 million people who suffer from a heart attack in any given year
- 8 million people who have adult-onset (non-insulin-dependent) diabetes
- 95,000 people who discover they have colon cancer each year

- 250,000 people who suffer from hip fractures each year
- 50 million people who have high blood pressure
- More than 60 million overweight adults, that is, those with a body mass index (BMI) of 25–29.9.

These ailments are well documented. But here's another view to consider: how climate change forces a new balancing act of preventive care activities.

The overweight are encouraged to exercise, and because monthly payments for gym memberships might not fit their budget, they might choose to walk outdoors, exercise at a local park, or ride a bike. But a changing climate means it's too hot outside for safe exercise or the air quality is prohibitive. How can public health stakeholders continue to enable healthy lifestyle choices when some options are removed? It's the type of thinking in which stakeholders should engage when looking at their options and the challenges facing their organization.

In this illustration, a health threat risk assessment would show the relatively low-cost intervention of increased physical activity if adopted by large numbers of the community. It would show the general health benefit. Any resulting reductions in clinical care from the health benefit may measurably reduce healthcare facility energy use and provide some mitigation of GHG production by clinical facilities. It would also take into account the factors that make implementing such a plan more difficult. The changing climate is just one of these factors.

GHG Mitigation Opportunities: Leveraging Partnerships and Technology

Several mitigation responses concern managers. Decision makers and stakeholders, in conjunction with public health managers, can address some of the public health outcomes that relate to climate change mitigation through 2025.

As a group, they generally involve viewing reduction opportunities as the sum of partnerships, technology, and critical thinking, as seen in the following recommendations:

- Use assessment tools to gauge areas of risk and opportunity.
- Increase the use of telehealth.
- Promote actions that reduce the impact of urban heat islands.
- Make energy use in medical facilities more efficient.

- Consider how GHG reduction activities impact disease transmission and prevention.

Assessments in Every Toolkit

Recommendation: *Expand the use of HIAs in public health programs. Promote and use HIAs to assess, identify, and recommend changes to proposed projects and policies to promote health and reduce illness requiring clinical care and the GHG emissions associated with patient diagnosis and treatment.*

CHALLENGE

HIAs are similar to the environmental impact assessments required by the National Environmental Policy Act but differ in that their use is not mandatory. Despite this, state and local governments have begun to show more interest in requiring HIAs. Unfortunately, up until this point, only a few jurisdictions have widely implemented them. A further concern is that most existing HIAs do not consider climate change—certainly not in regard to its relationship to health.

HIAs are comprehensive tools for identifying projects or policies where objective assessments and recommendations would be helpful. They primarily serve to show how to increase positive health outcomes while decreasing potential health threats and related negative outcomes. The HIA evaluates the potential effect on human health of a proposed project or policy while defining probable health outcomes. It identifies the affected community and offers recommendations for controlling or eliminating adverse outcomes. Its reach is broad: an HIA can be used for housing, transportation, air and water quality, safety, green space, nutrition, public services, or any other topical area with the potential to influence health.

SOLUTION

Mitigation Potential	★★
Operational Impacts	★★★
Financial Impacts	★★
Feasibility and Timing	★★

Because the HIA process fosters community involvement, the prime sponsor of HIAs should remain state and local governments—at least in the near term. With local application under way (allowing stakeholders to address local and regional characteristics), the national level should lead the development of standards for HIA tools. These would provide local jurisdictions with a consistent and validated framework.

Establishing a uniform national approach will also allow for better analysis of compiled data for health indicators and outcomes, a key step toward evidence-based recommendations resulting from the HIAs.

By following the recommendations of an HIA, communities could potentially see decreases in the rates of illnesses and be able to identify future needs for clinical care services and facilities. The potential benefits that have GHG reduction qualities include fewer vehicle trips to healthcare providers, less energy use by healthcare equipment, and a reduced urgency to build new, or continuously operate current, clinical care facilities. HIAs can guide policy that incentivizes behavioral changes and cultural shifts.

EXAMPLE

Human Impact Partners, a California nonprofit, has facilitated several successful local HIAs. In Humboldt County, CA, an HIA performed in conjunction with county leaders assessed the county's general plan.[26] From this effort came a recommendation that the county support focused growth in areas that had existing utilities and support infrastructure.

The assessment suggested that by promoting growth in these areas, residents would remain closer to shopping, schools, and public transportation. The likely impact is fewer people driving, choosing instead to bike or walk. In this case, some health and environmental benefits are clear—less driving not only directly reduces energy use, but promotes healthy lifestyles. Projected improvements in obesity, cardiovascular disease, and diabetes means fewer people driving to healthcare facilities and less required use of healthcare facilities for treating illness.

Human Impact Partners also performed another HIA, this time with the Public Health Department of San Francisco. In this case, it found that a major reason many people come to work even when they're sick is that about half of the entire US workforce does not have paid sick days.[27] This HIA suggested the support of a proposed federal law that would require all companies with 15 or more employees to give their workers 7 sick days per year.

The assessment found that paid sick leave encourages workers and their children to stay home, which would reduce the spread of communicable disease. It cited similar legislation in Massachusetts, showing that workers without paid sick days used emergency room services 15 percent more than those with sick leave. If a community can reduce the spread of communicable diseases along with unnecessary emergency room visits, it can reduce the overall energy used by cutting patient and staff vehicle trips. Further reductions would be found in eliminating the need for energy-intense clinical treatment and laboratory support, additional benefits of this mitigation strategy.

Embrace Telehealth

Recommendation: *Increase use of telehealth techniques to provide public health services, preventive care, and related community education.*

Use telehealth techniques

CHALLENGE[28,29]

The healthcare system is still based on one-on-one, personal interaction. It relies on the patient's meeting and interacting with the healthcare provider in person to receive care. As a result, we still rely on large, fully functioning clinical facilities. People travel to these facilities, sometimes long distances. The system has a significant energy cost and that equates to more GHG emissions that spur climate change.

We also live in an era of great technology for promoting and sustaining human interaction; these systems have enormous potential to improve the healthcare system and reduce the current GHG burden of healthcare. Telehealth uses technology to exchange electronic information and provide long-distance clinical healthcare. It allows patients and healthcare professionals to engage in health-related education, public health services, and health administration—all from a distance (or even from the comfort of home). Technologies used include live videoconferencing, store-and-forward imaging, the Internet, streaming media, and terrestrial and wireless communications.

Much of the current usage of telehealth focuses less on primary or secondary prevention and more on clinical medicine and tertiary public health services, including the following:

- Specialist services assisting primary care providers (PCPs) with diagnoses
- PCP-to-patient consultations via audio, still, or live images to develop a diagnosis and treatment plan
- Remote patient monitoring to collect and send data for interpretation, replacing or supplementing the use of visiting healthcare professionals
- Medical education for health professionals and targeted groups in remote locations, reducing the need for the presenter or audience to travel

- Self-care health information to community groups or individuals.

Unfortunately, despite early adoption, especially among clinicians, telehealth still lags behind for a variety of reasons. As a relatively new delivery system of care, it suffers from incompatible technology. Also, older age groups are less comfortable with using technology (for example, 45 to 54 year olds use social networks at a 95 percent clip, but those 54 and above are at just 9 percent).[30]

Compensation of physicians for the care they provide is also a challenge. Many benefit plans, including those offered by the federal government, do not uniformly cover telemedicine services. Under the current system, face-to-face visits are required for the physician to be paid—a critical hurdle for any system to succeed.

SOLUTION

Mitigation Potential	★★
Operational Impacts	★★
Financial Impacts	★★
Feasibility and Timing	★★★

A key starting point is addressing the policy issues associated with telehealth and telemedicine. Many health insurance carriers follow the lead of the Centers for Medicare and Medicaid Services when determining physician reimbursement. By adopting policy that designates telehealth activities as uniformly covered, many, if not all, major health insurance providers will follow suit, allowing physicians to adopt these activities immediately.

To increase the effectiveness of telehealth, public health providers must increase the types and availability of primary and secondary prevention services, and the public sector needs to provide access to the technology that makes telehealth possible.

Telehealth in the United States presently includes about 200 telemedicine networks linking

- tertiary care hospitals,
- rural and suburban clinics and health centers,
- point-to-point private networks,
- clinical care to single-line home video systems,
- clinical monitoring via normal phone lines,
- web-based follow-up,
- education, and
- outreach services.

Fortunately, we've seen the rise of the most promising vehicle for telehealth: social networking. As more people have gained wide access to the Internet through not just personal computers, but also smart phones, tablets, and even their televisions, mass numbers have populated the various social networking groups, including Facebook, Twitter, and LinkedIn. Facebook, for example, gained its 50 millionth user in 2007; by 2010, it had reached 400 million.[31] More and more, these social networks are integrating video and chat functionality, not just in a standard browser environment, but also for mobile devices. As a result, the infrastructure is being created within systems that are already in use. (The "Information" chapter discusses the role of technology in climate change mitigation and adaptation.)

Stakeholders need to commit to infrastructure, offering users facilities where they can access telehealth services. Kiosks or "telehealth conference rooms" can be established in public places such as community centers, libraries, and health centers. The extent will vary by community, but should consider the availability of video and Internet technology, which are critical for people to take advantage of some telehealth offerings. Telehealth technologies located in community facilities, which are usually in a central location, will give patients ready access to the services, in some cases by walking, biking, or public transportation.

In a letter in the *British Medical Journal*, Richard E. Scott of the Global e-Health Research and Training Programme, Health Innovation and Information Technology Centre, University of Calgary, furnished unpublished data from a recent study estimating that about 36 percent of the more than 32.2 million annual home nurse visits across Canada could be performed virtually. These virtual visits could eliminate more than 75,000 miles of travel and more than 33,000 tons of GHG emissions each year.[32]

Sunscreen Alert!

Increasing telehealth services to achieve primary and secondary prevention will reduce vehicle use for both the patient and many public health providers. It will also reduce illnesses that require more intensive clinical and tertiary preventive care.[j] Use of text messaging or social and professional networking as the communication medium may offer quick and widely acceptable access to a wide range of providers and users for a number of notifications, warnings, and simple educational services.

[j] Tertiary care occurs in situations where the disease or illness is already present.

EXAMPLE

Researchers at the Center for Connected Health assessed the effect of text-message reminders delivered via cell phone on adherence to sunscreen application via short message service (SMS) texting technology.[33] Researchers found that, at the end of 6 weeks, adherence rates for the reminder group, 56 percent, were almost double that of the control group that did not receive reminder texts, 30 percent. The study suggests that existing SMS technology offers a low-cost and effective method of improving treatment adherence.[34]

Address Urban Heat Islands

Reduce heat islands

Recommendation: *Public health managers should advocate community action to reduce the adverse personal and environmental health effects of urban heat islands.*

CHALLENGE

People flock to cities because of job opportunities, cultural offerings, and other benefits of urban living; they also find that built-up heat islands are simply warmer than nearby rural areas.[35] These warmer temperatures are quite noticeable: the annual average air temperature of a city with 1 million people or more can be warmer by 1.8 to 5.4°F than the surrounding areas. In the evening, the difference can be as high as 22°F. Heat islands affect communities by increasing summertime peak energy demand because of attempts to cool livable spaces—and air conditioning costs money. Because the warmer temperatures promote ground-level ozone, people must contend with air pollution (and GHG emissions) leading to heat-related illness and mortality. This added heat also affects runoff, influencing water quality.

Mitigation Potential	★★★
Operational Impacts	★★★
Financial Impacts	★★
Feasibility and Timing	★★★

SOLUTION

Public health stakeholders must support development, adoption, and enforcement of regulations and standards, public services, and educational programs.[36] A number of activities can help do so, including

- increasing tree and vegetative cover to provide more green space and parks and urban gardening;
- allowing the installation of green roofs to reduce heat islands;
- encouraging cool reflective roofs and cool pavements;
- identifying restricted building zones (possibly through the use of the previously discussed HIAs); and

- minimizing motor vehicle use and promoting optimized mass transit to serve population centers and encouraging walking and biking through infrastructure development.[k]

These actions could not only reduce direct heat-induced illnesses but decrease the exposures to ozone, volatile and non-volatile contaminants, and hazardous particulates that cause cancer and non-cancer illnesses and aggravate asthma symptoms. This reduction in illness will result in fewer patient visits to healthcare facilities and fewer vehicle miles.

EXAMPLES

Since its completion in 1911, Chicago's city hall has been the hub of civic activity, housing the office of the mayor, city clerk and treasurer, some city departments, and aldermen of Chicago's various wards. Today, it also shows how civic facilities can help alleviate the impact of urban heat islands. Since 2001, the Chicago city hall has had a green roof, a part of former Mayor Richard M. Daley's Urban Heat Island Initiative. The roof is home to more than 20,000 plants, shrubs, grasses, vines, and trees. The city expects to save more than 9,270 kilowatt-hours (kWh) per year of electricity and nearly 740 million British thermal units (Btu) per year of natural gas for heating. This energy savings translates into about $5,000 annually, which will increase if energy prices rise. Chicago's green roof is also a test bed for different types of rooftop garden systems, success rates of native and non-native vegetation, and reductions in stormwater runoff. The city hall example reduces GHG emissions while saving money and has increased general visibility of green roofs while increasing community awareness of their use.[37] Today, Chicago is a leader in green roof projects, with more than 400 new green roof projects in development, a total of 7 million square feet of green roofs constructed or under way—more than all other US cities combined.[38]

An example of an industrial project benefiting from a green roof can be found in Dearborn, MI, where the Ford Motor Company's Dearborn Truck Plant at Rouge Center has a 454,000-square-foot living roof.[39] The 10.4-acre project is the largest living roof in the world and includes a groundbreaking rainwater reclamation system. Although its primary function is to collect and filter rainfall as part of a natural stormwater management system, it also insulates the building. The roof uses sedum planted into a four-layer mat, which helps reduce the

[k] One further possibility, albeit a politically sensitive one, is to put in place urban automobile congestion charges.

heat buildup compared with the alternative (more than 10 acres of regular roofing materials). Ford says it reduces heating and cooling costs by 5 percent.

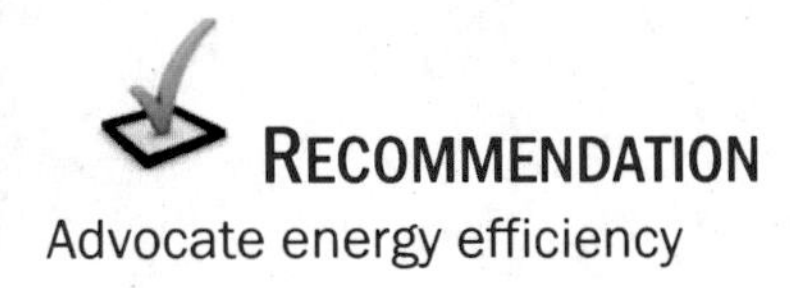

Reduce Healthcare's Energy Consumption

Recommendation: *Public health stakeholders should advocate and support public health interventions that directly or indirectly reduce energy use by hospitals, clinics, and other patient care facilities.*

CHALLENGE

The American healthcare system is a significant user of energy and an equally large emitter of GHG. Healthcare buildings account for more than 9 percent of all commercial energy consumption, using 1,094 trillion Btu of electricity, natural gas, fuel oil, and district steam or hot water (Figure 2-2).[1] These buildings are the fourth highest consumer of total energy of all building types.[40]

Figure 2-2. Significant Energy Use Burden from Healthcare Operations

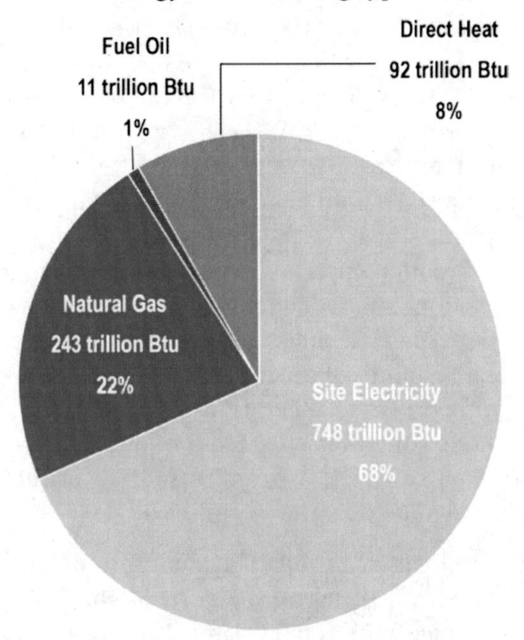

The largest energy use within healthcare buildings is site electricity (electricity consumed within buildings), totaling 748 trillion Btu (248 trillion Btu directly on site + 500 trillion Btu in energy wasted during generation and transmission).

[1] Anyone serious about grilling knows their Btu, but for the layperson, 1 Btu is the amount of energy required to heat 1 pound of 39°F water (about a pint) 1°F at 1 atmosphere of pressure.

The breakdown is astounding. Healthcare facilities consume an average of 4.6 billion Btu per building each year, with an energy use intensity (EUI)[m] of 187,700 Btu per square foot.[41] Inpatient facilities use 59.4 billion Btu per building, 249,200 Btu per square foot; in outpatient facilities, these numbers are 0.98 billion Btu and 94,600 Btu per square foot. Not surprisingly, inpatient care has the second highest intensity among all commercial building types, using energy almost three times as intensively as outpatient healthcare facilities (Figure 2-3). Any actions, therefore, that reduce patient loads or the amount of energy used per patient will significantly reduce emissions in the long term.

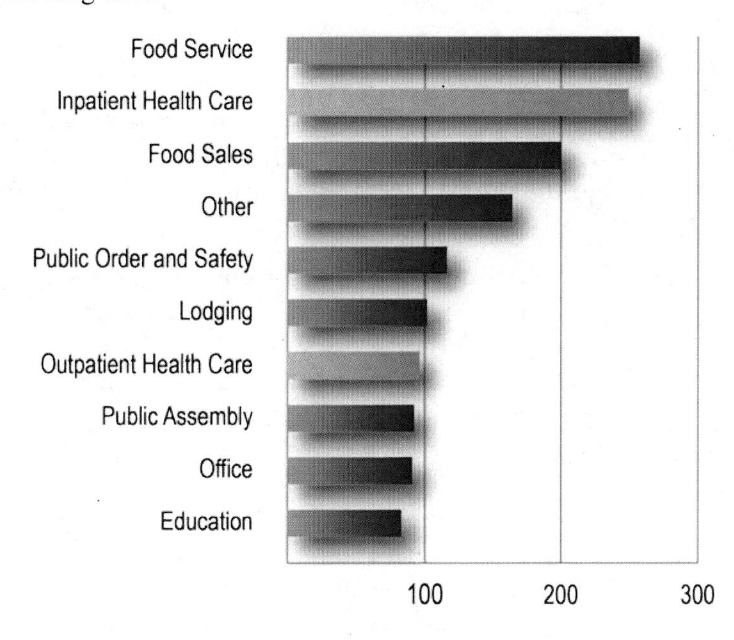

Figure 2-3. Healthcare Energy Use Is Significant Compared with Other Sectors

GHG production is equally high because, once an illness occurs, modern clinical care generally requires frequent transportation to see a healthcare provider. Medical facilities are expected to operate around the clock, every day of the year. Today's care involves the application of high-tech, high-energy diagnostic and treatment equipment. All this adds up, directly or indirectly, to GHG produc-

[m] EUI is a measure of building energy use per unit area. It differs from energy intensity, a metric to gauge an entire nation's energy consumption in the context of its gross domestic product.

tion. The potential sources of GHG from healthcare facilities are wide ranging,[n] and include

- heat and power buildings (CO_2, nitrogen oxide, sulfur dioxide, volatile organic compounds) and air conditioning and refrigeration units (HCFC);
- transportation (CO_2, nitrogen oxide);
- emergency generators (CO_2, nitrogen oxide) and particulates;
- anesthesia (nitrous oxide, desflurane, isoflurane, sevflurane);
- hospital laboratories (volatile organic compounds); and
- medical waste incinerators (CO_2, nitrogen oxide, oxide, sulfur dioxide, volatile organic compounds, particulates).[42]

SOLUTION

Mitigation Potential	★★
Operational Impacts	★★★
Financial Impacts	★★
Feasibility and Timing	★★

Inroads in decreased energy consumption can be made by reducing the energy needs of the healthcare sector, by focusing the timing and type of public health campaigns, and by reducing the need for clinical services.

Using public health initiatives to help drive construction and use of fewer, smaller, and more energy-efficient hospitals has great long-term potential to reduce healthcare facility energy demand. Stakeholders can work with public health managers to advocate and implement actions that reduce energy use directly or indirectly. These activities should also help the United States reduce GHG emissions while protecting our health. Opportunities include increasing the number of available public health programs, embracing ENERGY STAR processes and techniques, and designing and constructing energy-efficient facilities. They can support initiatives to reprocess single-use medical equipment, a step that can help control GHGs while saving resources and funds.

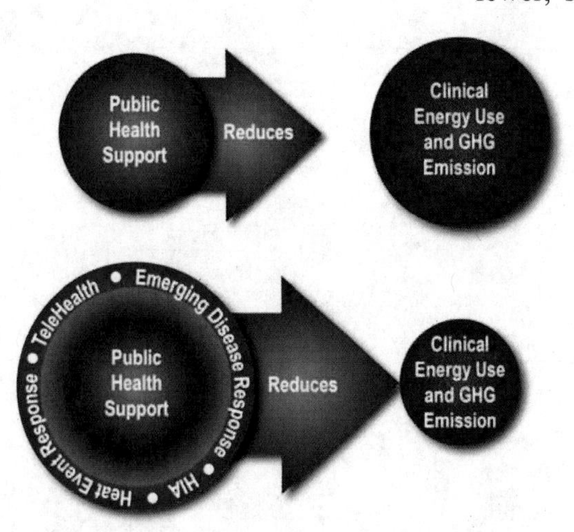

While public health reduces the need for clinical care, adding new public health support activities further reduces clinical care and associated energy use and GHG production.

Better timing of public health interventions can affect disease rates and, accordingly, attendant GHG emissions. These interventions can start before the onset of illness through screenings or efforts to address existing illness to keep it from getting worse. Evidence suggests you can reduce clinical disease and the associated use of facilities that produce GHG emissions by following three stages of prevention:

---

[n] See Table 1-1 for details of these GHGs.

- Primary (such as vaccines and sun screens)
- Secondary (such as regular checks of weight, blood pressure, and blood sugar)
- Tertiary (such as stroke rehab and diabetics' circulatory system treatment).

Many community-wide service interventions are effective and especially desirable because they transparently become part of the way we live. Think of how drinking water treatment has controlled waterborne gastrointestinal disease and protected teeth from dental cavities or how food additives (including vitamin D in milk and iodized salt) have promoted health. Stakeholder support initiatives to provide services on a variety of care levels—including one to one (patient/healthcare provider, such as sexually transmitted disease clinics), many to one (patient groups to healthcare provider, such as tobacco-use cessation training), and indirect community-wide (built into infrastructure, such as air pollution controls)—offer an opportunity for further gains in preventive health.

Preventive actions can help populations altogether avoid, or at least reduce, the extent of many illnesses, including those caused by pathogens, deficiencies, physiological conditions, or the surrounding environment. For example, the preventive activities discussed above have reduced the need for both outpatient doctor's visits and inpatient hospital stays. Embracing existing and new public health interventions will reduce the need to seek clinical care and its associated energy use—thus supporting mitigation and further improving overall personal and community health.

EXAMPLES

New York-Presbyterian Hospital (NYPH) is the first healthcare system to earn the ENERGY STAR Partner of the Year Award three times: twice for Excellence in Energy Management (2005 and 2006) and a third time in the Sustained Excellence category (2007). Through energy management, NYPH will save an estimated $1.77 million annually, more than it saved in 2004 and 2005 combined.

It saved this money through a series of sweeping updates, including a comprehensive heating, ventilation, and air conditioning retrofit, replacing 5,000 light fixtures, installing a high-efficiency chiller, and replacing more than 400 aging window air conditioning units with a central high-efficiency cooling loop. It also educated its staff on energy-reducing behavioral changes, some of them as simple as asking the housekeeping staff to turn off copy machines during nighttime rounds.

Stryker Sustainability Solutions (formerly Ascent) is a company supporting 1,800 hospitals that follow Federal Drug Administration (FDA) guidelines for reprocessing single-use medical devices rather than throw them away. It reports the annual elimination of several million pounds of waste (Table 2-1[43]) and direct supply cost savings of hundreds of millions of dollars. Participating hospitals use FDA published clearances to identify the devices each supporting company can reprocess. Currently, that can include catheters, cables, surgical instruments, forceps, and cuffs, among others. The number of participating hospitals is increasing as they discover the monetary savings and decreased waste generation that reprocessing affords.

Table 2-1. Reprocessing Medical Equipment Eliminates Waste

Year	Waste Avoided (lb)
2008	4,300,000
2009	5,300,000
2010	6,600,000

Think GHG Reduction—Within Context

Recommendation: *Stakeholders should recognize that simple GHG reduction doesn't always represent best practices for public health and work to balance actions to reduce GHG emissions with public health considerations.*

CHALLENGE

Not every plan that involves GHG reductions has positive health impacts. Programs for producing non-carbon energy, which include wind, solar, or hydroelectric dams that use less water during operations, are good for protecting local water supplies. However, hydroelectric dams change the ecology in gathered water above the dam, possibly providing an opportunity for bacteria and fungus to colonize and grow, and nuclear energy generation reduces GHG emissions from power plants but produces large amounts of warm water from cooling towers, which can provide environments for bacteria growth and affect stream ecology.

The usage of this produced energy has similar entanglements. Facility development planners and operations personnel constantly make decisions that attempt to integrate energy conservation into the construction or use of facilities. These decisions may affect the occupants of these facilities, but data and models are frequently lacking, leading to uncertainty for those responsible for balancing the wide range of effects, positive and negative, on populations and the environment.

SOLUTION

Public health programs must become integrated into every step in the development process for public facilities, including construction reviews, to ensure energy controls are balanced with possible undesirable health impacts. Often these aspects will be subtle and possibly difficult for non-health professionals to recognize. For instance, building energy use improvements that limit outside air introduction and use 100 percent recirculation of already tempered air frequently cause CO_2 buildup and humidity and mold problems.

Mitigation Potential	★★
Operational Impacts	★★
Financial Impacts	★★
Feasibility and Timing	★★

EXAMPLE

The late 2000s saw the rise of a new means for extracting natural gas from large deposits of shale. Shale gas was once just 2 percent of the US total energy production, but within a few years, it has quickly risen to 30 percent.[44]

This was an exciting development when news first reached public forums—the sheer amount of natural gas trapped in shale in North America alone has the ability to significantly alter US dependence on foreign oil and do so with an energy source that emits less GHG. President Barack Obama, in fact, declared that as much as a century's worth of shale gas was available in the United States. A government subcommittee studying the process noted that shale gas has enormous potential for economic and environmental benefits.

Action to take advantage of this new source of clean energy has been deliberate out of consideration for health impacts. Shale gas extraction is made possible by a process known informally as "fracking," or hydraulic fracturing. This process involves injecting highly pressurized fluid, such as water, creating new channels in the rock that can increase the extraction rates and ultimate recovery of fossil fuels. For shale gas, the typical formula of choice is a slurry of water, particles, and chemical additives.

The Department of Energy (DOE) is reviewing the process, including forming a Secretary of Energy Advisory Board (SEAB) Natural Gas Subcommittee, which has held public meetings and received reports on benefits and concerns. DOE researchers are considering possible pollution of drinking water from methane and chemicals used in fracturing fluids, air pollution, community disruption during shale gas production, and cumulative adverse impacts that intensive shale production can have on communities and ecosystems.

GHG Adaptation Opportunities

Adaptation actions for public health differ from those for mitigation in that they take place closer to the 2025 time frame we reference, but it remains critical that planners and stakeholders begin the necessary steps for implementing these activities immediately.

As a whole, they involve leveraging existing systems in ways that emphasize GHG reduction:

- Better planning for heat events
- Surveillance and early warning systems for health-related events
- Geographic organization of public health support and services for optimal response
- Preparing for contamination of drinking water and food caused by climate change–related events
- Recognizing and preparing for the impacts of climate change on mental health.

Change How We Deal with Heat Events

RECOMMENDATION
Anticipate acute events

Recommendation: *Define effective approaches for individual action and community support during heat and related air pollution events that adversely affect health.*

CHALLENGE

Heat is the number one weather-related killer—not hurricanes, tornadoes, or floods.[45] On average, more than 1,500 people in the United States die each year from excessive heat.[o] The *Intergovernmental Panel on Climate Change (IPCC) Fourth Assessment Report* summarizes the heat and weather effects of climate change:

> Since 1950, the number of heat waves has increased and widespread increases have occurred in the numbers of warm nights. The extent of regions affected by droughts has also increased as precipitation over land

[o] If you add up the average number of annual deaths from tornadoes, hurricanes, floods, and lightning over the past 30 years, you'd still fall short of the number of heat-related deaths.

> has marginally decreased while evaporation has increased due to warmer conditions. Generally, numbers of heavy daily precipitation events that lead to flooding have increased, but not everywhere. Tropical storm and hurricane frequencies vary considerably from year to year, but evidence suggests substantial increases in intensity and duration since the 1970s.[46]

A 2006 survey of people age 65 and older in four major North American cities showed that nearly all knew about impending heat events but remained unaware of the actions necessary to prevent heat illness.[47] Almost all had air conditioning but more than a third did not turn it on or run it for very long due to economic factors. Others actually may have increased risk by using fans to push air much hotter than body temperature over themselves. This shows a serious gap between the reach of public alerts and their effectiveness.

Heat waves are a high risk to the public because they are likely and have significant health consequences. People need to be prepared for the effects of these heat events, which include permanent disability and long-term medical care due to heat stroke and, of course, death. Yet, beyond their mentions in the media—which devotes regular, though relatively short attention to these events—many communities are unprepared to respond to heat waves and pollution events related to a rise in temperatures.

SOLUTION

Adaptation Potential	★★★
Operational Impacts	★★
Financial Impacts	★★
Feasibility and Timing	★★

A more robust effort to educate and inform the public can provide a strong basis for adapting to heat events. Existing literature can serve as the foundation for this effort. For example, CDC, the US Environmental Protection Agency (EPA), National Oceanic and Atmospheric Administration (NOAA), and Department of Homeland Security (DHS) Federal Emergency Management Agency (FEMA) jointly publish the *Excessive Heat Events Guidebook*,[48] a good starting point. This guidebook details preparedness and response to heat waves and summarizes comprehensive response programs that public health managers should develop.

Because of the severe heat wave experience of 2003 (previously discussed in relation to France), the World Health Organization (WHO) published guidance for public health responses for protection of at-risk populations.[49]

Continuous monitoring of each response activity is necessary to ensure they continue to match current community conditions. For example, home cooling in the mid-1960s was relatively rare; today, nearly all homes in the United States have installed some sort of cooling equipment.[50] With the wide availability of home cooling, local governments need to focus heat wave relief on those with no access to cooled buildings or the high-risk elderly who do not use their air conditioners because of economic concerns.

EXAMPLE[51]

In 1991, 20 people died in Philadelphia because of heat exposure, leading it to set up response policies and procedures. The city continues to refine policies each year, using lessons learned from each heat event, giving Philadelphia flexibility and institutional knowledge for tailoring its response to the predicted level of heat.

Its wide-ranging plan combines leveraged local partnerships with empowering individual citizens to act on behalf of their family and neighbors. Options for actions used include the following:

- Notification of local news media of heat events by the Philadelphia health commissioner and encouragement to broadcast information on how to minimize exposure to heat during the event
- Buddy system advocacy, reinforced through media coverage, including recommendations for friends, relatives, neighbors, and block captain volunteers to check on local high-risk residents (sick and older individuals) throughout the heat event
- Activating a HEATLINE phone system to take calls on heat-related questions
- Signage displaying the HEATLINE number, including an electronic billboard on top of the downtown Philadelphia Electric Company building visible over a large area
- Health department staff visits to homes identified from calls received on the HEATLINE
- Partnering with utility companies that service the city to reach agreements that they would not shut off electrical and water service for non-payment when the health department issued a high heat warning
- Increased emergency medical service staffing on duty for the duration of the high heat period, including fire department and emergency medical services personnel
- Increased outreach to the homeless to minimize their heat exposure, including the extension of shelter hours to include days
- Extending hours of operation at cooling shelters and senior refuges.

HEATLINE is no longer the exclusive domain of the dog days of summer. According to the Philadelphia Corporation for Aging's HEATLINE call center director, Chris Gallagher, the May 31, 2011, activation of Philadelphia's HEATLINE represents the earliest activation date in the history of the program.

What was the result of these efforts? A 2004 study analyzed the impact of publicizing heat events within the context of excess heat event–related deaths from 1995 to 1998.[52] The study found the system saved an estimated 117 lives over that period. It also looked into the program's cost-benefit framework, discovering these actions added only $10,000 in costs per heat event advisory day—primarily for the wages of the extra emergency medical staff.

Embrace Early Warning—For All Events

Recommendation: *Develop, deploy, and update surveillance and early warning systems for extreme weather, air pollution events, and vector-, food-, and waterborne diseases.*

Prepare warning systems

CHALLENGE

We do a very good job predicting, tracking, and warning people about things like storms, but coverage of other events that affect public health shows a major gap. For example, no uniform system is in place to provide local residents with sufficient warnings on vector-, food-, and waterborne diseases.

In the early 1990s, the response to the first cases of West Nile virus was relatively slow. As a result, the disease managed to gain a foothold in the native bird population, which now serves as a reservoir for the disease, allowing it to spread across the country.

Information isn't the problem: US and international partners are already researching and coordinating efforts regarding diseases. What we lack is a standard notification system for when these diseases emerge. Mandatory reporting of these illnesses to create early warning of new disease incursions is important, not only to begin treatment early but to control exposure to at-risk populations.

SOLUTION

Adaptation Potential	★★★
Operational Impacts	★★
Financial Impacts	★
Feasibility and Timing	★★

Government managers and stakeholders at all levels need to advocate integrating early warning systems for extreme weather, air pollution events, and vector-, food-, and waterborne diseases to make them consistent in delivery of potential hazard messages and personal safety steps.

The media has been a strong ally in these efforts, a relationship that should continue to be leveraged. Early warning of impending tornadoes, floods, high winds, bad air days, and heat waves is becoming commonplace in our urban areas via TV and radio news reports. This relationship can be expanded to include warnings on approaching events related to health that don't fit in with a station's weather coverage. By presenting information from credible sources in the same way you'd alert the media of an approaching storm, you can get the message out about vector-, food-, and waterborne diseases.

In addition to using the media, some local jurisdictions are also now using automated alert notification from their emergency management centers directly to residences to warn citizens about impending dangerous weather events. These should be expanded to account for other areas of concern.

Education messaging should help the public understand what's at stake and include tangible actions they can take. For example, although encouraging use of air conditioning might seem counter to standard environmental messaging because it increases energy use and GHG emissions, in the case of a local surge in West Nile virus cases, air conditioning can also protect users from vector contact. If this information were given to the public, it could act to protect itself.

Finally, a strategy without a standard way of implementing it is difficult to manage. Standardizing notification systems for all warnings will add legitimacy to the effort while presenting people with something familiar each time the system is deployed. Standardized warnings are familiar no matter where you are. Like road signs, people will recognize the warning's meaning regardless of the source.

However, these standardized warnings must be clear and concise. The terror alert system unveiled after September 11 was criticized for being vague and confusing.[53] A warning system based on a combination of symbols and colors should be flexible enough for varying circumstances, but simple enough not to confound the target audience.

EXAMPLES

The San Francisco BioWatch initiative, which has shown the ability to identify and track illnesses, is now being used as the basis for developing a regional response to disease outbreak. BioWatch is a federal government program for detecting the release of infectious agents into the air as part of a terrorist attack on major American cities. In San Francisco, the program collects and tests outdoor air samples from specialized air sampling devices mounted on outdoor monitors throughout the city. A confirmed positive result would initiate a major response from local, regional, state, and federal agencies. Depending on the biological threat, community clinicians may be asked to assist with local response.

The San Francisco Department of Public Health is now engaging regional partners to form a response through a BioWatch Advisory Committee. Such a program could be adapted to track and alert authorities of climate change–related disease outbreaks, air pollution, or other health events. It could even be tracked down to the neighborhood or building (or section of building) where the disease has been identified, allowing prevention strategies to begin at the source—and begin *immediately*. This capacity has been demonstrated during recent flu seasons.

Stage Health Services to Ensure They Get to Where They're Needed

Recommendation: *Organize public health support and services to better deploy vaccines and treatment regimens to public health providers where they're needed, considering changing regional geographic margins of vector-, food-, and waterborne and zoonotic diseases.*

CHALLENGE

Disease threat is the proverbial X factor. It's not contained by borders, and the things that do a good job keeping it localized—carriers and climate, for example—will be impacted by a changing environment. Vector-, food-, and waterborne and zoonotic diseases are not currently significant causes of illness or death in the United States, but have been during several periods in our history. Stakeholders must rely on lessons learned to ensure these diseases are held in check.

Only dedicated efforts at primary prevention—especially through indirect, community-level services—allowed us to greatly reduce their incidence; unfortunately, future incidence of these diseases can increase rapidly. For many preventable communicable diseases, the US population has had no exposure, so many people have little or no personal resistance. Many of these diseases occur more frequently in warmer areas, which are likely to expand with climate change. Failure to recognize the worldwide reservoirs of these diseases that can arrive in the United States via personal or business travel and via air, sea, or land shipment of import items can lead to infections that will spread to people without immunity in the presence of the right causative agents and vectors.

Public health programs, especially those in regions becoming warmer, will need to be vigilant and maintain the ability to react quickly.

Public health managers must prepare to predict, identify, and respond to disease outbreaks to contain them and produce as little illness as possible.

Adaptation Potential ★★
Operational Impacts ★
Financial Impacts ★
Feasibility and Timing ★★

SOLUTION

Public managers must engage partners to create robust tracking and response capabilities. They can take several steps to address regional changes and remain flexible for response purposes:

- Maintain surveillance, once it has begun, for emerging vector-borne, waterborne, and food-related disease infections in an area. Develop or enhance health and disease data collection for weather and air pollution events. Have surveillance teams incorporate predictive data for these diseases on the basis of temperatures, precipitation, and vector populations.
- Ensure that strategic information from the surveillance and data collection reaches the appropriate local operators.
- Deploy rapid diagnostic tests for vector-borne, waterborne, and food-related diseases to clinics and healthcare providers.
- Develop and implement effective integrated disease vector control techniques, targeting the vectors and ecology of the local area. Establish land-use policies to help control disease vector habitats to help in this effort and adjust them to allow for this type of action.
- Develop best practices for individual disease vector avoidance and communicate them to the public to prepare local populations to perform personal countermeasures.
- Develop *and enforce* food safety, drinking water safety, and clean air requirements.
- Develop, communicate, and enforce restrictions on the placement of sources of infection near food or water supplies.

Stakeholders and managers must also advocate funding to support these activities. Federal or state public health programs need to fund and support the following:

- Development of rapid diagnostic tools for vector-, food-, and waterborne diseases
- Development of vaccines and other clinical preventive measures for these diseases
- Expansion of professional development and training activities to increase the knowledge of the public health and clinical care communities—specifically where diseases and health conditions are at risk of change along with the climate.

EXAMPLE

We have already been slow to act in stemming a disease's spread. When the West Nile virus emerged in New York City in 1999, recognizing the threat and determining that the outbreak was in the local bird population took a few weeks. By then, it had spread to surrounding areas. Regional authorities eventually used aerial insecticide spraying to reduce mosquito populations in the area, but the West Nile virus continued to spread outside the spraying zones.

Today, the virus is present in bird populations across the country, and cases appear every year during mosquito season. Since first appearing in 1999, West Nile has reached the entire continental United States through human, bird, or mosquito activity (Figure 2-4). Local mosquito control appears to lower incidence, but areas that have not normally controlled mosquitoes because of low mosquito populations or funding constraints are at higher risk each year.

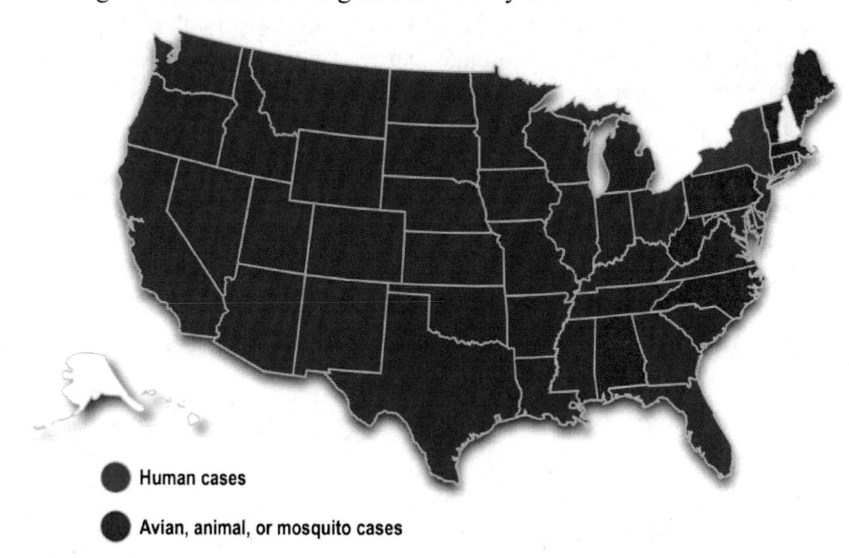

Figure 2-4. Since 1999, West Nile Virus Has Spread Across the United States

Prepare for Food and Water Contamination

Recommendation: *Identify and address the potential for drinking water and food contamination from flooding, sea level rise, or drought due to climate change.*

CHALLENGE

As the world heats up, atmospheric water vapor increases, and we're seeing the effects of more water falling from the sky. It differs regionally, but overall, average precipitation in the United States has increased by 5 percent over the last 50 years, and the amount of rain that falls during the heaviest downpours has increased by approximately 20 percent over the past century.[54]

Obviously, this can produce localized flooding (and direct injury and death from drowning). The indirect effects—from uncontrolled sewage treatment plant runoff, polluted drinking water, and contaminated food crops and shellfish—are also serious. They lead to gastrointestinal disease from widespread exposure to disease-causing bacteria and viral organisms in our water and food.

All stages of potable water production, availability, and use have public health concerns, especially related to volume and quality. Utilities that place potable water production facilities close to surface water features sometimes do so without considering the risk of flooding during heavy rain or the effect of upstream runoff or storm surges. This isn't intentional—most projects consider historical records for precipitation during the planning and development phases. The problem is that climate change carries the possibility of record-shattering weather events.

In areas of climate change–induced drought, low water flows can increase the concentration of chemical or biological contamination to levels that a water treatment plant cannot handle. Although increasing or protecting wetlands can help guard against storm damage and provide filters to clean surface waters, these wetlands can also serve as reservoirs of infectious agents, becoming a source of downstream contamination and placing an additional burden on water treatment activities.

SOLUTION

Adaptation Potential	★★
Operational Impacts	★★
Financial Impacts	★★
Feasibility and Timing	★★★

Public health managers must identify facilities at risk of flooding from stream overflow, levee failure (accidental or intentional), or sea level rise, especially if they are drinking water production and sewage treatment plants. Stakeholders should advocate the relocation of those at risk to areas that are less prone to flooding (before beginning major upgrades or additions to lessen the financial burden).

Facilities that remain at risk of flooding are inevitable, if only because of the resources and process required to relocate them. Public health managers should coordinate with the facility managers to ensure they are protected from minor floods. They should also prepare plans to secure alternative sources of potable water while responding to drinking water contamination events or sewage exposed during major flooding. The EPA has developed pre- and post-hurricane checklists for water and wastewater facilities that can be immediately adapted and exercised on a local level.[55]

In 2011, the city of Omaha took proactive steps to protect its wastewater treatment plant from the rising waters of the Missouri River. The resulting levee protects $100 million worth of city assets at the plant against flood stages up to 42 feet.

In the event of flooding, public health managers must have a plan for collecting and testing for contamination in land food crops, shellfish, and other seafood. Any such plan must emphasize to surrounding communities the hazards of consuming contaminated foods. They should also educate the public on personal responses that will help protect the community at large.

Public health managers should also identify water treatment plants at risk from supply source contamination by chemicals concentrated by drought or from infectious agents in surface water. Stakeholders should identify the potential chemical or biological contaminants and prepare and institute targeted sampling plans, including alerting operators and the public of positive results.

EXAMPLES

In EPA's *Security Information Collaboratives: A Guide for Water Utilities*,[56] the agency reports on the collaboration between a utility, Milwaukee Water Works, and the City of Milwaukee Health Department. After an outbreak of cryptosporidiosis in April 1993, and continuing to this day, the mayor directs a formal collaboration of stakeholders: officials from the water utility, the public works department, and the health department, as well as representatives of local and state government agencies. The Wisconsin Department of Natural Resources (the state's water quality regulatory agency) and local sewerage districts are also involved. City officials recognize that routine communication between the water utility and public health department is essential for ensuring the prevention of future waterborne disease outbreaks. A working committee meets every month and is on call for situations that emerge. The regular meetings include an update on water quality, reports on other water treatment projects (including capital improvement construction projects), and an update from the health department; this update covers disease trends, laboratory capacity, and environmental matters such as watershed influences, source water quality, and surface water testing and monitoring.

Stressing Mental Health

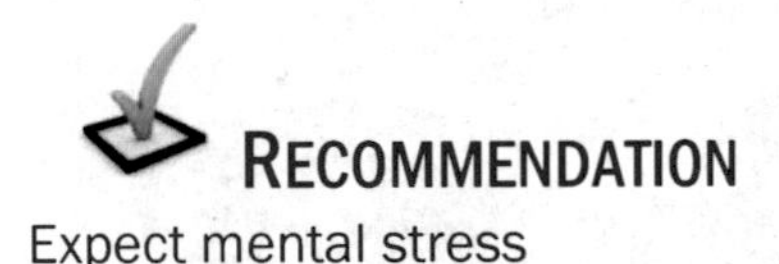

RECOMMENDATION
Expect mental stress

Recommendation: *Prepare for community-wide mental health issues related to stress from climate change outcomes.*

CHALLENGE

Mental illness is pervasive in society, especially in the United States.[57] A 2005 study found that one in four Americans meets the criteria for having a mental illness, and these ailments disrupt the ability of many to function day to day.[58] Like all times of change and discomfort, the adverse effects of climate change will undermine mental health within a community, unsettling social, economic, and environmental aspects that either exacerbate existing disorders or lead to the creation of new ones.

A 2011 study by Australian researchers found that several years of continued catastrophic weather events on that continent have resulted in anxiety and insecurity in children,[59] the likes of which have not been seen since the Cold War.[p] These events are also cited as contributing to an overall rise in depression, anxiety, substance abuse, suicide, and self-harm, with one in five in a community prone to suffer from extreme stress, emotional injury, or despair.

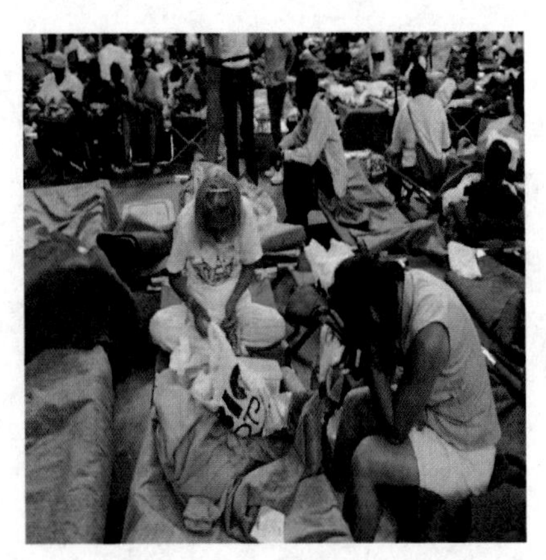

Hurricane Katrina survivors at the Red Cross shelter inside the Astrodome in Houston.

Of concern are the increased anxiety, emotional stress, and physiological responses brought on in the wake of extreme weather events, prolonged heat or cold periods, and air pollution episodes. Secondary events, such as the elimination of access to adequate food and water and job loss, are already documented stressors.

This is especially worrisome in already vulnerable populations, where the addition of mental health stressors can add to problems of existing medical conditions, including cardiovascular disease and diabetes. Stress leads to anxiety disorders, depression, sleep difficulty, social avoidance, and drug and alcohol abuse, resulting in large groups who need care.

---

[p] Australia has been devastated by a number of extreme weather events and natural disasters, including floods, cyclones, drought, and fires.

A final consideration is the long recovery time associated with extreme weather events—we already discussed displacement from drought and hurricanes—and for areas of migration. This requires consideration as to how long it can take to regain a normal life. Relocating means having to find work, obtain adequate food and water, and get relief from prolonged heat and cold. Many times, the movement is from rural to urban locations, which can result in overcrowding and increases in the transmission of communicable diseases. We saw this in Haiti after the earthquake of 2010, where cases of cholera were documented in significant numbers.

The exposure to disease and continued inability to access safe food and water increase stress and mental health issues, for example, the doubling of mental illness that occurred post–Hurricane Katrina in adults from two census divisions affected by the hurricane.[60] Public managers need to understand that these factors related to the effects of climate change will increase levels of personal and community stress over the next 10 to 15 years.

SOLUTION

Managers need to ensure adequate community health services with enough capacity to handle a wide array of mental health challenges. Managers in urban locations should consider how they will assist a population explosion due to migrating groups arriving in a community—and how their needs stress community resources already taxed by climate change effects.

Adaptation Potential ★★
Operational Impacts ★
Financial Impacts ★
Feasibility and Timing ★★

Managers need to develop a communication strategy tailored to the population they serve that emphasizes the expected physical and mental health effects. Stakeholders can help develop a robust plan for educating the public on the services and resources available to address those effects. Similarly, managers must assist their constituents in physically preparing for possible utility outages and evacuations during severe weather events.

EXAMPLES

A 2008 study on the impact of climate change by the Western Australia Department of Health presented a scenario in which farmers encountered the direct and indirect health impacts of a temperature increase in the year 2030,[61] one of which was an increase in mental health issues.

Under this scenario, they found several realistic concerns related to mental health, including psychological stress of loss of home and sense of community, traumatic stress from experiencing an extreme event, and an absence of feeling safe from surrounding civil unrest.

In response, the Department of Health developed a comprehensive plan for adapting to the mental health impacts brought on by climate change. It offers specific policy recommendations to address concerns in dislocation, community, lifestyle, and economic activity, including the following:

- Legislative and regulatory reform
- Public education and communication
- Surveillance and monitoring
- Ecosystem intervention
- Infrastructure development
- Health intervention.

In its findings, it noted a general lack of research into the social effects of climate change.

Final Thoughts

Climate change is altering our environment in ways that will affect human health for decades. Some change is due to our use of technology and machines to make life better. While technology is partially to blame, it also offers an opportunity for action. We can use technology and science to increase our understanding of the interrelationships of climate change, proposed responses, and how human health relates to mitigation and adaptation techniques.

Government managers and stakeholders have a responsibility to further define the issues and develop policies and plans that ensure future generations will continue to enjoy improving health. Over the next 15 years, we must undertake initiatives to identify the greatest climate change–related risks to the population. We need to prepare to implement policies and programs to reduce the probability of harm to that population. The mitigation and adaptation challenges and possible solutions described above are a starting point for tailoring local responses to the international challenges of climate change with regard to health.

---

[1] Edward W. Maibach and others, "Climate Change and Local Public Health in the United States: Preparedness, Programs and Perceptions of Local Pub-

lic Health Department Directors," *PLoS ONE* 2008; 3(7): e2838. doi:10.1371/journal.pone.0002838.
[2] Anthony McMichael and others, eds., World Health Organization (WHO), *Climate Change and Human Health—Risks and Responses* (Geneva: WHO, 2003).
[3] Grayson Vincent and Victoria Velkoff, US Department of Commerce (DOC), US Census Bureau, *The Next Four Decades, The Older Population in the United States: 2010 to 2050, Population Estimates and Projections*, P25-1138 (May 2010).
[4] OSHA standards and guidance on preventing illness from heat stress can be found at http://www.osha.gov/SLTC/heatstress/ (website as of October 12, 2011).
[5] Mercedes Medina-Ramon and Joel Schwartz, "Temperature, Temperature Extremes and Mortality: A Study of Acclimatization and Effect Modification in 50 US Cities," *Occupational and Environmental Medicine*, Vol. 64, No. 12 (June 2007), p. 827–833.
[6] National Drought Mitigation Center, "Drought in the Dust Bowl Years," *Drought Basics*, http://www.drought.unl.edu/DroughtBasics/DustBowl/DroughtintheDustBowlYears.aspx.
[7] Donald Worster, *Dust Bowl: The Southern Plains in the 1930s* (New York: Oxford University Press, 1979).
[8] Martin Parry and others, Working Group II to the Fourth Assessment Report of the Intergovernmental Panel on Climate Change (IPCC), *Climate Change 2007: Impacts, Adaptation and Vulnerability: Contribution of Working Group II to the Fourth Assessment Report of the Intergovernmental Panel on Climate Change* (Cambridge: Cambridge University Press, 2007), para 8.2.3.
[9] Ibid., para 8.2.6.
[10] Kristie Ebi and others, US Environmental Protection Agency (EPA), US Climate Change Science Program and the Subcommittee on Global Change Research, *Analyses of the Effects of Global Change on Human Health and Welfare and Human Systems* (Washington, DC: US EPA 2008), para 2.3.5.
[11] US EPA, "Fine Particle ($PM_{2.5}$) Designations," *Frequent Questions*, http://www.epa.gov/pmdesignations/faq.htm.
[12] Arden Pope and others, "Lung Cancer, cardiopulmonary mortality, and long-term exposure to fine particulate air pollution," *American Medical Association*, Vol. 287, No. 9 (March 2002), p. 1132–1141.
[13] Environmental Health Perspectives and the National Institute of Environmental Health Sciences, *A Human Health Perspective on Climate Change: A Report Outlining the Research Needs on the Human Health Effects of Climate Change* (Research Triangle Park, NC: 2010).
[14] James McCarthy and others, Working Group II to the Third Assessment Report of the IPCC, *Climate Change 2001: Impacts, Adaptation, and Vulnerability: Contribution of Working Group II to the Third Assessment Report of the Intergovernmental Panel on Climate Change* (Cambridge: Cambridge University Press, 2001), para 7.2.2.3.3.

[15] US EPA, *Heat Island Impacts*, http://www.epa.gov/heatisld/impacts/index.htm.
[16] Lewis Ziska and others, "Recent warming by latitude associated with increased length of ragweed pollen season in central North America," *Proceedings of the National Academies of Science (PNAS)*, Vol. 108, No. 10 (Washington, DC: PNAS, 2011).
[17] Anthony McMichael and others, eds., *Climate Change and Human Health—Risks and Responses* (Geneva: WHO, 2003), pp. 11–13 and 159–180.
[18] Martin Parry and others, *Climate Change 2007: Impacts, Adaptation and Vulnerability: Contribution of Working Group II to the Fourth Assessment Report of the Intergovernmental Panel on Climate Change*, para 8.2.2.
[19] D. Michal Freedman and others, "Prospective study of serum Vitamin D and cancer mortality in the United States," *Journal of the National Cancer Institute*, Vol. 99, No. 21 (2007), pp. 1594–1602.
[20] Jason Samenow, "Holy hot river! Potomac water temperature highest on record last week," *Capital Weather Gang*, http://www.washingtonpost.com/blogs/capital-weather-gang/post/holy-hot-river-potomac-water-temperature-highest-on-record-last-week/2011/07/27/gIQAwcVJdI_blog.html.
[21] Martin Parry and others, *Climate Change 2007: Impacts, Adaptation and Vulnerability: Contribution of Working Group II to the Fourth Assessment Report of the Intergovernmental Panel on Climate Change*, para 8.2.5.
[22] Ibid., para 8.2.4.
[23] US Department of Health and Human Services (DHHS), Centers for Disease Control and Prevention (CDC), *National Center for Emerging and Zoonotic Infectious Diseases*, http://www.cdc.gov/ncezid/.
[24] Suzanne Austin Alchon, *A Pest in the Land: New World Epidemics in a Global Perspective* (Albuquerque: University of New Mexico Press, 2003), p. 21.
[25] US DHHS, CDC, National Center for Chronic Disease Prevention and Health Promotion, "The Link Between Physical Activity and Morbidity and Mortality," *Physical Activity and Health: A Report of the Surgeon General* (Atlanta, GA: US DHHS, 1996).
[26] Health Impact Project, *Case Study: Humboldt County, CA*, http://www.healthimpactproject.org/resources/case-study-humboldt-county-california.
[27] Health Impact Project, *Case Study: Healthy Families Act: Paid Sick Day*, http://www.healthimpactproject.org/resources/case-study-healthy-families-act-paid-sick-days.
[28] US DHHS, "Telehealth," *Health Resources and Services Administration, Rural Health*, http://www.hrsa.gov/ruralhealth/about/telehealth/.
[29] American Telemedicine Association, *Telemedicine Defined*, http://www.americantelemed.org/i4a/pages/index.cfm?pageid=3333.
[30] Tom Webster, "The Social Habit – Frequent Social Networkers," *Perspectives, News & Opinions from the Researchers at Edison*, June 17, 2010.

[31] Muhammad Saleem, "Visualizing 6 Years of Facebook," *Mashable Social Media*, February 10, 2010.
[32] Anthony Smith, Victor Patterson, and Richard Scott, "Reducing your carbon footprint: How telemedicine helps," *British Medical Journal*, November 24, 2007.
[33] Center for Connected Health, "Sunscreen Adherence," *mHealth*, http://www.connected-health.org/programs/mhealth/center-for-connected-health-initiatives/sunscreen-adherence.aspx.
[34] April Armstrong and others, "Text-Message Reminders to Improve Sunscreen Use: A Randomized, Controlled Trial Using Electronic Monitoring, *Archives of Dermatology*, Vol. 145, No. 11 (2009), p. 1230–1236.
[35] US EPA, *Heat Island Effect*, http://www.epa.gov/heatisld/index.htm.
[36] US EPA, *Reducing Urban Heat Islands: Compendium of Strategies* (October 2008).
[37] City of Chicago, "Green Buildings, Roofs and Homes," *Department of Environment*, http://www.cityofchicago.org/city/en/depts/doe/provdrs/green.html.
[38] City of Chicago, "City Hall Rooftop Garden," *Green Roofs*, http://explorechicago.org/city/en/about_the_city/green_chicago/Green_Roofs_.html.
[39] The Henry Ford, "The Living Roof," *Ford Rouge Factory: LEED*, http://www.hfmgv.org/rouge/leedlivingroof.aspx.
[40] US Department of Energy (DOE), Energy Information Administration (EIA), "Table C1: Total Energy Consumption by Major Fuel," *2003 Commercial Buildings Energy Consumption Survey (CBECS) Detailed Tables* (released September 2008). We estimated district heat energy use as major fuels minus the sum of all other major energy sources.
[41] See US Department of Energy (DOE), Energy Information Administration (EIA), "Table E2: Major Fuel Consumption (Btu) Intensities by End Use for Non-Mall Buildings, *2003 Commercial Buildings Energy Consumption Survey (CBECS) Detailed Tables* (released September 2008).
[42] Practice Greenhealth, "Best Practices in Carbon Mitigation," *Energy, Water & Climate: Climate*, http://practicegreenhealth.org/topics/energy-water-and-climate/climate/best-practices-carbon-mitigation.
[43] Lars Thording, Senior Director, Marketing & Public Affairs, Stryker Sustainability Solutions, personal communication with the author, October 19, 2011.
[44] US DOE, Secretary of Energy Advisory Board (SEAB) Natural Gas Subcommittee, *The SEAB Shale Gas Production Subcommittee Ninety-Day Report* (Washington, DC: August 2011).
[45] US DOC, National Oceanic and Atmospheric Administration (NOAA), *Heat Wave: A Major Summer Killer,* http://www.noaawatch.gov/themes/heat.php.
[46] Martin Parry and others, *Climate Change 2007: Impacts, Adaptation and Vulnerability: Contribution of Working Group II to the Fourth Assessment Report of the Intergovernmental Panel on Climate Change*, p. 1000.

[47] Scott Sheridan, "A Survey of Public Perception and Response to Heat Warnings across Four North American Cities: An Evaluation of Municipal Effectiveness," *International Journal of Biometeorology*, Vol. 52 (September 2006), p. 3–15.
[48] US EPA, Office of Atmospheric Programs, *Excessive Heat Events Guidebook*, EPA 430-B-06-005, June 2006.
[49] WHO Europe, *Improving Public Health Responses to Extreme Weather /Heat-Waves – EuroHEAT Meeting Report: Bonn, Germany, March 22–23, 2007* (Copenhagen: 2008).
[50] The latest data from the Energy Information Administration (EIA) shows that 84 percent of all homes have air conditioning. See US DOE, EIA, "Table HC2.7, Air Conditioning Usage Indicators by Type of Housing Unit, 2005," *2005 Residential Energy Consumption Survey: Preliminary Housing Characteristics*.
[51] US EPA, Office of Atmospheric Programs, *Excessive Heat Events Guidebook*, EPA 430-B-06-005, June 2006.
[52] Kristie Ebi and others, "Heat watch/warning systems save lives: Estimated Costs and Benefits for Philadelphia," *American Meteorological Society*, Vol. 85, No. 8 (August 2004), p. 1067–1073.
[53]Scott Neuman, " New Terror Alert System Aims For Clarity, Not Color," NPR, April 20, 2011.
[54] US Global Change Research Program, "National Climate Change," *Global Climate Change Impacts in the US*, http://www.globalchange.gov/publications/reports/scientific-assessments/us-impacts/full-report/national-climate-change.
[55] US EPA, "Suggested pre-hurricane activities for water and wastewater facilities" and "Suggested post-hurricane activities for water and wastewater facilities," *Emergency Activities for Water and Wastewater Facilities*, http://water.epa.gov/infrastructure/watersecurity/emergencyinfo/.
[56] US EPA, Office of Research and Development, *Security Information Collaboratives: A Guide for Water Utilities*, EPA/625/R-05/002, May 2005.
[57] Rick Weiss, "Study: US Leads in Mental Illness, Lags in Treatment," *The Washington Post,* June 7, 2005.
[58] Harvard School of Medicine, *National Comorbidity Survey (NCS) and National Comorbidity Survey Replication (NCS-R)*, http://www.hcp.med.harvard.edu/ncs/.
[59] The Climate Institute, *A Climate of Suffering: The Real Costs of Living with Inaction on Climate Change* (Melbourne & Sydney: The Climate Institute, 2011).
[60] Ronald Kessler and others, "Hurricane Katrina Community Advisory Group, Mental Illness and Suicidality after Hurricane Katrina," *Bulletin of the World Health Organization*, Vol. 84 (August 2006), p. 930–939.
[61] Jeff Spickett, Helen Brown, and Dianne Katscherian, *Health Impacts of Climate Change: Adaptation Strategies for Western Australia* (Perth: Department of Health, 2008).

Information

As the storm cleared the area, the city, quieted by the event, reawakened. After the evacuations, not many were left to ride out the storm, but it was time to clean up. One lingering effect of the wind, rain, and surging waves, however, was that most businesses remained closed—especially those that relied on network access to function. The bay area had been hardest hit by the storm, yet one tiny Internet start-up near the bay was still up and running. That company, which developed software designed to help data centers improve their energy efficiency, was back in business immediately. Early on, it had invested in a mobile data center for its work. Now, as the city's business community took stock of the damage to its networks, this start-up was already back at work—its network hub, after all, was located far away from the damaging weather.

Now Transmitting

Communication and access to information dominate who we are and how we operate personally, professionally, and socially. How and why we communicate form the basis of everything we do, and the effectiveness of this communication can make or break any initiative, large or small.

Communication has always been driven by technology—the printing press kick-started the Renaissance and is a verified "agent of change." Today, we operate in a world dominated by another equally transformational communication device: the Internet.

The baseline for communication technology in the United States today is access to a computer, specifically one that's Internet enabled.

Many people have Internet access through mobile devices, and smart phones are increasingly saturating the market. Tablet computers, like Apple's iPad, are finding their way into business and education and are bridging the gap between the laptop computer and the smart phone. E-mail and text messaging are household terms, as is social networking, a communication method for many, at least on a casual basis.

All of this communication is made possible by information and communications technologies (ICTs). We use the term ICT because it represents a more holistic view of how technology is used in enterprise today. For example, social media is as much about a new way of communicating as it is about new technology.

Today, ICT pervades nearly every aspect of the world around us. The pace at which new ICT capabilities are being introduced is phenomenal. Organizations that succeed in adapting ICTs to their missions are ones who embrace the field as "emerging technology," weighing each new option for what it brings to the table and how worthy of investment it is.

That element of emergence is important. Today's "geek speak" is tomorrow's regular vernacular. Consider the following: cloud computing, mashups, web services, social media, context-aware computing, apps (mobile applications), semantic web, and geographic information system (GIS). Some of these terms have already reached our day-to-day conversations, and if the rest haven't by the time you're reading this, they soon will. Each of these terms represents a significant piece of the ICT universe.

Most ICTs leverage the Internet, a global information and communications platform. ICTs are becoming integral parts of our daily lives, and they, along with their descendants, will profoundly affect the way we live and work.

This is already evident in how inextricably linked ICTs are to our personal and professional lives. They shape how we present ourselves and our work on a daily basis. For example, few organizations today operate without e-mail. The first e-mail was sent in 1971, and by 2010, the number of e-mails had grown to almost 300 million *per day*. Following in the footsteps of e-mail is social media, which is now an ICT-enabled service ubiquitous in modern culture.

Appetite for Consumption

In August 2011, we visited Facebook more than 162 million times,[1] and we tweeted more than 6 billion times. By the time you read this, these numbers will be much larger. This behavior reflects our growing appetite for ICT, but it comes at a price. The energy required to satisfy our hunger for technology is enormous, and the resultant carbon emissions, while hidden from view, are equally tremendous.

As ICT became a central part of modern enterprise, the green computing movement helped promote business models that focus on the efficient use of computing resources. This focus allows organizations to minimize these environmental impacts and reduce costs.

Large organizations face major hurdles on the road to understanding the best ways to reduce energy consumption across their asset portfolio. Such efforts require extensive information management capabilities. Only when sufficient data are gathered and aggregated can expert analyses be performed. From the resulting data, intelligent decisions may be made, whether strategic, tactical, or policy related. The effective use of information technology (IT) makes such an endeavor possible, not only allowing data to be aggregated but also to be used in a meaningful way.

ICTs also have the potential to curtail our habits. They can enable energy use that is data driven and smart. They can substantially aid the mitigation of and adaptation to our changing climate.

New York City (NYC), for example, has established a Sustainable Energy Property Tracking System to help city decision makers compare various facilities to identify the ones that use resources most efficiently. The information garnered from this system allows the city to prioritize energy-efficiency investment decisions (Figure 3-1).

ICTs are expected to have profound impacts on an organization's energy consumption and carbon emissions in the years ahead. This will happen in a multitude of ways. Stakeholders need answers to a few simple questions.

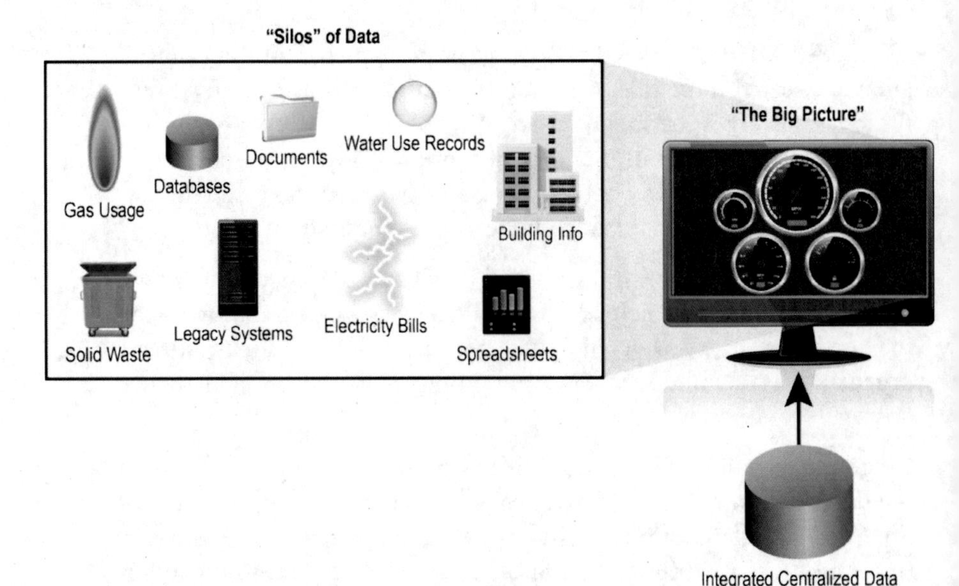

Figure 3-1. New York City's Challenge of Integrating Energy Data

- How can fusing data from otherwise disparate sources help my organization make smart energy decisions to reduce greenhouse gas (GHG) emissions?
- Can mobile computing and social media help climate change mitigation and adaption efforts?
- Can virtual offices and teleworking lower my organization's energy needs (and costs!) and carbon footprint?[a]
- What systems and applications already illustrate the use of IT to help understand, plan for, and mitigate climate change?
- How can I optimize the use of my ICT resources to improve efficiency and reduce the overall ICT carbon footprint?

What *Is* the Carbon Impact of the ICT Industry?

Consider two examples of ICT and its relationship to carbon emissions. The first example is global, dealing with the ICT industry as a whole.

In 2008, The Climate Group, a nonprofit organization, along with the Global e-Sustainability Initiative, issued a study highlighting the significant and rapidly growing footprint of the ICT industry.[2] The study

[a] A virtual office is a nonphysical workspace (or a shared, hoteling space) in which interconnected computer and communication services are available to workers.

predicted that by 2020, emissions from ICT sources will increase by 600 million metric tons of CO_2 equivalent because of rapid economic expansion in places like India, China, and elsewhere. This increase is more than *70 percent* (Table 3-1).

Table 3-1. Projected Increases in World and ICT GHG Emissions, 2007-2020[3]

World	+ 72.3%
Server farms/data centers	+ 121.6%
Telecoms' infrastructure and devices	+ 16.6%
PCs and peripherals	+ 100.2%

ICT emissions growth is driven by the fast rise of server farms and data centers. Also, as personal computers and peripherals spread worldwide, even in developing countries, emissions from equipment are expected to double by 2020.

Quick Search, Lasting Impact

The other example of how ICT relates to carbon emissions is the opposite of global—in fact, it's as local as you could imagine, since you probably participated in this activity very recently.

But for the sake of not assuming, imagine you're at a restaurant with friends. A song is playing on the juke box and a debate is in full swing over the year this particular pop song came out. Your companion pulls out his phone and Googles the answer. The exchange lasts no more than 45 seconds, but have you ever wondered what the carbon footprint of that Google search is?

On average, Americans perform one Google search per day (though if you're reading this book, you likely average much more than that). Google, via its blog,[4] states that one Google search is equivalent to about 0.2 grams of CO_2 and that a thousand Google searches produce as much CO_2 as driving the average car for 0.6 miles. If you do the math, considering searches in a single month—August 2011, for example—which amounted to 14.6 billion (according to digital business analytics firm comScore),[5] the effect is equivalent to 8.7 million car miles, or what 645 drivers would otherwise produce on their own in *an entire year*.[6]

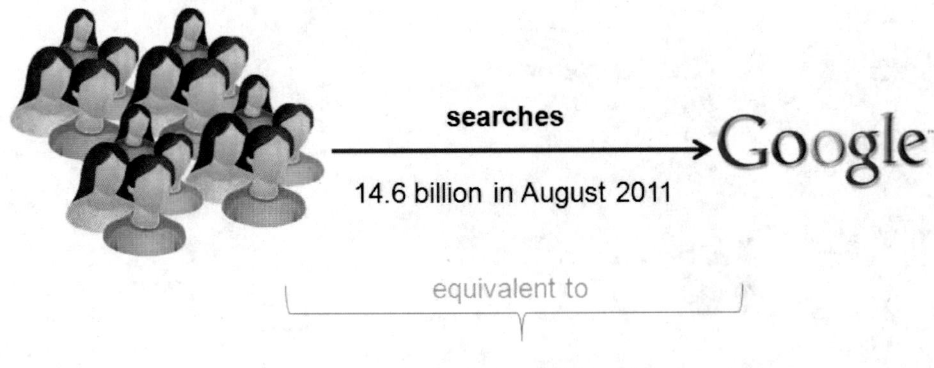

Part of this effect is a result of what Google provides: a variety of possibilities, and quickly. To deliver results to its users almost instantly, Google maintains vast data centers around the world, each packed with powerful computers. These computers, in addition to needing huge amounts of energy that produces lots of CO_2, emit a great deal of heat. As a result, these data centers need significant amounts of cooling power, which uses even more energy.

Google is perhaps the most widely used search engine, but it is by no means the only one. And the Internet search business is, of course, just one segment of an increasingly vast array of businesses relying on data centers. The carbon emissions associated with just this part of the ICT landscape—data centers and cloud computing—is beyond significant. (Both are discussed at length later.)

Fortunately, ICT yields opportunities to mitigate and adapt to climate change and offers a means for altering the relationship between economic growth and energy consumption. The relatively strong economy associated with web-based commerce and services, mobile applications, and social media are proof of this. The carbon footprint associated with the ICT industry is comparatively low—especially next to industrial activity such as aluminum or steel production. Low, however, does not equate to zero. The explosive growth of ICT makes figuring out a reduction strategy a necessity.

The ability to process and manage the large amounts of data that will help facilitate reduction strategies requires complex software and big databases (and processes to manage them). These, in turn, require an extensive physical computing infrastructure to function. The reliance on computing power throughout every sector of the economy means that the purchase, use, and disposal of computing devices of all sorts can have a significant impact on carbon emissions.

The term "green computing" has gained momentum over the past several years through initiatives such as the Federal Data Center Consolidation Initiative.[7] In 2010, the American Recovery and Reinvestment Act, by allocating more than $90 billion for investment in green initiatives (such as renewable energy, smart grids, and energy efficiency), helped raise public awareness of the importance of reducing energy consumption and emissions. It also began linking this effort to economic gains.

ICT and Mitigation

In the area of climate change mitigation, ICTs will have the most impact through their ability to enable prompt decisions and strategies based on data-driven analysis. Well-performing information systems will allow stakeholders to manage their organizations more precisely and will enable the achievement of goals that once would have been viewed as too difficult to even attempt.

Plainly put, better access to data will allow informed decisions with respect to carbon emissions. It is critical, however, to consider and weigh the energy used and carbon emissions associated with this increased reliance on computing.

The opportunities for GHG mitigation include the following:

- Explore the virtues of virtual.
- Become lean to be green.
- See whether elastic computing fits.
- Promote green buildings with data-driven energy use.
- Prepare for smart things and big data.

Explore the Virtues of Virtual

Recommendation: *Pursue a total carbon emissions reduction by increasing your organization's remote-location work and identifying the technical solutions appropriate for such an expansion.*

CHALLENGE

Businesses and agencies are under constant pressure to find ways to increase productivity while reducing costs. Many organizations are also looking for ways to decrease their environmental impact, specifically their GHG emissions. At the core, we're all simply looking for ways to decrease our energy bills.

Technology now lets us be more mobile, but the work is also different and relies less on being watched and more on being connected. Telework fits that model.

—Martha N. Johnson, GSA Administrator, speaking on October 7, 2010, at Telework Exchange Fall Town Hall Meeting in Washington, DC

Historically, businesses operated with an employee commuting to the workplace, generally located in a shop, lab, or office. These workers stored records in paper form and then commuted home again. This is an engrained system full of inefficiencies with many associated emissions.

Take just people's time, for example. Real hourly wages in the United States continue to rise, implying that time is becoming an ever more valuable resource and its loss ever more wasteful. At the same time, road congestion is a constant, if not growing, problem, particularly during employee commute hours. In fact, large amounts of time are wasted dealing with traffic just getting to and from the workplace.

According to one source, traffic congestion cost the United States about $115 billion in 2009, as the average urban driver spent about 4 days in traffic.[8] The challenge is how to use ICT to overcome this waste of people's time and productivity—while also ensuring that productivity at a remote/home office remains high.

Similarly, the real-estate market in the United States constantly fluctuates, but the usual pattern is for land and building costs to rise over time. This rise occurs because our population continues to expand and more and more people join the labor force. Organizations must find ways to reduce their reliance on floor and parking space.

The Internet is now the primary communications medium between knowledge workers (people valued for specialized subject matter expertise such as scientists and teachers), who routinely share ideas and content (such as documents) instantaneously with others located elsewhere. In the coming years, this trend will profoundly affect commuting patterns and the nature of how teams work together and share information.

Further, technologies continue to advance. The mobile web is replacing desktop and browser-centric computing models, providing a universal platform for application use and content delivery. As this occurs, using physical space to assemble coworkers becomes less and less critical.

Even if organizations embrace remote work environments, employees engaged in telecommuting, file sharing, and electronic filing might still come to the office frequently and continue storing hard copies of certain documents or books.

SOLUTION

Mitigation Potential	★★★
Operational Impacts	★
Financial Impacts	★★★
Feasibility and Timing	★★★

As of January 2010, 55 percent of American adults were able to connect to the Internet wirelessly, and statistics indicate that number is steadily rising. This fact, along with new technology, presents a golden opportunity for organizations to embrace remote workplaces.

Advances in technology have given organizations new and reliable ways to access and develop paper-free content. Businesses are able to connect people virtually, reducing emissions associated with document creation and transportation, including commuting.

In some cases, paper records and commuting may continue to be essential. But to a surprisingly great degree, organizations are finding they can save resources—office, parking, and filing space and employee time—through the use of virtual technology.

Today, an employee's office can be reproduced on his or her computer no matter where the employee happens to be. Files are available instantly from servers and can be accessed and worked on (sometimes collaboratively) at any time. This eliminates the need for paper. Even if employees are not working from home, they can be in satellite offices smaller and cheaper than a single, centrally located building. This arrangement reduces office space and commuting-related wear and tear. Meetings between remotely located workers are made possible through visualization technologies that enable participants to see the same things at the same time.

Teleconferencing can reduce travel-related emissions.

These meetings and conferences can benefit from the use of teleconferencing, videoconferencing, and webinars, which make travel less necessary without hindering the exchange of information within an organization. It's also less expensive.

The ICT options for remote workplaces are numerous: Voice over Internet Protocol (VoIP), mobile devices, social media, videoconferencing, file-sharing software, and GIS, among others, all fit into the puzzle. These options offer enough variety that organizations can tailor what they use to best meet their needs.

Virtual offices, and the workforce mobility they create, can significantly reduce an organization's energy consumption profile by slashing commuting and other transportation, lessening the need for physical office space, and decreasing workplace energy requirements.

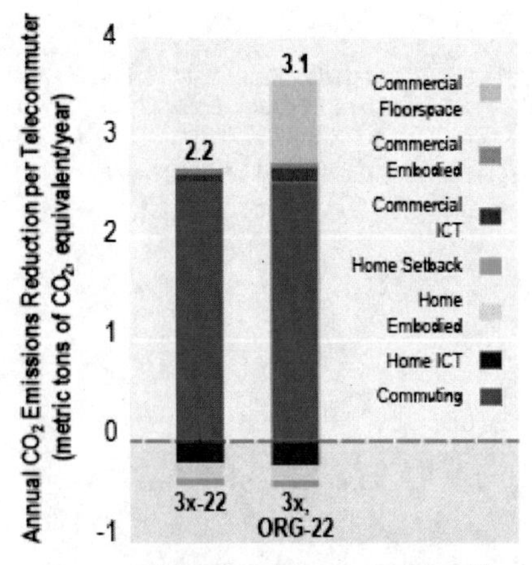

Elimination of commercial floor space accounts for an increasingly larger portion of potential CO_2 reductions as commute distance decreases.

This example shows how telecommuting instead of a 22-mile commute reduces GHG emissions (items below the zero line). This effect is even larger if the organization reallocates space and reduces office footprint in response to fewer in-office workers (ORG-22 on chart).

But do virtual offices and telecommuting make economic sense? An organization must consider not only its mission, but also its stakeholders and unique business requirements. In addition to evaluating technical solutions, it must consider the expected return on investment. As a rule, the savings should outweigh the total cost of ownership (TCO) within an acceptable time frame—typically 3 to 5 years.

The areas where savings are traditionally found include labor time, travel, physical infrastructure, and energy consumption. The TCO, on the other hand, includes implementation, maintenance, and periodic upgrades of technical enablers. Increasingly, net carbon load is also considered in these calculations as an unintended benefit of virtualization.

One positive factor is the value that many employees assign to having the freedom to telecommute at least part of the time and the preference for positions that offer this flexibility. Organizations then benefit from offering this feature because it helps them secure higher-quality, and sometimes less-expensive, employees than they might otherwise attract.

An organization's success is largely measured by two components: the quality of its work and its labor expense. The deployment of ICT to facilitate remote work can give it a distinct competitive advantage. But how do employees regard telecommuting? Approximately 80 percent of *Fortune Magazine*'s "Best 100 Companies to Work For" reportedly offer telecommuting as an option.[9] At these companies, 72 percent of workers say that they would be more likely to choose a job with flexible working arrangements than one without them. In fact, 37 percent of those surveyed mentioned telecommuting as a desired option, and some said they would take a pay cut in order to be able to work from home.[10]

The emergence of virtual workforce systems as a replacement for traditional on-site systems has raised questions about their relative costs and benefits—specifically energy consumption and GHG emissions. Generally speaking, the rise in the number of computing devices (and the industrial base needed to create and maintain them) will increase carbon emissions. However, this rise will be balanced by the reductions in office space and employee travel.

If ICT managers implement these technologies efficiently, their organizations should see an increase in worker productivity, a reduction in administrative costs, and a significant decrease in carbon contribution.

Consider the data measuring the effects of ICT-enabled telecommuting on energy consumption and GHG emissions. An analysis of federal commuter patterns shows that, on average, a federal employee produces nearly 4 metric tons of carbon emissions just by traveling to and from the workplace. A 2007 study commissioned by the Consumer Electronics Association, meanwhile, found that telecommuting reduced the nation's CO_2 emissions by 14 million tons, the equivalent of removing 2 million vehicles from the road.[11]

Although virtual offices and workforce mobility have the potential to reduce organizational energy consumption and GHG emissions, the effects will vary with each organization and location. Managers will need to break down the effects into components such as workplace energy requirements, physical office space utilization, and commuter transportation needs. They also need to understand how changes in each of these components will impact their organization's carbon footprint.

EXAMPLES

Today, more than 19,000 Sun Microsystems employees (approximately half the workforce) telecommute at least part-time. In 2008, Sun Microsystems (which in 2010 became part of Oracle Corporation) released the results of a study it performed comparing the energy requirements associated with traditional office commuting and telecommuting.[12] Sun's study found that employees saved more than $1,700 per year in gasoline by working at home an average of 2.5 days a week. Energy consumption at offices was two times that of home office equipment.

Commuting, meanwhile, represented more than 98 percent of each employee's work-related carbon footprint, compared with the less than 1.7 percent of total carbon emissions that resulted from powering office equipment.

Telecommuting savings are being realized overseas as well. A 2009 joint study by World Wildlife Fund-China and China Mobile showed that low-carbon solutions had enabled China's telecommunications sector to cut 48.5 million tons of carbon emissions in 2008 and 58.2 million tons in 2009. These solutions included telecommuting, electronic data interfaces, and more efficient logistics.

Become Lean to Be Green

Increase operating efficiency

Recommendation: *Use enterprise architecture (EA) as a means of identifying redundant or inefficient business processes, including IT applications and technologies, and rationalize them to reduce energy consumption and GHG emissions.*

CHALLENGE

Organizations today face increased pressure to identify opportunities for internal cost savings. At the same time, these organizations are seeking ways to mitigate the environmental impact of their business operations. Meanwhile, they must also plan for growth and organize their resources to accomplish these goals smoothly and efficiently.

Many organizations are required to develop and execute sustainability performance plans that communicate energy requirements across a broad range of business functions. These plans cover GHG management, pollution and waste management, sustainable buildings and communities, and fleet and transportation.

SOLUTION

Mitigation Potential	★★★
Operational Impacts	★
Financial Impacts	★
Feasibility and Timing	★★★

EA is an integrated, comprehensive framework for managing an organization's IT portfolio and its relationship to the core business. EA involves establishing a baseline for an organization's current state, where it wants to go, and what processes will be needed to bridge the gap. The process requires a holistic approach to technology systems and embraces new technology and opportunities. It involves vision, planning, and execution and often allows an organization to cut costs by discovering inefficiencies and using the savings to fund innovation. With an effective EA, an organization can answer questions such as, "How will outsourcing inventory management affect my IT organization?"

Today, EA is used as a tool to balance growth with energy efficiency and reduce carbon emissions. Because organizations are complex systems—composed of people, processes, information, applications, and infrastructure—they need EA to provide an analytic framework that is both comprehensive and structured. This framework helps an organization reduce inefficiencies (even those related to ICTs) and create a system that uscs lcss energy with fewer emissions. EA can show the impacts of near- and long-term changes to such systems. GHG emissions reductions can then be planned strategically and in the context of streamlining core business functions.

EA also makes it possible to model an organization's layers. These layers include strategy, performance, business, information, applications, and infrastructure, and each is directly tied to EA. Because each EA layer is modeled holistically, their interrelationships are carefully documented. This serves as a starting point from which to devise a more efficient overall structure.

GHG emissions-reductions strategies can then be developed intelligently, first by documenting the current ("as-is") operational environment and then the desired ("to-be") future state. EA helps reveal the gaps that can be bridged by implementing strategic and tactical projects. The strategy for moving the enterprise from the baseline to the target architecture is set down in a sequencing plan, which includes the projects needed to execute the transition.

To illustrate this point, consider how an organization could leverage EA tools, methods, and frameworks to achieve GHG emissions-reduction goals: first, by documenting existing architecture and then by documenting target architecture while conducting a gap analysis.

Document the current architecture

Every EA project begins by documenting the current environment, including the EA layers and their interrelationships. The organization identifies business functions and documents processes to show how these functions are performed.

It then identifies how information flows between participating elements and the systems employed, using interface diagrams to show the flow between various systems—a road map of sorts. It documents the infrastructure on which applications and organizational operations depend, records the performance metrics associated with each of the layers, and notes any progress against them.

The result is a set of interrelationships between performance metrics, business functions, organizational units, applications, data entities, and infrastructure. These interrelationships can then be viewed through a series of two- or three-dimensional matrices, which highlight any redundant or outdated processes, applications, and technologies. Figure 3-2 shows commonly used EA models and their relationships.[13]

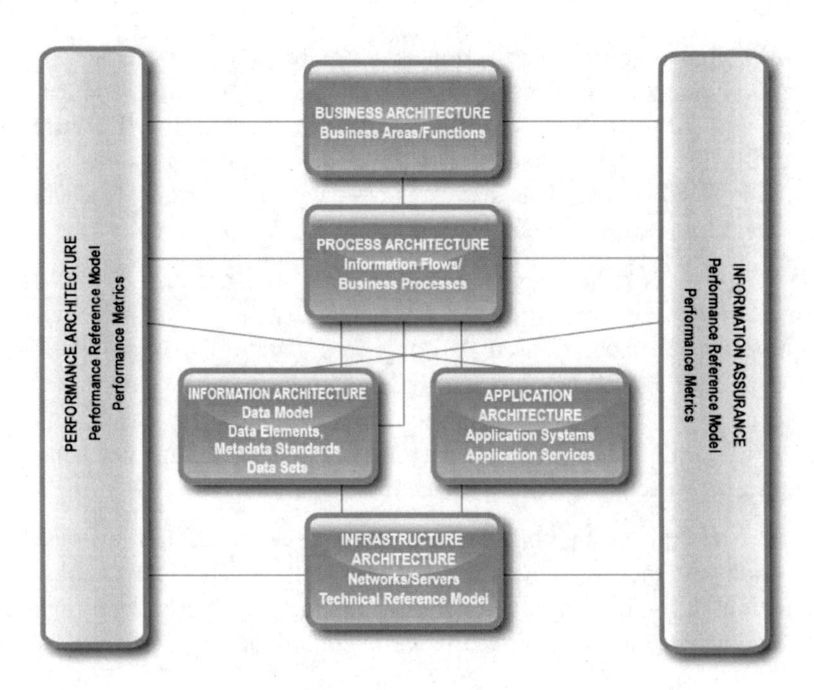

Figure 3-2. EA Looks across the Organization to Achieve Efficiencies

Different EA tools can be used to capture this information through different types of modeling methods. The goal remains the same: telling the story of the organization's performance, business, information, application, and technology architecture. This narrative must be comprehensive and holistic to identify ways in which its structural features can be improved. For example, in Figure 3-2, the link between business architecture and performance architecture shows the need to ensure that the business performs only those tasks that are necessary.

Documenting the target architecture and analyzing gaps

An organization's target architecture must align with its strategic goals and objectives. In this example, to reach its GHG emissions-reduction goals, the organization needs to develop an architecture target in line with those goals. This alignment allows it to identify and document how these goals will impact different aspects of the enterprise. Metrics continue to be a must, allowing any progress in meeting reduction goals and objectives to be measured.

How the organization will cope with these impacts involves answering several questions:

- *What business functions are most responsible for the organization's GHG emissions?* How can GHG emissions reduction be built into core business areas? What parts of the organization must be involved to make this happen? For instance, if an objective is to implement a telework initiative that spans the entire enterprise, will the ICT unit operate alone, or will it benefit from having the human resources staff work in conjunction with its efforts?

- *Will existing business functions need to be modified or new ones developed?* How will business processes need to change? For instance, if a GHG-reduction division is to be created, will new functions have to be added and new processes developed to reflect how they are to be executed?
- *What new information exchanges between people and systems will be necessary?* For instance, if the organization intends to report its GHG emissions, what's the process for defining and tracking the documentation of this new information?
- *Are any current systems available that can support changes in organizational strategy and operations?* For instance, if an organization finds that existing systems do not fully capture and report the information needed to achieve reduction goals, what new systems might need to be developed?
- *What changes in the technology infrastructure might be required to ensure the organization remains aligned with its reduction goals?* For instance, an organization's technologies are often documented as part of a technical reference model (TRM) that identifies the current, tactical, and strategic standards of the infrastructure. Will the organization have to establish new tactical and strategic technology standards within its TRM?

EA has the answers to these questions. When a change to one part of an organization impacts others, EA helps identify where the gaps are between current and target systems and processes. For instance, if a new telework initiative is to be implemented as part of an activity to meet reduction goals, an EA gap analysis might reveal a major hurdle: the organization's employees cannot access a set of systems from their telework sites because those systems are not web enabled. Along those lines, the gap analysis may reveal that no current system or process is in place to track the capture of GHG emissions related to an organization's operations.

The final step, implementation, is accomplished by attacking any identified gaps. EA can help establish the project, or set of projects, that should be prioritized by the organization's leadership. EA can also help specify timelines and milestones, and develop metrics for tracking progress.

EXAMPLES

The Federal Data Center Consolidation Initiative (FDCCI) is an ambitious effort launched in 2010 to guide federal agencies toward a goal of closing 800 of the government's 2,094 data centers by 2015.

The reason for this initiative is well documented; in 2006, federal servers and data centers consumed more than 6 billion kWh of electricity. That figure could already have doubled, becoming the equivalent of the carbon emissions of 10,000 cars in a single year.

FDCCI's goals are as follows:

- Promote the use of "green IT" by reducing the overall energy and real-estate footprint of government data centers.
- Reduce the cost of data center hardware, software, and operations.
- Increase the overall IT security posture of the government.
- Increase the use of more efficient computing platforms and technologies.[14]

This initiative is an opportunity for agencies to reduce their environmental impact by targeting underutilized and smaller data centers for consolidation. To do so, the FDCCI is using EA to help agencies map their existing systems and technologies as they conduct their baseline inventory assessments.

Specifically, they're using Federal Enterprise Architecture (FEA) reference models. Mapping to the FEA TRM facilitates the common identification of consolidation opportunities. Although still early in the process, EA is proving instrumental in helping government agencies meet their consolidation goals.

This effort is expected to save $3 billion annually through efficiencies gained by eliminating the energy consumption, maintenance, and management of these shuttered data centers—without hurting performance or sacrificing user experience. Instead, the intent is to gain IT efficiencies across agencies and foster greater collaboration.[15]

In the private sector, companies have used EA effectively to reduce ICT-related emissions while saving money, according to reports from IBM.[16] For example, a telecommunications provider saved 53 percent in ICT costs over 5 years by following an EA road map and consolidating and retiring overlapping and obsolete systems. Another organization, this time a major financial institution, saved $4.2 million in annual ICT costs by decommissioning excess servers identified through application of EA.

See Whether Elastic Computing Fits

Recommendation: *Investigate adaptable computing platforms (such as cloud computing and microservers) and how they can decrease the carbon emissions associated with the use of ICT within the enterprise.*

CHALLENGE

Servers are the actual, physical *things* that fuel today's tech world. They allow information sharing and collaboration. Without servers, and the activities they enable, advances such as teleworking would be impossible for any organization. But servers require vast data centers, and, if every company looking for remote desktop access were to create their own personal data center, the GHG toll would be catastrophic.

Cloud computing is based on the premise that computing is more efficient when concentrated in large computer centers capable of serving many clients at once—the "cloud"—providing *economies of scale* not available to smaller, individual data centers. It's the ICT equivalent of buying in bulk to get a better price.

The economies of scale offered by cloud computing also give organizations the ability to easily increase or decrease their use (hence, the term elastic computing) to meet their computing needs. Cloud computing also can involve computer virtualization, in which multiple software-based virtual machines, each with different operating systems, run on the same physical machine but in isolation, side by side.

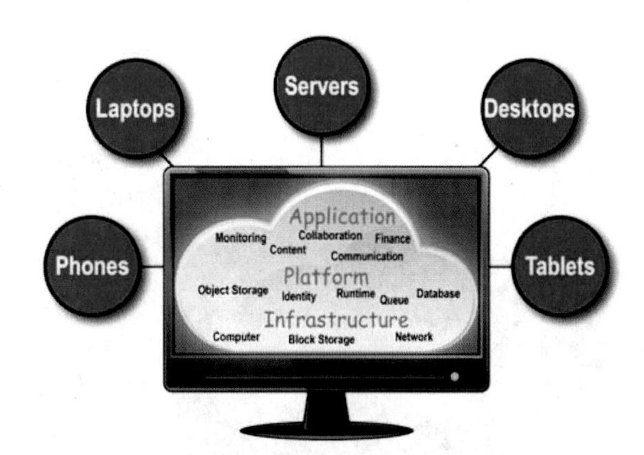

Cloud computing affects all aspects of your computing experience and provides new challenges and opportunities for GHG reduction.

However, server virtualization isn't a cure-all for solving all data center problems, and large organizations must investigate alternative solutions.

From a climate change mitigation perspective, the movement to cloud-based ICT services can significantly impact the way computing infrastructure is supported and how its energy use may be attributed to any one organization. So while data centers offer big opportunities for carbon emissions reduction, managers must first figure out how to go about making the right decisions in a complex, fast-changing, and vendor-rich environment.

SOLUTION

Mitigation Potential	★★★
Operational Impacts	★★
Financial Impacts	★★
Feasibility and Timing	★★

You might already operate in the cloud without realizing it, especially if your personal e-mail is provided by Hotmail, Gmail, or Yahoo.

In the late 2000s, when cloud computing finally arrived on a grand scale, it represented a legitimate option for ICT to transform how technology infrastructure is delivered and managed. Cloud computing reduces up-front data center needs and reduces energy needs to maintain that infrastructure.

Embracing cloud computing can help reduce energy consumption and GHG emissions in two primary ways. First, the scale economies in ICT infrastructure imply that computers will be used more efficiently and therefore require less energy to run. These savings can be added to those gained through computer consolidation. Second, more efficient use of energy at cloud centers is increasingly possible, which also saves money and GHG emissions.[17]

How can this be so? The story with current "legacy" data centers is that the power required to cool the facility is as great as, or greater than, the power needed to run the equipment; in other words, at least half of the power consumed in the data center is used to cool it. This doesn't have to be the case, for a couple of reasons.

Data centers have historically been highly controlled environments with temperature settings of 68–72°F and 45–55 percent relative humidity (RH). But this environment is as much a legacy of where computers came from as anything else. The reason for such stringent humidity requirements is that the first computers ran on paper punch cards, and 45–55 percent RH was what was recommended to ensure that paper punch cards didn't warp and get stuck in the computers' card readers. Today, that's clearly not the case; modern hardware supports a much broader RH range (0–90 percent RH) with no measurable change in failure rates.

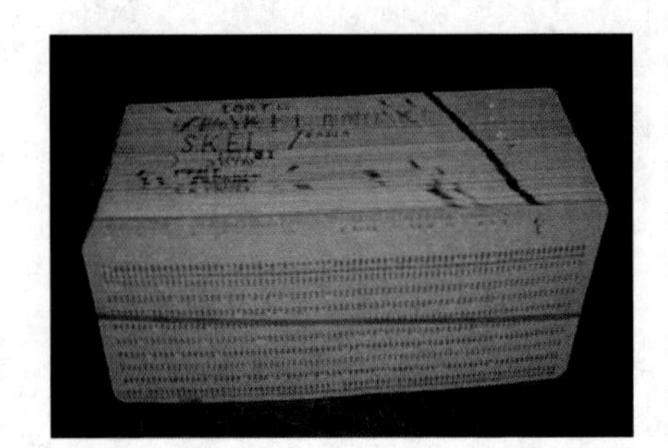

Punch cards, circa 1969.

Data center operators also are challenging the historic 68–72°F cooling requirements. Many providers now operate their data centers year round at 75°F, with short spikes of temperatures to 80–90°F easily tolerated. However, one particularly massive user of server farms, the federal government, has continued to detail in their contract proposal process the strict traditional requirements that data centers operate at 68–72°F with 45–55 percent RH. These requirements are outdated.

With these stringent specifications out of the way, data centers are free to use filtered air from the outside for cooling over a larger range of climate conditions. Outside air is free, and this "free" cooling system greatly reduces the power requirements of running a data center.

To best support the move to using outside air, new data centers are being located in cooler climates having more average cool days per year. Google followed the model long used by nuclear power plants in leveraging cool moving water for its temperature-reducing abilities; they positioned one of their major data centers along the Columbia River in Oregon to leverage free cool air (along with the inexpensive hydro-sourced electricity the area offers). Other Google server farms use sea water for cooling.[18]

Google's data center in The Dalles, Oregon, is located right on the Columbia River. It runs on hydroelectric power.

Using these (and other) techniques, some large data center operators have been able to reduce their data center cooling power usage from approximately 100 percent of server power usage to around 10 percent.

Still, many organizations will require preparation before they're able to completely embrace cloud computing. Many of the issues that need to be addressed go beyond the technical aspects of the cloud computing environment itself because, in many ways, cloud computing represents more of a business model shift than a technology shift.

For example, organizations will likely have to address how they will maintain their business processes, support their culture, and protect their privacy. As they begin to think about how they can leverage cloud computing to support their business needs, they will need to consider these and other elements to prepare for a smooth transition and realize the intended benefits.

The other promising development follows the long tradition of technology improving itself by getting smaller. Major players—including Intel, Dell, Advanced Micro Devices, and ARM Holdings—are competing with smaller start-ups (SeaMicro, for example) to produce a relatively new type of processor that promises to use less energy, resulting in lower carbon emissions (and costs).

These are known as microservers, and they are basically smaller versions of their larger, power-hungry cousins. In SeaMicro's case, the company used "Atom" chips—tiny, relatively low-powered chips traditionally used in inexpensive "netbook" laptop computers—and repurposed them for use in data centers.[19]

Microservers have a smaller physical computing footprint and are optimized for small, repetitive tasks such as processing e-mail. They are efficient in that they typically share power and cooling systems as well as storage and network connections—all of which results in reduced overall power consumption.

Microservers, though a possible solution to the data center efficiency problem, are a classic example of how the path you choose must meet the mission of your organization. For example (as of 2011), Facebook has initiated the use of microservers, something that sets them apart from counterparts such as Google. Facebook is interested in microserver technology because the company believes that managing its social content requires brief, but frequent, data connections for which microservers are perfect.[b] The company found that putting Facebook technology into virtual machines didn't provide the financial gains it expected, and the cost of dealing with hardware failures was too high.[20]

Alternatively, with Google, it's a question of what an organization needs. In Google's role as a search engine, it thought that "brawny" processors win over "wimpy" ones because in warehouse-scale systems, throughput (the average rate of successful message delivery) is more important than single-threaded peak performance because no single processor can handle the full workload.[21]

EXAMPLES

The Department of Health and Human Services uses cloud computing to manage customer relationships as well as specific projects. The organization has begun implementing cloud computing–based systems for customer relationship management and project management. Its regional extension centers will use these systems to track, manage, and report on initiatives to encourage the adoption of health ICT across the nation.[22] A recent Brookings Institution study estimates

[b] In the film, *The Social Network*, Facebook founder Mark Zuckerberg (played by actor Jesse Eisenberg) emphatically declared, "Okay, let me tell you the difference between Facebook and everyone else, we don't crash EVER! If those servers are down for even a day, our entire reputation is irreversibly destroyed!"

that government agencies that adopt cloud computing save 25 to 50 percent—potentially a multi-billion-dollar endeavor.[23]

These examples lack estimated GHG savings. However, the Carbon Disclosure Project has ballpark numbers for a different application:

- If a typical food and beverage firm transitions its human resources application from dedicated ICT to a *private internal* cloud, it can reduce CO_2 emissions by 25,000 metric tons over 5 years.
- On the other hand, if the same food and beverage firm transitions its its human resources application from dedicated ICT to a *public* cloud, it can reduce CO_2 emissions by 30,000 metric tons over 5 years.[24]

A report by the Technology CEO Council revealed that Applied Materials, a company that produces capital equipment (including semiconductors), was able to reduce its electrical consumption by 34 percent after consolidating 27 server rooms in California into one super-energy-efficient data center.[25] This case likely required an up-front investment, but doing so probably reduced the firm's operating costs as well as its energy usage and GHG; this is the type of thing that must be weighed ahead of time.

Promote Green Buildings with Data-Driven Energy Use

Recommendation: *Use integrated data sources to facilitate empirical, quantitative analyses, such as identification of the most and least energy-efficient buildings on a campus on the basis of hard data.*

Inform efficiency decisions

CHALLENGE

According to the US Energy Information Administration , 40 percent of US energy consumption can be attributed to the nation's residential and commercial sectors. Such consumption comes almost entirely from buildings. The industrial sector accounts for another 32 percent, some of which is attributable to buildings as well. In larger organizations, the data used to associate this energy use with each building are recorded in various, dissimilar information systems. Examples include systems that separately, and not always equally, record consumption metrics for electricity, gas, water, and fuel oil.

Typically, these systems are not integrated, making it impractical (often impossible) to holistically analyze which buildings use energy most intensively. Without this information, an organization can't possibly discern the structures most in need of energy audits or refurbishment.

Over the coming years, large organizations will need to measure the GHG inventories of their enterprise and supply chains and determine how enterprise carbon accounting systems can help (see the "Supply" chapter). Many products have been designed to coordinate data in this area, but selecting the right solution is likely to be a challenge.

Mitigation Potential	★★
Operational Impacts	★★
Financial Impacts	★★★
Feasibility and Timing	★★★

SOLUTION

The ability to identify inefficiencies in the largest sources of energy consumption promises to significantly impact climate change mitigation. The same is true for carbon emissions.

Over the past several years, a new class of information systems has emerged under the banner "enterprise carbon accounting" (ECA). These systems (typically vendor-specific products) offer a consolidated set of services for an organization to collect, integrate, and report enterprise and supply chain GHG inventories for Scopes 1, 2, and 3 emissions.[c]

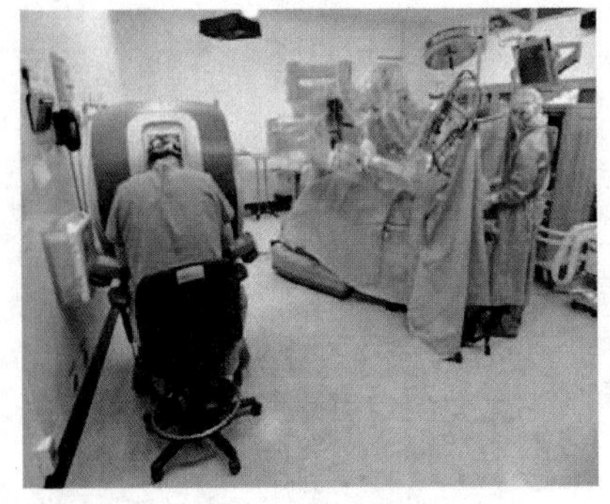

Linking surgery schedules and HVAC settings reduces energy use.

With this in mind, integration is a must for data-driven decision making, along with a baseline of standards for such data. ICT networks provide the means to integrate various sources of otherwise disparate data to make smart, comprehensive operational and strategic decisions on energy.

Integrating data collection systems and their compiled information is an important step toward a comprehensive view of energy use and GHG emissions. Semantic data integration is an important part of this process because it establishes a baseline for *language*. Semantics enables consistent terms and meanings across organizations.

Data standards and service-oriented architectures in ICT help make this possible by allowing organizations to better understand their energy use and take steps to reduce it. Many organizations have turned to ECA solutions (like software for socialized supply chain accounting) to better manage energy and carbon data at the enterprise level.

Of course, energy use data integration is only a necessary first step. Many factors influence building energy use, and proper analysis will

---

[c] Scope 1 refers to all direct GHG emissions; Scope 2 to indirect GHG emissions from consumption of purchased electricity, heat, or steam; and Scope 3 to other indirect emissions, such as transport-related activities in vehicles not owned or controlled by your organization. (See the "Action" chapter.)

require more complete information. For example, even if two seemingly identical buildings use the same amount of energy, they may not be equally energy efficient. One might contain a swimming pool or have fewer tenants. Viewing building energy use in the context of these secondary factors is crucial in developing informed physical infrastructure and energy strategies (see the "Structure" chapter).

EXAMPLES

NYC's Mayor Michael Bloomberg used Earth Day 2007 as a forum to unveil PlaNYC, a comprehensive sustainability initiative designed to reduce the city's greenhouse gas footprint. The goal was to decrease citywide carbon emissions by 30 percent below 2005 levels and slash carbon emissions from government operations 20 percent below 2006 levels by 2017.

Mayor Bloomberg unveils his ambitious PlaNYC initiative.

Given that the operation of buildings accounts for about 40 percent of carbon emissions and energy consumption in the United States, the mayor's office decided to investigate the feasibility of developing a new citywide information system to aggregate data from multiple sources to give city decision makers a comprehensive view of their building portfolio. Once in hand, this portfolio view could be used to examine current and planned projects, energy cost and consumption (for gas, steam, electricity, and fuel oil), and—in particular—the impacts these projects and rates of energy consumption have on the city's carbon footprint.

The city's building portfolio analysis demonstrated that each separate data component revealed a piece of the puzzle, but there was still no comprehensive view. Several NYC laws mandate that agencies collect and manage information on energy and water consumption, cost, efficiency, and building systems, but only at the individual facility level. The city needed a way to

- provide building performance benchmarks,
- more precisely monitor energy consumption,
- clarify the costs and benefits of new capital projects,

- monitor the status of ongoing projects, and
- identify opportunities to improve financial performance.

By centralizing data across buildings and energy sources, city decision makers were able to paint a "big picture" of the city's building performance they could then use to measure and manage energy usage and carbon emissions more effectively. The idea was to get beyond data silos for water use, electricity, physical building information, and gas usage and instead integrate and harmonize them.

The University of California at Santa Barbara has installed a campus-wide energy information system (EIS) and has begun measuring energy savings directly attributable to operational changes. According to a published report, the installed system, including an annual maintenance and support fee, cost the university $341,500. The first-year savings were $295,000.[26] At this savings rate, payback on the EIS was to be realized in 1.2 years.

Another school, Oberlin College in Ohio, has instituted a campus-wide electricity- and water-use information system in which usage is provided in real time to each residence hall on campus, along with the environmental and financial consequences of such usage. For some residence halls, the system is broken out by wing or floor and by kitchen unit. Oberlin encourages residence halls to compete with one another to reduce energy consumption but leaves it to the residents of each hall to develop strategies to do so.[27] This is an example of technology's enabling an incentivized cultural shift—even if that incentive is bragging rights.

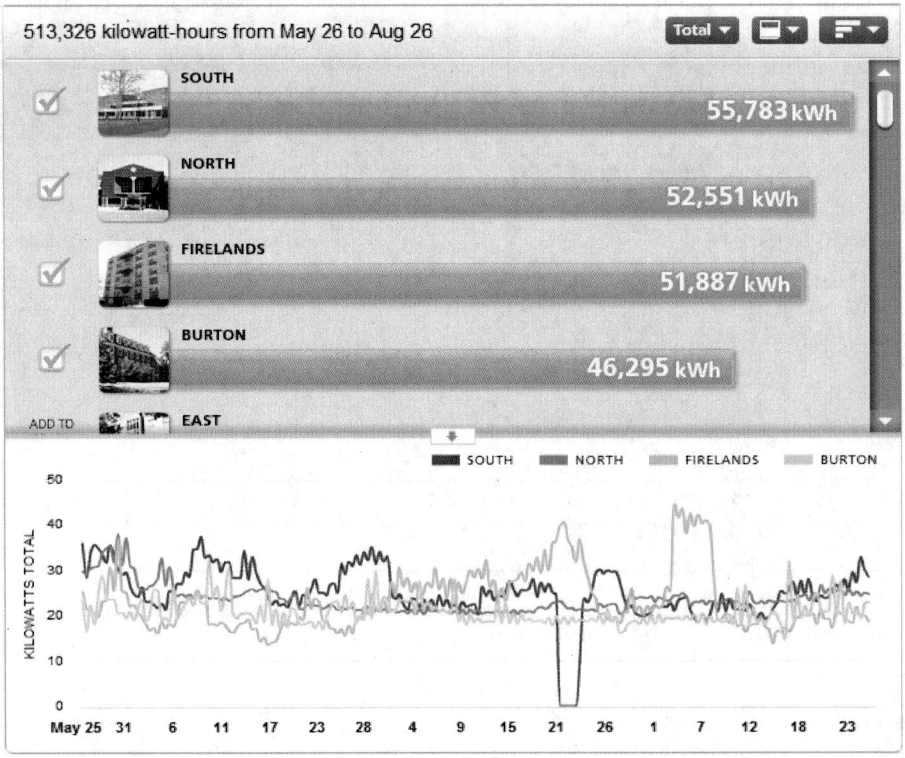

This dashboard based on real-time data allows students at Oberlin to monitor dorm energy use.

Prepare for Smart Things and Big Data

Recommendation: *Stakeholders must prepare their organization for the huge volumes of data that will be generated by the monitoring of physical objects—and the competitive advantage these data can provide—and seek opportunities to use these data to facilitate efforts to reduce GHG emissions.*

CHALLENGE

The Internet is evolving. When the World Wide Web invaded mainstream culture in the mid-1990s, it comprised mainly static, hyperlinked pages of text and images, a phase now referred to as Web 1.0.

Users soon demanded more, and a new platform (Web 2.0) allowed people to participate in the creation of content on the web (think YouTube). This evolution will continue, and new generations of the web (such as "Web 3.0" and the "Internet of Things") are now being developed.

Vast amounts of data have never been collected, from objects and non-human living things, for example. Can we make use of these data, and how can we separate what is useful from what is not? Few organizations have yet considered how to integrate the "smart" objects contained in these data sources into their enterprise planning, let alone set up ways to access, process, and store the massive volumes of data that will be made available. Analysis is an afterthought.

A new term, "big data," has become a shorthand reference to the massive volumes of data that will result from these increasing levels of interconnectedness. Smarter data storage and data analytics solutions will be required to make sense of it all.

Already, the social web produces volumes of data readily accessible to us all, data previously exclusive to niche research areas. For example, as of August 2011, Twitter is home to 5 billion tweets per week, a number that is steadily increasing. Intelligent use of this emerging data source is one challenge; using it to reduce GHG emissions is another.

The Chevy Volt is a connected source of data. It can transmit real-time updates on its operation status to a smart phone.

SOLUTION

If the Internet is connectivity largely revolving around people, then the Internet of Things is the same, but for … *things*. Consider the notion that everyday objects in our world have and can be a source of

Mitigation Potential	★★
Operational Impacts	★
Financial Impacts	★
Feasibility and Timing	★

data, and that data can be channeled into an interconnected global network.

These data, which come from natural and human systems and physical objects, have historically never been captured. But, since these objects are increasingly instrumented with sensors and the Internet can help with data retrieval, we will increasingly have access to those data.

The Internet of Things provides us with the ability to collect information from countless geographically dispersed sensors that are linked to real objects. This ability will create a flood of data that will need to first be merged, culled, and analyzed, but once that's done, they can be used to help mitigate GHGs and accomplish other important goals.

ICT managers should consider preparing their enterprises for the potential influx and integrate smart devices into their asset portfolios, including physical infrastructure systems and subsystems and their employees' personal communication devices. If done properly, this integration will give a system access to new, previously unavailable data. For example, automated lighting systems could capture occupancy levels and promote space utilization optimization.

The benefits are likely varied and numerous, including real-time ordering and inventory management or smart equipment that monitors and reports on its own performance.

To harness the potential of the Internet of Things, organizations should track the progress and price of emerging and evolving technologies in this area. They should recognize the real possibility of sourcing data enclaves needed to capture and report on them in real time. Organizations that move early to develop these capabilities stand to benefit over those that don't. Moreover, the increasing number of Internet-enabled devices (and people) promises to produce a torrent of data.

Another promising, relatively recent emergent technology doesn't harness data so much as provide access to augmented reality. Football fans are already familiar with augmented reality when they watch a game where a bright yellow first-down line appears on their TV, which is nowhere to be found on the actual field.

One emerging form of this technology is "quick response" (QR) codes. QR codes are 2-dimensional bar codes that anyone with a smart phone and an app for reading QR codes can scan and be directed to any web destination. They are related to the Internet of

Things in that they are intended to augment a physical object in the real world (such as a magazine advertisement, street sign, or business card) with instant additional data from the Internet. QR codes are used in construction to link to technical information, job-site signage, and emergency contact information.[28]

That said, organizations will find no shame in not doing it all themselves. Many that attempt to manage these data using on-site IT infrastructure (servers in closets) will find that doing so may put them at a competitive disadvantage to those who adopt a more scalable cloud computing solution. The latter is much more cost competitive.

EXAMPLES

While the Internet of Things is still nascent in every regard, a number of prominent organizations are beginning to deploy these kinds of capabilities.

One example is the 2011 Chevy Volt, which has its own unique address on the Internet at http://gm-volt.com/. Currently, GM plans to use Internet connectivity as a mechanism to push software updates to the car, which has a whopping 10 million lines of code (more than the Boeing 787 Dreamliner jet, which has 8 million).

Using software, GM engineers have fused the vehicle's various propulsion systems, such as the electric motor, battery, and power electronics. Traditionally, they were coupled to a power train mechanically, but in the Volt, they're now coupled electronically.

What's truly revolutionary are the electronics capabilities built into the Volt. For example, the vehicle can track its own GHG emissions and transmit them to a central location such as the Environmental Protection Agency (EPA)—or the vehicle owner's smart phone. If all US vehicles did so, EPA would have a far better estimate of transportation GHGs than it does today.

Another example is the Soil Climate Analysis Network (SCAN), a comprehensive, nationwide soil moisture and climate information system that provides data to support natural resource assessments and conservation activities. SCAN uses Pachube.com, an online service that allows users to "store, share and discover real time sensor, energy and environment data from objects, devices and buildings around the world." Pachube ("patch-bay") serves as a web switchboard of sorts, using remote sensors to connect people to devices, applications, and the Internet of Things (Figure 3-3). These data can be visualized and fed to and merged with other data sources.

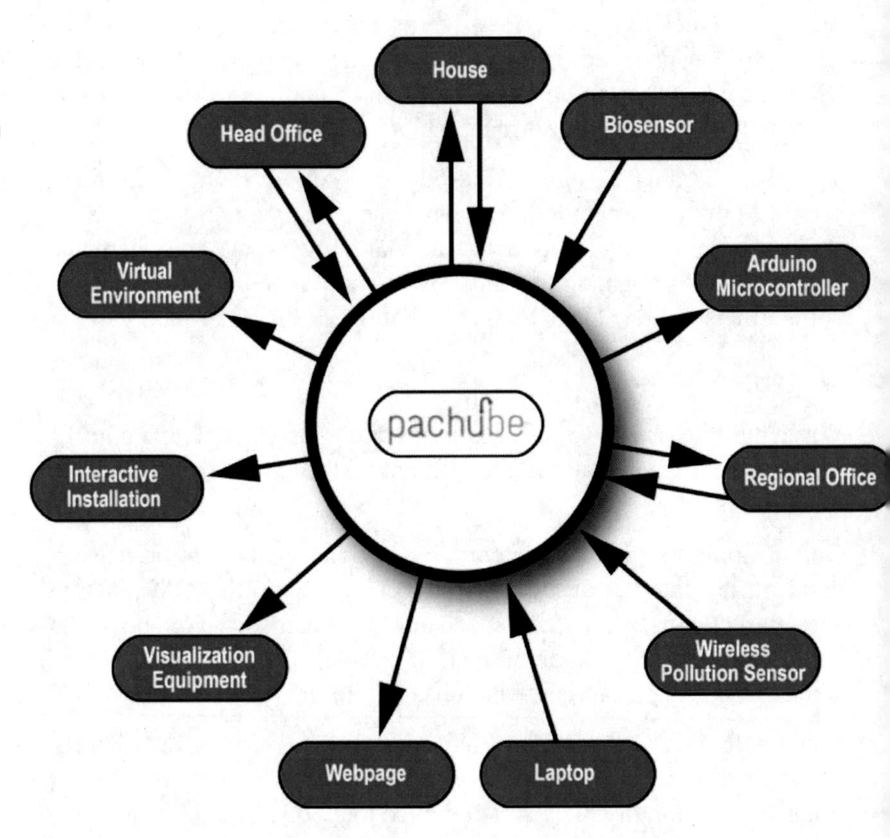

Figure 3-3. The Internet of Things Allows Objects to Communicate and Respond without Human Intervention

SCAN focuses on agricultural areas across the United States, monitoring soil temperature, water level, air temperature, and other climate-related variables. Global climate modelers can use the data to

- monitor drought and develop plans and policies for mitigation,
- compare climate model predictions to data on the ground, and
- monitor and predict changes in crop, range, and woodland productivity relative to soil moisture-temperature changes.

ICT and Adaptation

If ICT can help mitigate climate change by using energy and resources more efficiently, it can also help organizations adapt to a changing climate. The physical infrastructures of ICT (data centers and networks, for example) are at risk of adverse climate change consequences, for which organizations should be planning adaptation actions. That said, ICTs can play a significant role in facilitating climate change adaptation for the entire organization. Also, they are increasingly being used by people in mobile and social environments to better communicate during disasters and react to them.

Hurricane Katrina uprooted 1,000 wireless towers and knocked down 11,000 utility poles, costing the telecommunications sector as much as $600 million.

Organizations should adopt strategies for ICT use in adaptation, including the following:

- Protect your ICT resources.
- Use information from climate models to inform your risk management plans.
- Use emerging media to manage first response.
- Take advantage of crowdsourcing.
- Coordinate ICT adaptation planning with your organization's overall adaptation plans.

Protect Your ICT Resources

Recommendation: *Identify your organization's ICT exposure to climate change risk and its ability to continue operations and recover during and after a natural disaster. Consider selective investments to increase your organization's ICT resilience in the face of such disasters.*

Address climate risks

CHALLENGE

Today, ICTs are so critical, so engrained in nearly every human endeavor, and necessary to connect us all on a global scale, that protecting ICT infrastructure is not an option, it's a necessity. We should be humble in our ability to forecast the frequency and impact of natural disasters associated with climate change, but we cannot ignore the history we have to draw upon for information. Figure 3-4 shows the trend of natural disaster events reported and the numbers of victims resulting from those events between 1975 and 2008.

As the number of reported events trends upward, consider this caveat: this upward trend is at least partly due to increased communications capabilities. Still, a number of prominent organizations, including the American Chemical Society and the United Nations World Health Organization,[29,30] predict that this trend will continue upward with a rising number of severe weather events.

Figure 3-4. Trends in Natural Disasters and Victims

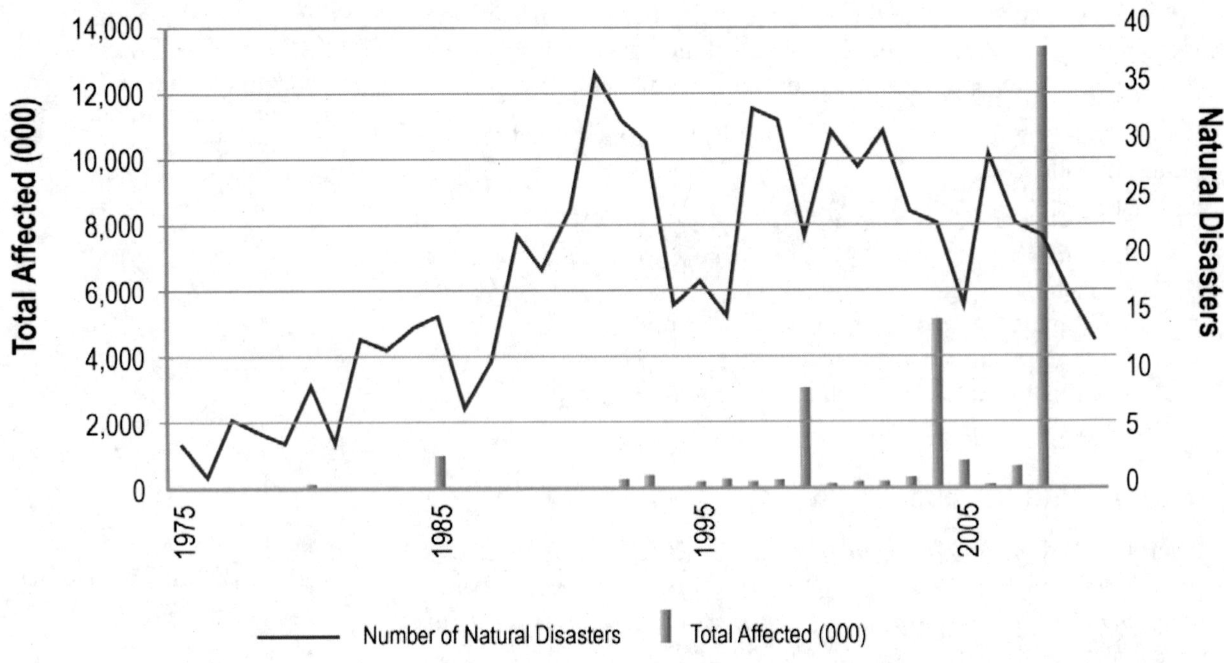

There is no argument, though, about the amount of uncertainty that comes with severe weather events. We don't know for sure where or when they will occur, or how intense they will be, and this uncertainty must be factored into any plans for protecting organizational ICT resources.

SOLUTION

Adaptation Potential	★★
Operational Impacts	★
Financial Impacts	★
Feasibility and Timing	★★★

Protecting an organization's ICT resources, including access to lines of communication and electric power, is one of the largest climate change adaptation responsibilities. Guaranteeing access involves first analyzing the risk and then using the results to drive organizational strategies. This is a common theme throughout this book, but it's particularly valid in regard to ICT assets.

What is worth protecting? At the core, every organization's ICT functions hinge on its people, equipment, the structures that house equipment, software, lines of communication, and access to electric power. Because all of them are critical to communication efforts, all should be examined for vulnerability to natural disasters. Sufficient damage from an adverse weather event could hurt your organization's ICT functionality and severely disrupt your ability to conduct business.

Because risk is a combination of the probability of an event, its impacts if it occurs, and the vulnerability of systems to such an event, stakeholders must look at what can and cannot be controlled. The probability of an event cannot be controlled, but resiliency can. Improving organizational resiliency, which means developing the ability to rebound from an event and maintain operations, can help minimize any damage.

Identifying the key people and key equipment in an organization's ICT network also involves the structures in which they are located, their exposure to climate-related events, or how vulnerable a severe event or long-term degradation will leave them (such as sea level rise near a key facility). Organizations must determine how to protect them to ensure continued operation and consider backup capabilities.

Any analysis should include the lines of communication the ICT network utilizes. It should incorporate the sources of power that allow these networks to operate.

Investments can certainly decrease the vulnerability of ICT assets to weather-related events, but you can't blow your entire budget in this area. Doing so would handcuff any organization, limiting its flexibility to adapt to the next evolution of ICTs. Lost is the ability to take advantage of other opportunities while facing other risks (because climate change isn't the only threat organizations face). Thus, stakeholders need to prioritize their investment plans. Organizations should still be able to identify a few things that offer solid returns in increasing ICT resilience—especially with respect to weather-related events and longer-term degradation.

The investments an organization should consider depend on its mission. One option might be to purchase access to ICT-related assets unlikely to be affected by any weather event with the potential to threaten the organization's facilities. The same can be said for a supplier's facilities. These assets would be held aside unless the first-string assets were disabled, but they would nonetheless be maintained for use should the need arise. This amounts to purchasing insurance, paying out relatively small sums as you go along to ensure access to ICT-related assets—assets that, under certain conditions, could be extremely valuable.

EXAMPLES

In 2006, Sun Microsystems announced "Project Blackbox," later known as the Sun Modular Datacenter. Sun's concept was a mobile data center in a shipping container. Several other technology companies jumped in to try to capitalize, including Google.[31] Sun's mobile data center design was remarkably simple and efficient, filling a standard 8- by 8- by 20-foot shipping container with as many as 280 servers. The container can be placed wherever there was access to a source of electricity, making it transportable and adaptable to any surrounding. It can even be used in conjunction with nearby clean energy sources such as solar panels or windmills.

In announcing its plans, Sun boasted that this model of data center could be equipped to operate on a rooftop, in a parking garage—even on another planet or moon.[32] It noted that the cost to make such a system operational would be just 1 percent of the price tag of a traditional data center.

Today, you can still purchase a mobile data center, but it's not central to the company's product strategy. Still, Oracle says it will build one to order for organizations looking to add this to their tech portfolio.

The opportunities presented by such a product would seemingly fit well within a climate change impact adaptation strategy. An organization utilizing a mobile data center could insulate itself from coming severe weather and ensure that the functionality of necessary servers is maintained even after a region is hit by such an event.

Use climate models

Use Information from Climate Models to Inform Your Risk Management Plans

Recommendation: *Use climate information to gain an understanding of how a broadly changing climate will affect your regions of interest and the risks, opportunities, and planning implications your organization faces.*

CHALLENGE

One daunting problem facing any ICT manager looking at adaptation to climate change is figuring out how to digest the extraordinary amount of relevant information that will become available over the coming decade. Equally challenging is how to apply this information to real solutions. On top of that is helping the rest of your organization do the same thing.

Organizations must find ways to absorb this continuing stream of information, sift through it for what's important, and use it responsibly. In the private sector, publicly traded companies must report their climate change risks to the Securities and Exchange Commission to demonstrate their ability to adapt to these challenges, which, in turn, places them on solid financial footing with their investors.

Of the information sources that threaten to overwhelm decision makers, the projections of climate change (created using complex climate models) are perhaps the most alien and most easily misused. Most of these projections offer crude details for the whole globe, though regional projections are much more complete, offering specific scenarios for any part of the globe—including where your organization operates or has interests.

The information itself has massive potential to help stakeholders creating ICT risk management plans. An organization that focuses on the supply chain and relies on Egyptian cotton needs to know how future growing conditions in that area might affect supply or price. Stakeholders, then, need to recognize when disruptions could become severe.

Those involved in ICT as it relates to national defense will look to climate change model scenarios as they plan missions and anticipate needed intervention for humanitarian and stabilizing missions.

The information is there and easily accessible. The key is to interpret and digest it.

Intelligent adaptation strategies require detailed information about many places and times, plus an understanding of the uncertainties and ranges of projections. This means someone, probably you and the ICT team you manage, will need to help key stakeholders access this vast amount of information and synthesize it from many sources. This information needs to be placed in a usable format for planners, and archives must be created for data used as the basis of important decisions. This archiving allows the information to be revisited, reevaluated, and kept current as challenges become more pressing with more sophisticated efforts. In short, your group must become "climate information users."

SOLUTION

Adaptation Potential	★★
Operational Impacts	★★★
Financial Impacts	★★★
Feasibility and Timing	★★

Climate information is steadily moving toward localized analysis, and ICT can help organizations access and understand it to better gauge threats to their facilities and activities.

The role of ICT in this effort will be to ensure an organization can access selected information, maintain uncompromised analysis records, and analyze and display results to potential users throughout the organization. These users include, of course, you.

Climate change information can help us envision how land use may change in the next decade and clarify how storm surge and sea level rise could threaten coastal facilities. As these various projections continue to improve, better information becomes available and previous results become outdated. ICT processes can help version control, where the latest data are presented and previous data are archived to allow comparison.

Still, because climate models are far too complex and expensive for organizations to use and operate, they must rely on modeling centers as resources. The National Oceanic and Atmospheric Administration (NOAA), for example, established the Environmental Modeling Center (EMC) to support the National Centers for Environmental Prediction. The EMC works to improve numerical weather, marine, and climate predictions using extensive data assimilation and modeling research. It is just one of dozens of unique modeling centers across the nation, some with very specialized focuses. Organizations should consider how well various centers' activities and data fit their needs.

Organizations must also learn how to identify experts who can be consulted for short-term engagements to identify and distill the proper projections and scenarios, and output the most useful data for a particular application. This field of expertise is expected to grow in the coming years.

Kennedy Space Center is on Merritt Island, 3 feet above sea level and at risk to sea level rise.

EXAMPLES

NASA recognized that its spaceports are vulnerable and, realizing it had no systematic approach to address this vulnerability, began site-specific risk characterizations.

More than two-thirds of NASA's developed property is within 16 feet of sea level, so it began hosting a series of interagency workshops at Kennedy Space Center to consider risk and how tools such as modeling can inform that effort. The Climate Adaptation Science Investigator (CASI) work group leads the effort, which acknowledges the four major climate modeling groups in the US infrastructure:

- NASA Goddard Space Flight Center (GSFC), MD
- NASA GSFC/Goddard Institute for Space Studies, NY
- NOAA Geophysical Fluid Dynamics Laboratory, NJ
- National Science Foundation National Center for Atmospheric Research, CO.

The work group intends to leverage these assets to best process the available information, using the most relevant data to develop an agency-wide, facility-specific plan for managing risk.

Use Emerging Media to Manage First Response

Recommendation: *Investigate strategies and tools to leverage social media to help manage crisis response efforts that may touch upon your organization's facilities or other assets.*

CHALLENGE

We know that severe weather events are likely to occur more frequently in the decades to come. But, because we still don't know their location or severity, the challenge to stakeholders in the private and public sectors is how to harness social media and mobile technologies to their advantage.

Mobile computing and social media now play an important role in the way we communicate and consume information and media. The relatively recent emergence of popular social media platforms such as YouTube, Facebook, and Twitter have revolutionized the way citizens communicate and engage with media, government, other organizations, and one another.

Social media platforms enable a new kind of "public information stream," an interaction that is the hallmark of Web 2.0, where engagement is as important as informing. If harnessed correctly, social media can give both individuals and organizations a very effective crisis-response tool.

Meanwhile, mobile device capabilities have improved significantly. They allow social media to be used nearly everywhere, connecting people and their experiences and perspectives in ways that weren't possible just a few years ago. However, harnessing meaningful, actionable data from these new public information streams requires careful planning and familiarity with a number of cutting-edge skills. Increasingly, technologies that process natural language and mine text are making it possible for computers to identify patterns of communication across vast quantities of social media content.

But how can emerging media be used as a tool? For example, how can this technology be harnessed to improve our ability to rapidly understand the magnitude and impacts of a weather-related event and shape our response? How can we determine where and how to respond?

SOLUTION

Adaptation Potential	★★
Operational Impacts	★★★
Financial Impacts	★★★
Feasibility and Timing	★★★

What's the first thing you should do in an emergency? Since 1968, the answer has been to call 911. Over the past several years, however, the emergence of social media (such as Twitter and Facebook) has introduced a radically different kind of crisis-response capability.

New, emerging media can complement emergency response efforts. Organizations can take advantage of these media sources to improve this response when it involves their own assets.

Emerging media have allowed us to move past a static publication life cycle and a hierarchical chain of command. Now, media content can be created by anyone, and through Web 2.0 connectivity, can be disseminated instantly and globally. It takes only a moment to send one-way or two-way asynchronous text messages to anyone on the planet via Twitter, and it's possible to do the same with video via YouTube.

The ability to participate in this information stream assumes, of course, that you have a smart phone (such as an iOS-powered iPhone or an Android-powered phone). But, these days, that's an increasingly safe assumption. Of the world's 7 billion people, more than a billion own smart phones (16 percent of the earth's population).[33] Smart phone penetration in the United States is even higher: as of May 2011, 77 million Americans owned a smart phone.[34]

Smart phone technology is in demand, and its capabilities are now the norm. Five years ago, you would've shelled out a lot of cash for a smart phone (on top of an expensive data plan with your carrier); today's market puts this technology in your hands for far less. So we can expect smart phone use to continue to expand.

With easy access to social media, people are embracing these platforms in droves. A 2011 Pew Institute study revealed that 65 percent of adult Internet users spend time on social networking sites—such as Facebook, Twitter, LinkedIn, and MySpace—up from just 5 percent in 2005.[35] Furthermore, at the time of the survey, 43 percent of participants said they'd used social media in the *last day*. So, people are using social media, and they are using it regularly. Investing in a social media strategy for managing first response, therefore, is a solid bet.

What's different about social media is the engagement factor. Consider traditional media and how people interact with it: they consume it. That's it. Social media offers the opportunity for give and take between sender and consumer.

Social media offers organizations the ability to create a valuable layer of information that would not be available otherwise, for the following two reasons:

- *Social information is widely scattered.* The location of "ground-truth" situational awareness cannot be known ahead of time and seldom will be captured automatically by video systems or other sensors. Humans with access to mobile technology can expand access to immediate, relevant information on the ground.
- *Mobile devices provide real-time images.* An increasingly large percentage of the population carries a smart phone, able to take pictures and access the Internet, so they can transmit ground-truth information to the world in real time. Until very recently, this was not possible.

Artist Eric Fischer makes art using the *what* and *where* of social media by tracking geotags, geographical identification metadata that's attached to a photo, tweet, or video. Geotags basically tell you where an individual was at the moment she or he engaged in the use of social media.

Fischer's work with social media platforms, particularly Twitter and Flickr (a photo-sharing service), paints a beautiful picture (Figure 3-5). His view of the continental United States shows how Twitter use is very much an East Coast phenomenon, certainly right up to the Mississippi River, where the social media habits shift.

We don't advocate that stakeholders embrace one medium while eschewing another. Rather, every good plan should take into account what is in use in a particular locality. Plans that fail to consider what works best for the people affected may not have the maximum impact; in the worst case, they'll fail altogether.

Figure 3-5. The Work of Artist Eric Fischer Shows How Social Media Usage Varies across the United States[36]

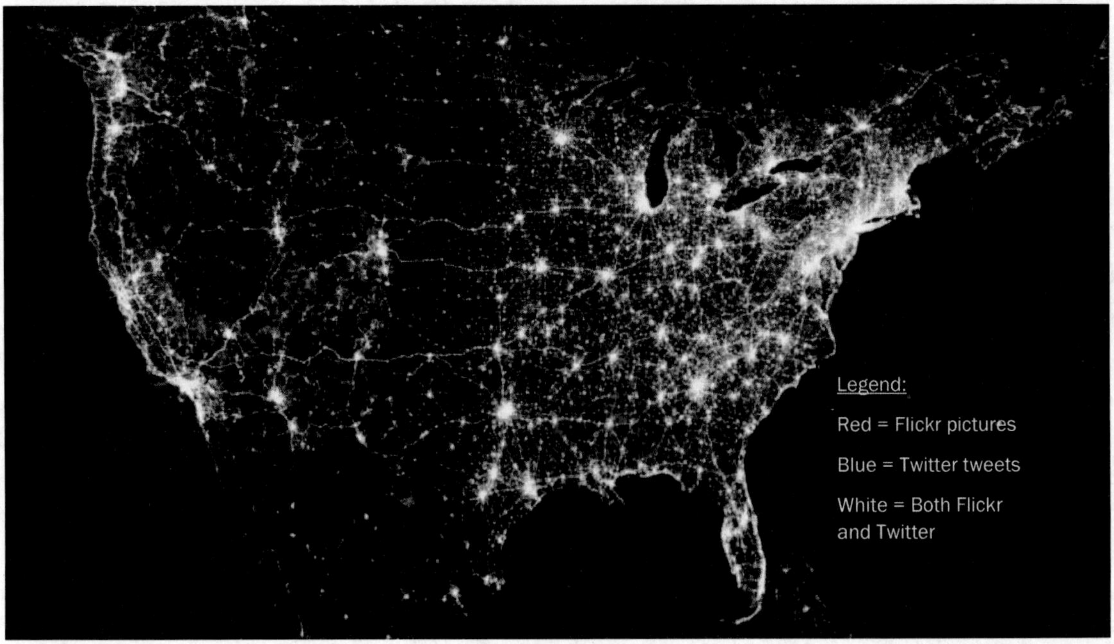

Legend:
Red = Flickr pictures
Blue = Twitter tweets
White = Both Flickr and Twitter

Social media and mobile computing already have had a positive impact on crisis response and disaster relief. Future disaster relief efforts will become more timely and accurate as social media venues continue to develop and expand.

Laying the groundwork that guarantees access to these networks in times of need is one way to adapt to climate change. Organizations can improve their own crisis response capabilities by planning to access the appropriate social networks and tap real-time situational information made available by people "on the ground."

EXAMPLES

In 2010, when a crippling blizzard walloped the American Northeast, Newark, NJ, was hit with a total snowfall of 24.2 inches.

Before the storm, Newark Mayor Cory Booker used Twitter to remind people to pay heed to road conditions and check on seniors. The city's tweets also demonstrated a suitable awareness of the coming storm,[37] offering tips for shoveling and reminding people that garbage collection would be pushed to the following week.

After the blizzard hit, Mayor Booker took to Twitter to engage and interact with those in need.[38] He personally responded to tweets from people who were snowed in, and then he and his staff took to the streets.

Booker found out where someone needed shoveling and showed up with a crew. If a street was blocked, he sent a tweet directing a plow to that neighborhood. A sample tweet from that day: "Just doug [sic] a car out on Springfield Ave and broke the cardinal rule: 'Lift with your Knees!!' I think I left part of my back back there."

Booker's efforts reinforce the notion that social media is communication in the truest sense—not just a one-way stream of information but an opportunity to leverage information from other sources—in this case, citizens on Twitter.

Take Advantage of Crowdsourcing

Recommendation: *Use the emerging capability of crowdsourcing to harness global, collective intelligence and identify new and innovative ideas to adapt to climate change. Deploy intelligent screening methods to sift through ideas and select those that apply to your organization's circumstances.*

Innovate through crowdsourcing

CHALLENGE

Due diligence. It's a phrase tossed around in every organization, not necessarily in a legal sense, but as a reminder that every avenue should be explored, every possibility exhausted. When something falls through the cracks, people wonder, "Didn't they perform their due diligence?"

As much as we might try to think of various ways our organizations can adapt to climate change, we are unlikely to think of everything. The existing and emerging technologies are too many, and there's too much uncertainty. Our knowledge can cover only so much at any given time. Thus, no matter how carefully we seek solutions for adapting to a changing climate, we likely have left out one or more good ideas that others have considered.

SOLUTION

Organizations need to understand that ensuring the best access to a collective intelligence and experience means tapping into emerging ICTs to borrow from others. This strategy will advance our knowledge and ability to adapt to climate change. Crowdsourcing is one means.

Adaptation Potential	★★★
Operational Impacts	★★★
Financial Impacts	★★★
Feasibility and Timing	★★★

Crowdsourcing involves an open call for input, especially in areas traditionally covered by individuals, not professionals. The term was coined by Jeff Howe in a 1996 *Wired* magazine article.[39] The goal is to tap into a community's collective experience and viewpoint to put more options on the table.

Crowdsourcing works because good ideas often come from a variety of sources. ICT can take advantage of collective intelligence to develop and implement new strategies to adapt to climate change. Over the past several years, crowdsourcing has emerged as one of the newest and most effective forms of bringing people together via the web.

In a climate change context, crowdsourcing would involve seeking knowledge from people in different countries, cultures, and climates about how best to adapt to climate change. Such a plan would embrace the fact that climate change is a global phenomenon and that other people are, and will be, facing similar problems in regard to protecting assets and maintaining activity in the face of weather-related events and long-term degradation.

Crowdsourcing enlarges the flow of creative ideas.

Crowdsourcing does two things: it gets people thinking about things that they should consider, and it leverages thoughts regarding events that have already happened. In some cases, the solutions are already out there, but we just didn't know about them.

Technology's role in crowdsourcing is the medium it provides. Crowdsourcing can take place via chat sites, on message boards, or even on specialized networks that deal with questions of this kind. The route chosen is less important than deciding this is the right way to attack a problem.

Organizations must decide whether crowdsourcing makes sense for their particular challenge. This medium has promise, but it has to be used judiciously. After all, putting out an open call for information on how to adapt to climate change may result in thousands of pieces of useless information, and sifting through these responses will take time and effort. Again, this is where the intelligent use of ICT comes in.

Organizations should start by setting some parameters that allow for information to be screened through ICT. Just as you use your personal intelligence to sift through ideas that you read or hear about, you employ an intelligent ICT process to flag or highlight ideas that are sent your way. Properly done, organizations should be able to see fairly quickly the ideas that can be of help and make swift use of them.

EXAMPLES

The Climate CoLab, a project sponsored by the Massachusetts Institute of Technology's Center for Collective Intelligence, employs crowdsourcing to extract ideas and proposals for dealing with climate change from the general public.[40] Individuals are encouraged to submit ideas alone or as members of teams. The center selects among the various submissions, and the winning proposals receive a prize and are featured in briefings for policymakers at the United Nations and Congress.

Some are more successful than others, but that's the nature of proposals. For example, a team of three researchers put forward a proposal for "carbon-negative 'biochar economies.'"[41] The proposal was in response to the website's 2011 contest, which asked entrants "How should the global economy evolve through 2100, given the risks of climate change?"

This particular proposal had 21 supporters and 17 comments, which offered an outside perspective of its feasibility and the challenges it faced. The discussion element led to the proposal's becoming a dynamic work, constantly evolving and hopefully improving. Its interaction is much more immediate and collaborative than any an article in a scientific journal could offer.

Plan adaptation cooperatively

Coordinate ICT Adaptation Planning with Your Organization's Overall Adaptation Plans

Recommendation: *Recognize that ICT and its associated assets are not the only concern of an organization and seek to add value to adaptation strategies by coordinating them with the organization's overall adaptation planning.*

CHALLENGE

The four previous adaptation recommendations are ways to protect an organization's ICT assets in the face of climate change or to use them to gain better insight into how to do so. Although ICT is important to any organization, it is not the only priority.

Any organization planning to adapt to climate change will also focus on protecting key physical assets and personnel while ensuring access to suppliers and financial capital. In all likelihood, ICT adaptation will be more valuable if coordinated with the organization's overall adaptation planning.

The challenge is twofold. First, to coordinate your ICT climate adaptation plans with those of the overall organization, you'll need access to those plans. Ideally, ICT stakeholders would be part of this planning, but even if they aren't, knowing the contents of the plans is critical. It's the only way to determine the scope of the plans covering, for example, weather-related emergency conditions.

Second, stakeholders need to know the relative emphasis the organization places on the things it needs to protect—again, key personnel, facilities, and equipment.

Planning needs to be flexible because ICT, more than any other area, features a variety of emerging technologies and options. A plan must leave room to incorporate a good solution when it reveals itself, even if that solution wasn't part of the original plan.

Adaptation Potential	★
Operational Impacts	★★
Financial Impacts	★
Feasibility and Timing	★★★

SOLUTION

The more information you can obtain on the resources vital to your organization's ongoing performance (and those less so), the better the job you can do in planning an ICT climate change adaptation strategy. This involves using ICT to facilitate risk management for climate change in your organization.

Earlier in this chapter, we focus on managing the risks of climate change to your organization's ICT assets: decreasing their vulnerability to severe weather events that might impair or outright destroy them. This risk management requires investments of various sorts, whether to "harden" sites or equipment, furnish backup resources, gain access to alternative sources, or something else.

The overall organization, however, takes a broader view of climate-related risks. They relate to *all* of the organization's assets in addition to any other source of risk to the organization's function and performance. Even with risks related only to climate change, the organization may have higher priorities than protecting the access and use of ICT assets.

For example, protecting key personnel at various locations might be an organization's highest priority. The possibility of injury, or even death, during a weather disaster might lead the organization to invest in ways to withdraw personnel or build physical protection into facilities that otherwise make this unnecessary.

Here's where an ICT manager comes in. By understanding this priority, improving the organization's overall plan is possible. For example, ICT could allow for tracking personnel by gaining information on their status during a disaster, keeping lines of communication open, and communicating the resources that are immediately available if they are injured or incapacitated. These measures would complement any direct physical protection of personnel and, in some cases, serve as a key part of that protection.

For example, an organization's management team might have incomplete information on a local disaster and withdraw personnel as a precautionary measure. Imagine, for instance, that ICTs provided reasons why a planned withdrawal during a disaster might expose personnel to even greater danger than staying put. ICT assets, by tracking and maintaining a real-time flow of information, identify emerging threats that can be factored into better decisions. The desired result is the same—greater personnel safety—but the constant flow of information provides a greater chance of success, even if it is simply telling the organization to stay the course.

ICT managers will be better positioned to act if they understand other organizational adaptation priorities (facilities, equipment, and access to key input suppliers and perhaps to financial capital). They can work through available technology to engage with suppliers and financial institutions. Because it's impossible to invest in everything, the work of an ICT manager to maximize the value of his or her own adaptation plans is just good business.

EXAMPLES

The idea of integrating an ICT risk analysis into a broader organizational risk assessment is relatively new, but it's similar to what Cisco has attempted with its Enterprise Risk Management (ERM) initiative and how this initiative fits into the organization's strategy as whole.

Cisco established its ERM team in 2004 with a goal of locating places to leverage risks, which are viewed by the organization as opportunities.[42] Cisco views effective ERM as a business enabler for meeting and exceeding corporate objectives.

The ERM team meets regularly with its executive sponsors and with Cisco's Risk Review Group, which consists of representatives of the IT, finance, human resources, and supply chain functions. Members of the ERM team are not full-time in ERM, but a multidisciplinary global group that represents insurance, operational risk, financial risk, and legal liability processes.

The focus on ERM at Cisco is directly connected to the corporate strategy in three areas:

1. Protect: "How do I reduce business risks?"
2. Optimize: "Is my current risk level in control?"
3. Grow: "How do I take more intelligent risks?"

By focusing on these three ERM areas, Cisco embeds cross-functional risk management into its core DNA.

The ERM process begins by determining risk priorities for the initiative. The team conducts interviews to support its risk analysis, covering multiple functional areas. In these interviews, it seeks not only to identify perceptions of key risks facing the company, but to gauge the probability, severity, and current effectiveness of managing the risk.

The interviews are followed by discussion and risk workshops with stakeholders where response to risk is a key topic. The group uses consistent metrics for assessing risk probabilities and severities company-wide.

The group then develops probability, severity, and management effectiveness scales, along with a risk inventory framework that can be applied across the organization. This framework includes specific external risk catalysts with internal concerns, which are broken into three areas: strategic, operational, and financial. Operational risks are further subdivided into operational processes, management information, human capital, integrity, and technology-related risks.

Cisco's ultimate goals are to raise awareness for ERM within the company and externally, while integrating already existent risk-management processes from the areas of investment management, strategic planning, and business development. The company believes that, at its core, ERM relates directly to adding strategic value.

Final Thoughts

One of the toughest things about technology is how quickly it changes. Seeing the new possibilities out there can be a great windfall to an organization. It can be equally maddening for those trying to plot a course on the basis of what they know now and ensure that any planning assumptions will remain valid.

Our society's ability to mitigate climate change will depend, in many ways, on our willingness to develop data-driven energy policies and make smart energy decisions. ICT is there to support such decisions, and vast quantities of information can, and must, be gathered, integrated, assessed, and rendered visually. These are big data, which, properly harnessed, can enable "big solutions."

Making sense of these big data will be critical in carbon-reduction strategies and tactics. As it matures, ICT is becoming so engrained in our lives that we have no choice but to look for ways to balance the equations where output and reductions make sense for the greater effort to curb climate change.

Because ICTs are the lynchpins for this, the ICT industry must evolve to create greener computing footprints, such as cloud computing and other green computing technologies. Managers in organizations of all sizes will need to consider how these advancements can reduce energy consumption and costs—in the near term and in the years ahead. And that long view is critical. An investment now doesn't ensure success later, but that's a case-by-case determination that should not be left out of the strategic planning process. Near-term costs may result in significant long-term gains.

Current economic conditions demand attention to where money is spent and where to get more for less. Organizations must become, and remain, efficient, nimble, and mission-focused in the eyes of stakeholders, investors, the public, and other watchful groups. The power of ICT to connect information and people more effectively can help organizations make smarter decisions to mitigate and adapt to climate change.

[1] http://www.comscore.com/Press_Events/Press_Releases/2011/9/comScore_Media_Metrix_Ranks_Top_50_U.S._Web_Properties_for_August_2011.
[2] The Climate Group, *SMART 2020: Enabling the low carbon economy in the information age* http://www.smart2020.org/_assets/files/03_Smart2020Report_lo_res.pdf.
[3] Greenpeace "Make IT Green: Cloud Computing and its Contribution to Climate Change", March 30, 2010 http://www.greenpeace.org/raw/content/usa/press-center/reports4/make-it-green-cloud-computing.pdf.
[4] Urs Holzle, "Powering a Google Search" *Google Official Blog*, January 12, 2009, http://googleblog.blogspot.com/2009/01/powering-google-search.html.
[5] "qSearch: A Comprehensive View of the Search Landscape" *comScore*, http://www.comscore.com/Products_Services/Product_Index/qSearch.
[6] Based on the Federal Highway Administration's statistics, the average driver in the United States is behind the wheel for 13,476 miles, http://www.fhwa.dot.gov/ohim/onh00/bar8.htm.
[7] "Federal Data Center Consolidation Initiative," CIO.gov, February 26, 2010, http://www.cio.gov/documents_details.cfm/uid/25A781B7-BDBE-6B59-F86D3F2751E5CB43/structure/Information%20Technology/category/Federal%20Data%20Center%20Consolidation%20Initiative.
[8] *Bloomberg Businessweek*, January 20, 2011.
[9] Kate Lister. "The Who, What, Where, and Why Not of Telecommuting." *Work Shifting,* June 26, 2011, accessed on July 12, 2011. http://www.workshifting.com/2011/06/the-who-what-where-and-why-not-of-telecommuting.html.
[10] Telework Research Network, see Note 4.
[11] TIAX LCC"The Energy and Greenhouse Gas Emissions Impact of Telecommuting and e-Commerce," July 2007.
[12] "Sun Microsystems Study Findsopen Work Program Saves Employees Time and Money, Decreases Carbon Output," *PRWeb*, June 9, 2008, http://www.prweb.com/releases/sun_microsystems/open_work/prweb1009224.htm.
[13] The figure is taken from LMI EA Practice Handbook, 2009.
[14] "Federal Data Center Consolidation Initiative FAQs," *CIO,gov*, June30,2010, http://www.cio.gov/documents_details.cfm/uid/68449CA0-BDBE-6B59-F29E2AC3CE863CFB/structure/Information%20Technology/category/Federal%20Data%20Center%20Consolidation%20Initiative.
[15] Steve Riley, "Q&A on Federal Data Center Consolidation Initiative," *Riverbed Blog*, September 13, 2011, http://blog.riverbed.com/2011/09/riverbed-technical-lead-steve-riley-qa-on-federal-data-center-consolidation-initiative.html.

[16] Martin Owen and Robert Shields, "Using Enterprise Architecture to Develop a Blueprint for Improving your Energy Efficiency and Reducing your Carbon Footprint," IBM, White Paper, December 2008.
[17] According to one recent study, US businesses potentially can save up to $12.3 billion and 85.7 million metric tons of carbon per year by 2020 through the use of cloud computing. See Carbon Disclosure Project, "Cloud Computing: The IT Solution for the 21st Century," July 20, 2011.
[18] Google Data Centers, http://www.google.com/about/datacenters/locations/index.html.
[19] Don Clark, "Server Startup SeaMicro First to use Intel Chip," *Wall Street Journal* Blog, http://blogs.wsj.com/digits/2011/02/28/server-startup-seamicro-first-to-use-new-intel-chip/.
[20] David Marshall, "Facebook gives the nod to Intel micro servers over virtualization," *Infoworld,* March 2011, http://www.infoworld.com/d/virtualization/facebook-gives-the-nod-intel-micro-servers-over-virtualization-109.
[21] Urs Holzle. "Brawny cores still beat wimpy cores, most of the time," http://static.googleusercontent.com/external_content/untrusted_dlcp/research.google.com/en/us/pubs/archive/36448.pdf.
[22] Healthimaging.com, "Acumen nabs ONC Cloud Computing Contract," February 16, 2010.
[23] Darrell M. West, "Saving Money Through Cloud Computing," Brookings Institution, April 7, 2010.
[24] *Cloud Computing—The IT Solution for the 21st Century*, The Carbon Disclosure Project, 211.https://www.cdproject.net/en-US/WhatWeDo/Pages/Cloud-Computing.aspx.
[25] "Green Information Technology," GAO-11-638, July 2011. The referenced report is Technology CEO Council, "One Trillion Reasons: How Commercial Best Practices to Maximize Productivity Can Save Taxpayer Money and Enhance Government Services," October 2010
[26] Naoya Motegi and others, "Case Studies of Energy Information Systems and Related Technology: Operational Practices, Costs, and Benefits," International Conference for Enhanced Building Operations, October 2003.
[27] Energy Use Dashboard, Oberlin College, http://www.oberlin.edu/dormenergy/.
[28] "5 Uses for QR Codes in Construction," *BuildingProductMarketing.com*, http://www.buildingproductmarketing.com/2010/06/5-uses-for-qr-codes-in-construction.html.
[29] American Chemical society, "Global Climate Change," http://portal.acs.org/portal/acs/corg/content?_nfpb=true&_pageLabel=PP_SUPERARTICLE&node_id=1907&use_sec=false&sec_url_var=region1&__uuid=0cbd57b5-5766-456d-800b-680b88c1c8bf.
[30] "Protecting Health from Climate Change," http://www.who.int/world-health-day/toolkit/report_web.pdf.
[31] Dan Farber, "Sun's Data Center in a Box," *ZDNet.com* October 17th, 2006, http://www.zdnet.com/blog/btl/suns-data-center-in-a-box/3790.

[32] Ryan Block, "Sun's Project Blackbox – Data Center in a Container," *Engadget,com* October 18, 2006,
http://www.engadget.com/2006/10/18/suns-project-blackbox-datacenter-in-a-container/.
[33] "Infographic: Moblie Statistics,Stats and Facts2011," *Digital Buzz Blog* April 4, 2011, http://www.digitalbuzzblog.com/2011-mobile-statistics-stats-facts-marketing-infographic/.
[34] "comScore Reports April 2011 U.S. Mobile Subscriber Market Share" *comScore* http://www.comscore.com/Press_Events/Press_Releases/2011/6/comScore_Reports_April_2011_US_Mobile_Subscriber_Market_Share
[35] Mary Madden and Kathryn Zickuhr, "65% of online Adults Use Social Networking Sites," *Pew Internet*, August 26, 2011,
http://www.pewinternet.org/Reports/2011/Social-Networking-Sites/Overview/Findings.aspx.
[36] Eric Fischer, North American detail map of Flickr and Twitter locations,
http://www.flickr.com/photos/walkingsf/5912385701/in/set-72157627140310742.
[37] City of Newark, "It's Official! Blizzard Warning Issued," *Twitter*, December 26, 2010,
http://twitter.com/#!/CityofNewarkNJ/status/19083011622445056.
[38] Sean Gregory, "Cory Booker: The Mayor of Twitter and Blizzard Superhero," *Time U.S.*, December 29, 2010,
http://www.time.com/time/nation/article/0,8599,2039945,00.html.
[39] Jeff Howe, "The Rise of Crowdsourcing, "*Wired*, June 2006,
http://www.wired.com/wired/archive/14.06/crowds.html.
[40] MIT Center of Collection Intelligence Climate CoLab,
http://climatecolab.org/.
[41] "Proposals: Carbon-negative 'biochar economies,'"
http://climatecolab.org/web/guest/plans/-/plans/contestId/4/planId/14637.
[42] Rob Rolfsen, Cisco Systmes, Inc., Speaker at NC State University ERM Roundtable, "ERM at Cisco: taking it to the Next Level," March 23, 2007,
http://www.poole.ncsu.edu/erm/index.php/articles/entry/rob-rolfson-roundtable/.

Land

The boy listened attentively as his father explained the farm's history. Two generations ago, the operation was of reasonable size: the family made enough from the wheat and salad lettuces they grew to live comfortably. However, a small, but significant, change in what the farm grew altered the family's fortunes. When rising sea levels changed the nature of the land around them, the boy's family tried growing barley in place of wheat and tomatoes and asparagus to replace lettuce. When the rising water became even saltier, the family was prepared; the farm was already growing crops that could flourish in this new environment. "That," his father told him, "is how we became the East Coast's largest producer of marsh mallow for bioenergy, not only oil from the seeds but gasohol from the leaves and stems."

Lay of the Land

The term "foundation" can be somewhat cliché, but the land on which we live is truly the foundation of our existence. The land provides us with food, shelter, and resources; these resources spur technology and progress. We're not alone in reaping these benefits; the land also offers safe haven for insects and pests that destroy food stocks and cause disease. So, considering how much we rely on the land around us, we can clearly see why much of the discussion of climate change has revolved around the land. The two are inseparable and affect each other in profound ways.

There's also an inescapable relationship between land and sea. Sea level rises, higher ocean temperatures, variable water supplies, and salt water intrusion have real, pronounced effects on land.

The management implications for protecting species, biological communities, and physical resources within finite land management boundaries in a rapidly changing climate are complex and without precedent.

—Jon Jarvis, Director, National Park Service, October 28, 2009

The land, and its influence over us, has been considered by great thinkers for thousands of years. Even in the earliest days of land use, man had an understanding of the bigger picture of human action. Plato, for one, wrote in 400 BC that deforestation could lead to soil erosion and the drying of springs.

Land use is the result of a number of concerted, conscious efforts to manage human activities that affect the land, including product use and its benefits. It also includes all the things gained from these activities. Land, and the way humans use it, affects the concentration of global greenhouse gas (GHG). Certain land-use activities contribute to the release of GHGs, while others help sequester these gases.

One consideration regarding the land actually stems from the atmosphere. The concentration of atmospheric carbon, along with other GHGs, is increasing. Put another way, the rate of CO_2 emission into the atmosphere (which is primarily from fossil fuel combustion) exceeds the rate at which natural filters such as trees take CO_2 away from the atmosphere through absorption. The amount of carbon movement is massive: gigatons—or billions of tons—of carbon (GtC).

This "carbon cycle" is a dynamic progression involving complex time scales and multidirectional flows that depend on chemical and biological processes. It includes the seasonal growth of plants, the formation of soils, the life cycle of algae, and even human metabolism.

Figure 4-1 illustrates this complexity, known as the global carbon flux, which includes considering global carbon reservoirs. Fossil fuel emissions from North America were 1,856 million tons per year as of 2003,[a] accounting for 27 percent of all fossil fuel emissions globally. Over the same period, North America[b] absorbed 505 million tons in carbon sinks—or approximately 30 percent of the continent's total fossil fuel emissions.[1]

The United States accounts for approximately 40 percent of the global carbon sink, mainly because of the country's unique historical land use, a side note to our nation's history. Following European colonization in the 17th and 18th centuries, large amounts of carbon were

---

[a] North America refers to Mexico, the United States, and Canada.

[b] North American carbon storage compartments (sinks that remove CO2 from the atmosphere), take in less carbon than released by burning fossil fuels—even while forests store more of the carbon than any other terrestrial sink.

released when large tracts of the eastern United States were converted from forests into agricultural lands.

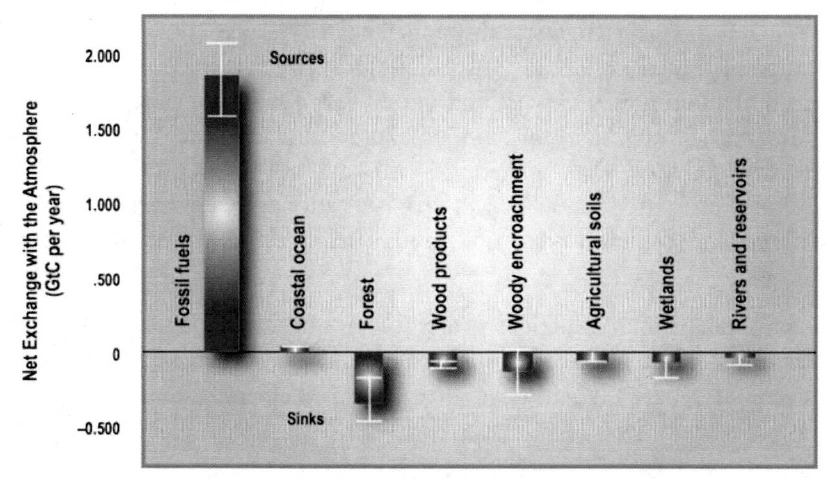

Figure 4-1. North American Carbon Sources and Sinks

This cycle has continued, but recently the agricultural focus of this converted land has been abandoned; as a result, up to half of current US carbon sink may be due to the subsequent and ongoing regrowth of eastern forests.[2] In fact, remember those emissions from the original settlements? Recapturing them (a one-time thing) may be one cause of the current sink rate.

Mitigation: Markets Abound

Mitigation of climate change will require reducing the amount of GHG in the portion of the carbon cycle that is the atmosphere. This is accomplished by reducing emissions, increasing sequestration, or both. Land use for mitigation not only provides a way to reduce emissions, but it also provides a sink for atmospheric carbon.

That said, it's a difficult process to measure accurately. As a result, land-use practices specifically for climate change mitigation can be difficult to prescribe.

Scientists can document the accumulation of carbon in some ecosystems extremely accurately; for example, the Forest Service measures carbon accumulation in trees at more than 200,000 locations across the country as part of their Forest Inventory and Analysis Program. However, such precise measurements are not always possible, certainly not for other important measures such as the amount of carbon that accumulates in the soil (which is measured much less accurately). Similarly, scientists have had mixed results with studies that explore how increased carbon and nitrogen in the atmosphere impact forest carbon sequestration rates over time.[3]

Climate change mitigation through land use is also challenging because of directives already in place that impact the way we use land in the United States. A vast and varied set of federal agencies—each pursuing its individual mission—make up (and complicate) the effort to manage lands. They manage federal lands for forest products, agricultural crop production, wildlife preservation, livestock grazing, resource and mineral extraction, and warfare training, among other things.

Land-management practices geared toward climate change mitigation *may* be consistent with other management goals, but when they are not, their benefits must be weighed against their impact on those other goals.

As you can see, even if the federal government were the sole landowner in the United States, managing land for climate change mitigation would be a complicated process. But the federal government, even as the nation's largest landowner, still owns just 30 percent of the total land area.[4] The remaining land is owned by state governments, local governments, Native American tribes, corporations, and private landowners—and each manages its lands independently. This patchwork of landownership and management goals makes comprehensive, effective carbon management much more difficult and expands the set of stakeholders substantially.

The combination of diverse land uses and various stakeholders means each entity must create a mitigation strategy. People deserving seats at the table include planners, regulators, managers of public and private land, and forest and water managers. Policymakers must be involved as well because much of the overall action needed to respond to climate change requires enhancements to current policy or, at least, new interpretations of existing laws and policy. And this, in itself, further widens the audience. In spite of the challenges presented by an enormous and diverse group of stakeholders, the potential for reducing atmospheric GHG through land-use strategies remains significant. Forestry and agriculture produce 30.9 percent of the GHG emissions from human activities (Figure 4-2). This is because of the reliance of these two sectors on fossil fuels to power equipment and transport products; they also release GHG because of the very nature of their activity of harvesting carbon-rich products, such as crops and wood, and generate a great deal of carbon-rich waste.

Figure 4-2. Proportion of GHG Emissions from Various Human Activities

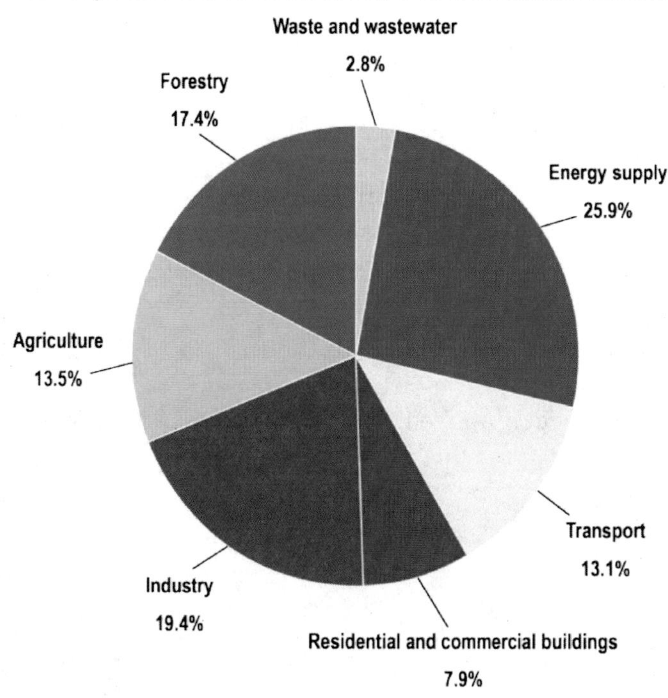

Combining savings in fossil fuel use in those land-use sectors with the land's innate ability to capture CO_2 can mitigate climate change. For example, trees in forests and urban settings store carbon in their biomass and account for a significant fraction of all CO_2 transitioning out of the atmosphere. A change, then, in forestry practices can impact that storage.

A number of land-management techniques or land uses can help reduce GHG emissions. Some may help pay their own way, while others may be less cost-effective unless a price mechanism for atmospheric GHG is present. Although these strategies lack easily quantifiable outcomes, they have clear payoffs in GHG reduction.

Our key mitigation recommendations for those with a vested interest in land use and management are as follows:

- Mitigate through local land use at the local regulation level.
- Manage forest lands to increase GHG sequestration.
- Use climate-friendly livestock practices.
- Adjust crop practices to maximize sequestration and minimize emission of GHG.
- Ensure market-based tradable offsets effectively reduce GHG emission.

Mitigate through Local Land Use at the Local Regulation Level

Take advantage of local land-use regulations

Recommendation: *Leverage local regulations as a way to mitigate while avoiding cumbersome political battles.*

CHALLENGE

Several obstacles can prevent effective mitigation of climate change through land use. Government can enact regulations that support mitigation efforts, but the higher the level of government, the more cumbersome this process can be.

Some practices widely held as beneficial in mitigating climate change—such as emissions reductions, cap-and-trade programs, and landscape protection—are only likely at certain levels of government (Table 4-1).

Table 4-1. Mitigation Activities That Can Be Accomplished at Various Levels of Government

Government	Primary	Indirect	Ancillary
International	Binding emission reductions	———	International conservation lands
National	Cap-and-trade program Emission reductions	Conservation land uses Land reform Banking and trading oversite services	Development of new biological materials Green highways Livestock research
State	Emission reductions State carbon banking and trading	Permitting local ordinance changes Conservation land reform Regulating CAFOs Promoting land application of carbon from agriculture and municipal waste	Outreach Livestock reform
Local	Landscape use rules Landscape protection ordinances Emission reductions	———	Retrofitting public buildings Encourage building soil carbon

As an example, binding emissions reductions can play a substantial role in mitigation, but enacting them would require international action to be truly effective. Such action would necessitate international treaties and consensus building, a difficult, lengthy process that is as likely to fail as it is to result in meaningful change.

In fact, no international body currently possesses the ability to accomplish such an initiative. None has the clout to develop and establish an international mitigation framework (or set of equivalencies) that can provide incentives in the market to change behavior with regard to climate change.

Things aren't any easier at the national level. US federal authority is generally limited to direct regulation of industry. Federal control of land use on private property is greatly limited. In fact, there isn't a single federal agency that can stipulate land-use practices. The rare cases where land use can be regulated involve off-the-property consequences—the Clean Water Act, for example. These cases preserve watersheds for their intended uses by enforcement on property, but only if the practices could result in a degradation of water quality off of the property. It's too narrow a scope to be effective in mitigation.

SOLUTION

Mitigation Potential	★★
Operational Impacts	★
Financial Impacts	★★
Feasibility and Timing	★★★

Local and regional governments are in the unique and enviable position of being able to adopt land-use regulations that could meaningfully contribute to climate change mitigation. This is true because local governments are an effective testing ground for policy regimes, more so because they act as a catalyst and proof of concept, which can influence national and international public opinion and practice.

One important local government mechanism for land-use regulation in the United States is zoning. Several state and municipal governments already have regulatory structures in place, such as building codes and zoning ordinances. There is ample opportunity to improve existing policies to better focus on reducing potential GHG emissions.

Plantings in a cloverleaf can be selected to increase carbon storage.

Zoning for mixed-use commercial/residential developments encourages residential construction directly adjacent to the business and commercial sites that serve nearby residents, reducing their need for GHG-emitting transportation. Similarly, a local government may use the building code and zoning ordinances to encourage conservation easements or other large-scale undeveloped lands, which help to remove GHGs from the atmosphere. Green spaces and plantings also help diminish the effect of urban heat islands (discussed in the "Health" chapter).

Small projects can add up to make a big difference. Examples abound of planters in the medians of roadways, where simple jersey barriers are employed and soil brought in from local composting facilities. Likewise, many cities have adopted tree guidelines stating that every inch of tree removed must be replaced by replanting a similar number of inches of new trees. These projects are effective at reducing heat islands as well as sequestering carbon. While this can only be accomplished on a local level, wide adoption of these simple measures can have a significant effect.

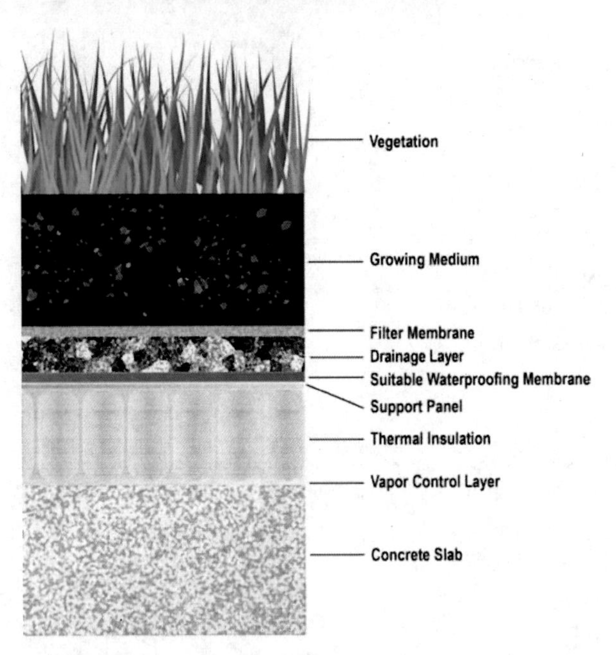

Elements of green roof construction.

One mitigation opportunity that can be regulated through zoning and building codes is the promotion of green rooftops (discussed in the "Health" chapter and significant to the "Structure" chapter). A green roof can be planted on almost any flat-roof surface. Most flat-roof engineering comprises a membrane and gravel (ballast), a construction with a life cycle of 10–15 years. Generally, at the time of replacement, an engineered roof can be switched to a green roof without an increase in the roof weight load that requires additional engineering support.

Plants selected for the roof are known to be healthy in the local climate and under harsh conditions, while sequestering small amounts of GHG. Green roofs are also known to ameliorate both building and ambient temperatures, reducing the energy needed for heating and cooling. Plants help protect the membrane from sun degradation, increasing its life span 15–20 years. This added benefit conserves on the energy needed to construct roofing material.

EXAMPLES

The city of North Vancouver, British Columbia, recently completed a comprehensive effort to consider how specific land use, building form, transportation, and infrastructure decisions contribute to the city's energy use and GHG emissions. A new development plan uses a standard means of measuring and comparing alternative land-use scenarios and evaluates them on the basis of GHG emissions.[5]

The resulting 100-Year Sustainability Vision incorporates green infrastructure investments, increased transportation and mobility options, and an increase in public spaces. In 2010, the city adopted a *Greenhouse Gas Emission Reduction* plan, which sets GHG reduction goals and an aggressive timeline for reaching them. To achieve their objectives, the city council approved programs to reduce energy use, add passive design elements in new buildings, encourage bicycle lanes, and require electric vehicle charging stations in multifamily dwellings.[6]

RECOMMENDATION

Sequester carbon in forests

Manage Forest Lands to Increase GHG Sequestration

Recommendation: *Target forest management to maximize GHG sequestration, including increasing overall forest cover.*

CHALLENGE

The United States has 749 million acres of forest land. The role of these forests in storing carbon is undeniable.

Forests store 84 percent of the world's aboveground carbon and 73 percent of its soil carbon. Forests have been routinely managed for many different outcomes, but only lately some forest managers have begun to manage their land for mitigation of GHG.

Today's management outcomes for forests vary greatly. Most forest managers practice techniques such as harvesting or extreme fire suppression. From there, things splinter substantially.

The National Park Service, for one, manages heritage resources on its lands;[7] conserves the scenery, natural and historic objects, and wildlife; and provides for enjoyment of these features by the American public.[8] The Forest Service, on the other hand, manages its forests to secure favorable water-flow conditions while producing a continuous supply of timber to meet the nation's needs.[9]

Meanwhile, the roster of agencies seeking different forest management outcomes includes the Bureaus of Land Management and Reclamation, the US Fish and Wildlife Service, and Department of Defense. Another group of stakeholders—private landowners—may be included as well. Most global deforestation, on the other hand, is the result of converting land for agricultural production.[10]

Some management strategies could decrease the forest's ability to sequester GHG or, worse, increase the output of GHG by the forest due to catastrophic events like wildfires. Adapting the assorted objectives of various agencies or private owners for GHG mitigation requires a set of techniques that can be implemented broadly.

SOLUTION

Mitigation Potential	★★
Operational Impacts	★★
Financial Impacts	★★
Feasibility and Timing	★★★

Numerous factors contribute to a forest's ability to sequester carbon from the atmosphere, including temperature, light exposure, species mix, presence of fire, and age of the forest. Unfortunately, the interplay of these factors, along with the expectation for increased atmospheric CO_2, makes crafting specific forest management recommendations extremely difficult. What works in a forest in the Pacific Northwest may not work in the Southeast.

Some straightforward and simple actions, however, can improve land cover sequestration with large payoffs. *Afforestation* is the establishment of a forest or stand of trees in areas where they did not previously exist; *reforestation*, on the other hand, involves planting an area that has been historically forested. Both are important steps toward mitigation because forests have the capacity to sequester more carbon from the atmosphere compared with grasslands, agricultural uses, or shrub-scrub environments.

Prescribed burns limit the chance of catastrophic wildfires.

Increasing the planned period for harvest in a managed forest and prescribing burns can drastically increase the CO_2 stored in a tree's biomass. Prescribed burning contributes to extending the period between timber harvests and allows wildfires to burn at lower temperatures, which causes less forest damage and releases less CO_2.[11] Wildfires often destroy large trees that store significant amounts of carbon. Prescribed fires, on the other hand, burn things that store less carbon—underbrush and small trees, for example. By clearing out the underbrush, controlled burns reduce the chance of subsequent high-severity wildfires. They protect large trees and allow more carbon to be sequestered in the forest.

EXAMPLES

Measurements taken in old-growth forests in Washington and Oregon show that the most effective way to harvest trees and conserve CO_2 at the same time is to increase the length of time between cutting trees.[12] Harvesting timber on a 100- or 120-year rotation, rather than a 40- to 60-year cycle, increases the total amount of CO_2 stored both in the trees and in the forest soil. Leaving some trees to become genuine old growth while harvesting a select number of surrounding trees on a century-long cycle would keep more CO_2 locked up. This practice would allow more timber to be harvested while storing more carbon in the landscape.[13]

Satellite observations and computer models show that fire emissions of CO_2 in the western United States can be reduced 18 to 25 percent by using prescribed burns.[14] In fact, certain forest systems might find those emissions reduced by as much as 60 percent.[15]

Use Climate-Friendly Livestock Practices

RECOMMENDATION

Reduce livestock emissions

Recommendation: *Reduce the use of fossil fuel in raising animals and adopt other livestock management practices considered "climate friendly."*

CHALLENGE

The demand for dietary protein from livestock is increasing. This has the negative GHG effect of promoting deforestation to provide pasture for livestock in developing nations. In addition, livestock production creates prodigious amounts of GHG, including methane, nitrous oxide, and CO_2. Animal waste and livestock production, meanwhile, account for nearly half of all GHG emissions for the agriculture segment and as much as 14.5 percent of all human GHG emissions.[16]

Gases associated with ruminant digestion (belching and flatulence) in livestock alone accounts for 19 percent of US total methane—even exhaling carries significant amounts of methane and other GHG out of the animal.[14] Additional GHG is released during the combustion of fossil fuels used in the production and transportation of feed and animal waste. Suggesting a reduction in the demand for meat and dairy products, which could potentially decrease GHG emissions associated with livestock operations, is not realistic. In fact, global demand for meat and dairy products, particularly in the developing world, is expected to increase.

SOLUTION

Mitigation Potential	★
Operational Impacts	★★
Financial Impacts	★★
Feasibility and Timing	★

Livestock managers can reduce methane emissions associated with ruminant digestion by altering what they feed their livestock. Certain feed can be managed for specific metabolic requirements of the animals consuming them. This type of management may increase animal productivity, thereby reducing the total methane emissions over an animal's lifetime. For example, distiller grain is high in protein, which promotes muscle growth and is easily digested.

Similarly, maintaining animal health and providing proper veterinary care promotes growth and leads to greater efficiency in animal production.[17]

Manure management practices can also significantly reduce methane produced from livestock manure. Two general options can diminish the methane emissions associated with livestock manure. The first involves using a dry manure management system instead of a liquid system. The other requires the recovery of released methane for use in energy production, thereby reducing demand for fossil fuels.

Technological improvements in meat and dairy production continue to improve overall efficiency as well. For example, a beef cow that can reach the appropriate weight of slaughter 20 days earlier than other types will release less methane over its lifetime. Through husbandry, selective breeding, and feed management techniques, livestock productivity can be increased and GHG emissions reduced.

EXAMPLES

The 80-year old Castelanelli Brothers Dairy, located on a 550-acre farm in Lodi, CA, recently expanded its operations to not only produce milk but to produce electricity. In addition to the 13,000 gallons of milk per day produced at the dairy, more than 4,000 kWh/day of electricity is generated from methane captured in a covered anaerobic lagoon.

By replacing its two uncovered anaerobic manure lagoons with a covered lagoon digester system, the Castelanelli Brothers Dairy has also reduced its annual methane release by approximately 20,000 tons of equivalent CO_2 per year. In addition to a reduction in GHG emissions, the dairy has benefited from a significant reduction in odor and an increase in net farm income as a result of the electricity generation.[18]

Adapt agricultural practices

Adjust Crop Practices to Maximize Sequestration and Minimize Emission of GHG

Recommendation: *Fine-tune agriculture practices to reflect the local environment with a focus on farming techniques that provide sequestration of GHG while minimizing emissions.*

CHALLENGE

Because agriculture is one of our largest and most important land uses, modern agricultural practices emphasize the commodity. The result is a system that consumes large amounts of resources. Fossil fuel input for agriculture is immense; it includes fuel for equipment operation, irrigation, and transport, along with fuel and feedstock for manufacturing fertilizers and pesticides.

The selection process for plants places value on high and immediate yields and emphasizes monoculture (where a single crop is grown over a large area). Wheat, rice, and corn, for example, cover half of the world's available croplands, even though more than 3,000 species of edible plants are known.

The widespread use of chemical fertilizers and constant soil disturbance, along with reduced sequestration, have a high GHG emissions cost.

Mitigation Potential	★★★
Operational Impacts	★★★
Financial Impacts	★★★
Feasibility and Timing	★★★

SOLUTION

Changes in agricultural practices could greatly reduce GHG levels given the large areas farmed and amount of fossil fuels used. These agricultural practices could be improved to better reflect the local environment. This is not the general practice.

Changes to modern agricultural techniques could reduce the need for fuel inputs (and their resulting GHG emissions), while sequestering large amounts of CO_2 and other GHGs. These approaches include the following:

- No-till agriculture, which uses less fuel, and therefore can reduce GHG while increasing the buildup of carbon in the soil
- Production of food locally, which decreases production of GHG that results from the long distances required to transport fruits and vegetables to market
- Perennial agriculture, which can act like no-till when it comes to fuel use and carbon preservation and allows further buildup of belowground living biomass (and carbon) in the perennial plants
- Adaptations to farming practices that can diminish the need for fertilizer
- Adaptations that incorporate organic matter into cultivated areas, such as poultry litter and animal manure, which can increase fertility and soil health, decreasing the need for fertilizer.

Agroforestry is an integrated approach to farming where crops and livestock coexist with perennial plants such as trees, grasses, and shrubs. Lately, perennial *crops* have been developed (most crops are annuals), including grasses for biofuels. Researchers have also developed perennial relatives of rice, sorghum, wheat, and sunflowers (oilseeds) that could provide alternatives to conventional annual crops.

Marginal land, degraded watersheds, and rangelands could be restored using plant materials suited to the soil type and slope. Because vegetation on the landscape holds more GHG than bare and eroding soils, restoring any degraded rangelands and watersheds helps reestablish their GHG-sequestering ability.

Another widely touted agricultural use for marginally or highly erodible land is to grow native grasses such as switch grass for the purpose of producing biomass as feedstock for fuel. The benefits of this approach are obvious: every gallon of fuel generated from renewable resources reduces the need for the release of GHG from fossil fuels. Restoring marginal soil also relieves pressure on good agricultural land for production of food and provides revenue from land that could otherwise not be affordably managed.

EXAMPLE

One perennial wheat strain developed and tested in Washington State produces almost 64 percent as much as conventional wheat varieties across several areas and 93 percent as much in Pullman, WA.[19] The strain has also shown resiliency, with confirmed field regrowth among the 4,000 perennial head rows planted at the Spillman Farm in Pullman. Trial plantings of the strain in other areas of the state, including Mt. Vernon, WA, have regrown for 4 consecutive years. The strain has also proved to be resistant to stripe rust, a small-grain wheat disease that is common in the state.

This strategy is significant because of what perennial cropping offers: reduced tilling, which, in turn, keeps carbon in the soil and cuts costs for fuel. Developing perennial strains for important commodity crops would offer a greater opportunity for sequestration of carbon in soils and, to a certain extent, in biomass—especially compared with traditional annual cropping systems. The prospects are also strong for marketing currently known perennial crops to build awareness and to cater to those who place a premium on sustainable food sources.

Today, the US Department of Agriculture (USDA) has several conservation tillage support programs. The Conservation Resource Enhancement Program (CREP) gives farmers and ranchers cash assistance for reclamation or credit for practicing good agricultural management. Presently, these programs are aimed at protecting watersheds, but they are very flexible. Texas, for example, has proposed using the CREP approach to develop a system that would provide ranchers and farmers with incentives in exchange for their help in controlling invasive species such as carrizo cane and salt cedar.

Ensure Market-Based Tradable Offsets Effectively Reduce GHG Emissions

Trade carbon credits

Recommendation: *Utilize verification and evaluation strategies to ensure that any market-based carbon credit actually reduces GHG from that under a baseline scenario.*

CHALLENGE

Some climate change scientists, economists, and policy professionals believe that a market-based carbon exchange is an important component of any effective climate change mitigation strategy. In a market, active participants earn and then sell carbon credits for taking steps that either reduce their operational emissions or sequester atmospheric GHG. Tradable credits represent an important opportunity to reduce emissions since they allow for reductions in the most cost-effective places.

A good example is tradable renewable energy credits. Power plants in several states have requirements for creating some percentage of their power from renewable energy. California power plants are required to produce 33 percent of their power from renewable sources by the year 2020.[20] Suppose high winds in Texas could be producing relatively cheap wind energy, yet capital limitations prevent wind farms from being constructed there. Tradable credits allow for a California power plant to invest in the Texas wind farm, thereby reducing overall emissions more cheaply than would have been possible through improvements in existing California plants alone. The economic incentives exist to encourage GHG reductions at the lowest price possible.

But a market-based approach also creates incentives to cheat. These instances involve either the exaggeration of an organization's actual emissions reductions (in some cases claiming credit for reductions that may have occurred anyway) or misrepresenting the long-term impact of a project. These practices are claimed to occur often.

For example, numerous climate activist groups charge that, since 2005, firms in developing countries have deliberately produced excess GHG pollution for the purpose of destroying it and then earning valuable carbon offset credits under the Kyoto Protocol's Clean Development Mechanism (CDM).[21]

In order for emissions to become tradable for the purpose of accomplishing meaningful reduction, carbon-trading systems must demonstrate that they reduce emissions. However, the most important part is that these reductions happen in addition to any reductions that may otherwise have occurred had the project not been undertaken.

SOLUTION

Mitigation Potential	★★★
Operational Impacts	★★★
Financial Impacts	★★★
Feasibility and Timing	★★★

Regulators, producers, and consumers can follow several key steps to ensure the carbon credits with which they are involved actually generate the emissions reductions claimed.

All projects for emissions reduction or offsets should adhere to two important rules before being initiated: first, the emissions-reduction technology of choice should be proven to demonstrate GHG emissions reductions; second, steps should be taken to ensure that subsidies are offered only in cases of true investment, as opposed to rewarding things that would have normally occurred in the absence of a carbon market.

Both of these components should be verified against a project baseline—a hypothetical state that, in the absence of the project (and holding all else constant), a reasonable prediction of the future behavior of actors involved can be made.[22]

A proposed project should reduce GHG emissions relative to this baseline scenario. It should answer a few key questions. Has the proposed technology been shown to reduce emissions as much as claimed? Will the reduced emissions be sustained over a suitably long period? How can reductions be verified? Would the emissions have been reduced even without investment in the project? Answering these questions should provide the first information needed to assess the value of an offset project.

The second rule requirement is known as "additionality," an important means for demonstrating that the project funding will result in emissions reductions beyond, or *in addition to*, those in a baseline scenario.[c] Projects to reduce emissions, when not financially limited, are assumed to occur in the baseline, so investments in these projects would yield no reductions beyond what is already expected to occur.

For example, consider an area of new growth and development. The planners of the growth need a power plant, and they seek to construct a low-GHG-emitting plant to support the area. But that power plant is needed regardless whether it's a low-GHG emitter or not; it's going to get built because of the need and a demand. This project, then, should not be incentivized. But the construction of a new power plant to replace one that's already functioning, albeit less green, could be incentivized at a level that would ensure a competitive rate of return. It's a difficult balancing act, yet one worth exploring and analyzing.

Finally, after a project is completed, it should be subject to third-party verification to ensure that emissions are actually being reduced consistent with expectations. The offset credit should be certified and tradable only after emissions reductions have been verified. Similarly, emissions levels should be monitored over time to ensure reductions are sustained.

---

[c] "Additionality" is a term normally used to explain the requirement that investments in GHG reductions should reduce emissions beyond what would otherwise normally occur.

EXAMPLES

In the early 2000s, the Kuyasa Urban Housing Development in Cape Town, South Africa, wanted to provide energy-efficiency improvements, including the installation of solar water heaters and insulation, to more than 2,300 homes in the development. Unfortunately, it lacked the financial resources to invest in this project. It instead decided to pursue certified emissions reduction credits through the CDM of the Kyoto Protocol. By 2005, the agency had developed a pilot project to verify the emissions reductions resulting from the efficiency upgrade.

The Kuyasa project also partnered with the Swiss not-for-profit The Gold Standard Registry, established under Swiss law to certify the quality of carbon credits. The Gold Standard Registry, using established benchmarks that go beyond established CDM criteria, independently certified emissions reductions in addition to what would have occurred in a baseline scenario and sustainable into the future.[23]

Adaptation: Managing What's Available

Adaptations of land use will be required due to climate change. All climate-sensitive systems of society and the natural environment will need to adapt to a changing climate or face diminished productivity, functioning, and health. These systems include agriculture, forestry, water resources, human health, coastal settlements, and natural ecosystems.

Discussions of climate change usually revolve around increases in temperature, rising sea levels, changes in water distribution and movement, and changed weather patterns. Within these broad categories, a number of specific changes will transform behavior. For example, climate change probably means warmer temperatures, but, more specifically, it also can mean one or more of these:

- Temperatures in the winter fail to drop low enough to kill agricultural and forest pests, resulting in the invasion of previously spared crops and forests from invading insect and weed pests.
- Milder winters, where snow doesn't materialize, and precipitation instead falls as rain, result in immediate water runoff rather than snowmelt later in the year. The immediate impact is winter floods and less water in summer when agriculture needs it.

- Longer frost-free periods allow invasive plant species, which could otherwise not set seed due to ground freeze, to spread into new areas.
- Rising sea levels result in coastal flooding and drive saline water further inland, affecting drinking and irrigation water as well as ecosystems not able to handle the new, saltier water.

Many types of land-use adaptation will be required in response to climate change, and some are already under way. For example, the Netherlands considers climate change in planning its coastal defense.

But what happens when entire populations find their lands impacted, especially the land on which they live? That's another area where land-use adaptation is needed. For example, an oft-cited National Oceanic and Atmospheric Administration report places the number of coastal inhabitants potentially affected by sea level rise at 53 percent of the US population by 2050.[24,d] For other nations—especially low-lying nations such as Bangladesh, the Netherlands, and the Maldives—even a small sea level rise would be catastrophic. The Intergovernmental Panel on Climate Change Working Group II Table 2 predicts that a 4°C increase in global temperatures would cause flooding due to ice melt and thermal expansion of nearly half a meter. More recent studies suggest the figure could be much higher.

That temperature increase would result in sea level rise that would force the relocation of more than 129 million humans worldwide (1,760,000 in North America alone) by 2050 (Table 4-2). Some of these people will be able to adapt within their own country due to large surrounding land mass and the relatively small number of people affected. The North American Atlantic coast, for example, could see the displacement of 909,000 people, but they have a larger space to move to than those on islands. Sea level rise in other areas will almost certainly result in emigration (see the "Security" chapter).

---

[d] This raises another point—whether the data we're collecting are actually the best and most accurate. A later analysis of these same data showed the need to fine-tune the geographic analysis to incorporate hyperlocal data. This has implications in the allocation of resources and certainly for some things discussed in the "Structure" chapter.

Table 4-2. Areas Where Additional People Will Be Flooded as a Result of Storm Surges and Sea Level Rise from an Increase of 4°C (using IPCC A1B Scenario)

Location	People Flooded
Atlantic Ocean small islands	67,000
Caribbean islands	1,097,000
Gulf states	737,000
North America Atlantic Ocean	909,000
North Ameria Pacific Ocean	851,000
Pacific Ocean small islands	529,000

Many existing approaches can be used as adaptations to climate change even though they were initially developed for other purposes. Recommended adaptation actions that land-use managers can adopt include the following:

- Develop an informed land-use policy for coastal systems.
- Manage changes in water availability with structured water-use management.
- Adapt forest management practices to climate change.
- Change agricultural practices to adapt to rising temperatures.
- Prepare for increased salinity in water sources.

Develop an Informed Land-Use Policy for Coastal Systems

Recommendation: *Support the development of a comprehensive planning process for areas vulnerable to sea level rise, flooding, or water supply salinization, including mapping and prioritizing areas for protection, land-use changes, or exclusion from costly engineering fixes.*

CHALLENGE

Adaptation planning to address climate change effects will involve multiple sectors and multiple legal regimes.[25] Unfortunately, most environmental laws, policies, and regulations only protect the status quo: repair the dam, repair the flood damage, replenish the beach, restore the wetland, etc. Most current land managers operate with a similar goal: maintain the range, maintain the forest, replant the commodity, reestablish habitability.

For sea level rise, the current approach is just not good enough. The sea level has risen along much of the US coastal areas over the last 50 years and will rise more in the future.[26] Most predictions point toward ½ to 1½ meters of sea level rise by the year 2100.[27]

The actual depth change at ports due to sea level rise will be small. However, the rising waters will move saturated soils inland and could easily require costly infrastructure changes to such things as parking areas, transshipment facilities, and rail yards (see the "Structure" chapter).

The resulting inundation will cause changes to fishing grounds, other important natural ecosystems, ports, arable farmlands, sources of irrigation and drinking water, and habitable land. The change in habitable land is perhaps the most financially devastating.

New approaches to land-use practices, regulations, and laws are necessary to adapt to the likely catastrophic effects of sea level rise. Maintenance and construction on flooded or unsuitable lands is extremely expensive. Such efforts are likely to fail if sea level rise exceeds what's predicted.

SOLUTION

Adaptation Potential	★★
Operational Impacts	★★
Financial Impacts	★
Feasibility and Timing	★

Lose your expectations of the land. Forget what you know it to be, and focus on what it can be. Sea level rise is substantial and impactful, and it forces the status quo out the window.

Sea level rise forces a "making lemonade out of lemons" approach to adaptation. For example, land saturated by rising water or that has become swampy or brackish (salty) due to sea level rise is no longer suitable for traditional habitation or farming. Salt water is especially unkind to most crops, habitats, or forests. It may, however, be suitable to reestablish ecosystems.

Again, though, the point is *lose your expectations of this land.* Attempting to adapt to these types of changes by trying to return to previous conditions may be impossibly expensive. On the other hand, re-purposing the land and changing the way it is used may be possible, and it may be beneficial to other goals such as the provision of ecosystem services.

A comprehensive planning process for areas known to be vulnerable to sea level rise, flooding, or water supply salinization can guide what is possible as adaptation activities become necessary. Through this process, planners map and prioritize areas for protection, those slated for changes in land use, and those excluded from costly engineering cures.

On a local level, land managers can best adapt to climate change and the required adaptations in land use through preparedness. Rather than reacting to imminent threats, they should prepare to suit the land to the new condition.

A climate change vulnerability assessment of managed land units provides two essential components to adaptation planning:

- Determining why species, uses, or users are vulnerable can help identify strategies more likely to succeed, helping avoid wasting precious resources on adaptation activities that are doomed to fail.
- Assessments allow you to manage expectations. Not all segments of society are equal in their capacity to adapt to climate change challenges. The same can be said of the resilience of many natural systems, especially when human needs for space and ecosystem services compete for space and for resources to adapt.

Because of the risk in maintenance and construction on flooded or unsuitable lands, abandoning the land and untenable land uses must be a consideration. It's better to know this in advance, something possible with comprehensive planning.

This sort of advance policy planning allows the public sector to support the needs of farmers and foresters, parties to whom land is an economic investment that must provide returns.

Policies that monetize certain required ecosystem services such as wetlands, nutrient management, and even carbon sequestration would help prepare land for the impacts of sea level rise.

EXAMPLES

Some states have established coastal area management regions with special zoning and other types of restrictions. The North Carolina Coastal Area Management Act was an early attempt at coastal land-use regulation, and although considered quite draconian at its inception, it has nonetheless proven its worth.

With so many miles of barrier islands and coastal areas, North Carolina had to find a way to avoid construction that imperiled other coastal inhabitants or fouled the waters. The legislation establishes certain building and development rules within the coastal zone. It prohibits building within a set distance of the high tide and even prohibits rebuilding if erosion has taken the buffer footage out of a property—furthermore, it offers no grandfather clause. The unforeseen benefit of this approach is that by prohibiting an armoring of the coastline to protect dwellings or land uses, the fringing ecosystems of the coastline can migrate inland during sea level rise.

In the Camargue region of France, the government has designated inundated land for conservation while adopting a multitasking policy that allows for its monetization. Industry, farming, ranching, and ecosystem services all coexist. Salt pans are harvested for sea salt. Rice is a commodity crop. Horses and cattle use the areas as range land, and vast tracts of wetlands provide ecosystem services. This supports the needs of farmers and foresters, to whom land is an economic investment that must provide returns.

Manage Changes in Water Availability with Structured Water-Use Management

Recommendation: *Establish better water management and retention practices that provide structure to help adapt to changes in water availability, including those that take effect when crops are grown.*

CHALLENGE

All plants have specific water requirements. The amount of rainfall or other moisture (snowmelt, dew set, and irrigation) heavily influences a cultivated plant's success and affects the ability of native plants to survive. Everybody understands the impact a drought has on plants and its role in crop failure. Climate change will disrupt the overall distribution of precipitation, resulting in arid conditions in some previously fertile locations.

Another predicted effect of climate change is a more erratic climate. Intermittent drought conditions can be expected. Texas, for example, has experienced a profound drought spanning several years, and scientists at Texas A&M anticipate a long, drawn-out stretch still to come—perhaps through 2020.

As some areas experience changeovers from snow to rain and earlier snowmelts that send this water downstream earlier in the year, this liquid nourishment may not be available when crops, forests, or other habitats need it most. This is an often misunderstood aspect of precipitation: the importance of timing. A plant has to be ready to take delivery in order to use water when it becomes available.

Milder temperatures can cause winter precipitation to fall as rain and run off throughout the winter. Normally, this water would be retained in snowpack. If it flows off of the land in winter, it is no longer available for agriculture or wild land plants in winter dormancy. Water use and retention will be problematic. The risk of crop failure will increase. The dry part of the summer could become even longer.

SOLUTION

Adaptation Potential	★★
Operational Impacts	★
Financial Impacts	★
Feasibility and Timing	★

Better water management and retention with dams, or distributed water retention like that provided by low-impact development, may be required. Changes in crop planting dates and irrigation schedules could be required so that plants can take advantage of earlier water availability.

The US Geological Survey has a long-term data set of flow measurements from virtually every watershed in the continental United States. Using rainfall, snowpack, and snowmelt data, one can identify watersheds for water-retention projects. Lakes, dams, levees, and wetlands storage areas can slow the water flow through the watershed, delaying it until crops need it.

Rehabilitation of forests in the upper watershed and marginal lands throughout the watershed has been shown to slow runoff and allow ground water recharge. Green water (that stored in plants) is nearly as valuable as blue water (surface water) for maintaining water in the landscape. Vegetative cover on landscape helps retain water, prevent erosion, and discourage the loss of water downstream.

We know a great deal about the water requirements for established crop species within the United States. It is common knowledge that many crop species respond to irrigation by increasing their yield. However, this yield increase is not always in line with the increased cost of irrigation. Corn is a good example of this concept. Understanding the true water-to-yield ratio for common crops, especially those grown for biofuels, allows land managers to make informed choices about planting location and irrigation water needs.

EXAMPLES

The USDA Agricultural Research Service (ARS) maintains several experimental watersheds and investigates their responsiveness to hydraulic alterations such as input, rainfall, and snowmelt. ARS confirmed that water can be stored and distributed quite well to meet agricultural needs and to maintain sustainable low levels in natural waterways.

The International Boundary and Water Commission, centered in North America's desert southwest and spanning the United States and Mexican border, is a binational body dating back to the 19th century. The commission's purpose is to deal with the water rights issues of both nations. Today, it exists almost solely for the regulation and delivery of water on both sides of the international border. It has a sophisticated system of sensors and dams, and operates under a sophisticated set of treaties and agreements. Water masters in each state "own" the water and distribute it according to need. Base flow is maintained in natural waterways. Through this work, the arid southwestern portion of Texas has emerged as one of the nation's fruit and vegetable baskets; today, its winter crop production rivals those of Florida and California's Central Valley.

ARS has also identified the absolute water requirements for crops as well as the optimum application time and rate. This program is a prime candidate for expansion. ARS studies with corn, for example, show that the expense of the added irrigation often does not equate to increased yield. In times of true low water, changes in crops from corn to switchgrass (which would then be used for fuel feedstock production) could be a way to maintain production on lands.

Switchgrass

Another emerging solution to change water availability is to target invasive, water-using weeds with biological control, something being done by the South African Working for Water (WfW) program. WfW seeks to control the invasive Australian acacia plant, the single biggest threat to the country's biodiversity. To assist in this endeavor, it has employed biological control: insects from the plant's native range in Australia. To date, 76 biocontrol agents have been released in South Africa against 40 weed species. The lower density of water-guzzling acacias allowed a documented increase in water captured in catchments.[28] Similar efforts by the ARS in the United States help control such water thieves as salt cedar and carrizo cane. (Incidentally, control of carrizo cane has national security implications, as illegal border crossings in the Southwest are often aided by the cane, which is tall and easy to hide in.)

Adapt Forest Management Practices to Climate Change

Recommendation: *Promote best management practices that help forests stay healthy, including actions that prevent disease and pest damage.*

Adapt forest management

CHALLENGE

Climate change will affect growing conditions essential to forest health. A healthy forest provides an important sink for GHG (Figure 4-2), supplies many forest products (such as lumber, fruit, and non-wood products), and furnishes other important ecosystem services (such as temperature mitigation, water-retention-erosion control, and habitat).

Climate changes will alter the physiology of plants in our forest ecosystems. Higher temperatures have been demonstrated to cause some trees to grow faster, especially higher-elevation trees such as bristlecone pines in the Southwest. Trees are also growing deeper into the tundra than in prior years due to warmer temperatures.[29]

Rampant growth can provide more fuel for wildfires while shading out lower-growing plants. Some weedy species of trees overtake native species; the fast-growing, difficult-to-kill, *Paulownia* tree, for example, grows faster at higher temperatures.

Increased CO_2 can also favor one species over another, including vines that choke out upright trees while severely affecting the ability of mature forests to sequester GHG.

Trees damaged by mountain pine beetle infestation.

Higher temperatures create water stress, which can result in reduced tree growth. Trees will grow more slowly and shorter, or sometimes simply stop growing altogether. Their ability to set and develop fruit and mast (acorns and other nuts) is curtailed. For example, the elevated temperatures cause stomata cells (which allow CO_2 water vapor and oxygen to move rapidly into and out of the leaf) under the leaves to close, slowing plant respiration.

Higher temperatures will also result in more water stress for forest plants. Water will transpire more quickly even if the amount of rainfall does not change.

Elevated temperatures may well release "sleeper" weeds. These are species that have previously been less aggressive in the landscape but that are more successful in elevated temperatures. According to the Commonwealth Scientific and Industrial Research Organisation, at least 41 species of weeds are likely to become pests to agriculture or forests as temperatures increase.

Insect pests can be expected to display different habits than previously noted. The mountain pine beetle was once generally confined to lower elevations in the Rocky Mountains, but warming winter temperatures have allowed this pest to overwinter at higher elevations. They've begun to infest lodgepole pine, which results in a fatality rate of nearly 100 percent.[30] Dead pines (and other trees) can contribute to the increased risk of wildfire, erosion, and flooding.

Adaptation Potential ★★★
Operational Impacts ★★★
Financial Impacts ★★★
Feasibility and Timing ★★★

SOLUTION

Employing best management practices in forestry will promote a healthy network of forests across the nation.

Management can help minimize disease and pest damage to ensure healthier trees in existing forests. Fewer incidences of pests and diseases will minimize the buildup of debris on the forest floor. The debris, if allowed to accumulate, plays a role in hotter fires, which cause more damage and burn over wider areas.

Managing forests to allow for more frequent "prescribed burns" can reduce fuel loads and minimize the damage of hotter fires. Perhaps the most important forest best management practice would be to stop deforestation where it is occurring.

EXAMPLES

The National Fire Plan is a program jointly developed by Congress and the Secretaries of the Interior and Agriculture, and implemented by the Forest Service, National Park Service, Bureau of Land Management, state fire services, and others.

The program features economic incentives for clearing debris and smaller plants to decrease wildland fuel load as well as to develop potential products from small-diameter lumber or non-forest products. A number of community assistance products are available, including educational materials aimed at the public, such as *Smart Yard Care: Big Rewards from Small Investments in Stewardship*,[31] and financial incentives for private landowners awarded the title *Firewise Communities* for fuel reduction projects.[32]

The United Nations Reducing Emissions from Deforestation and Forest Degradation (REDD) is an international program focused on developing methods for reducing deforestation pressure from outside the forest. REDD has employed the use of financial incentives, political pressure, and education to help stop deforestation.

Change Agricultural Practices to Adapt to Rising Temperatures

Recommendation: *Overhaul how agriculture is performed, including how and what farmers plant, with a focus on land practices and vegetation that can better withstand higher temperatures and longer growing seasons along with increased insect, pest, and disease pressure.*

RECOMMENDATION
Change agricultural practices

CHALLENGE

Weather conditions, temperature, and growing season length have profound impacts on crop growth and successful soil management (Table 4-3). However, some of these factors also affect the success of pests and diseases in plants.

Table 4-3. A Change of Only a Few Degrees Celsius Can Cause Significant Changes in Agriculture

Temperature Change	Impact
+1° to +2°	Some increase in yield Cold limitation alleviated Yield reduction in some latitudes (without adaptation) Seasonal increase in heat stress for livestock
+2° to +3°	Potential increase in yield due to CO_2 fertilization (but likely offset by other factors) Moderate production losses of pigs and confined cattle Increased heat stress Yields of all crops fall in low latitudes (without adaptation)
+3° to +5°	Maize and wheat yields fall regardless of adaptation in low latitudes High production losses of pigs and confined cattle Increased heat stress and mortality in livestock

A warmer climate is projected to increase the number of frost-free days, good news for fans of certain long-season crops, including artichokes and sweet potatoes. In general, small increases in temperature could produce increased yields for some crops. Warmer temperatures may increase the availability of arable land at northern latitudes.

However, elevated temperatures can also slow down the rate of CO_2 fixation by plants, and as temperatures rise, plant growth is even further stunted. Plants won't be able to flower or fertilize themselves, and other biochemical effects may occur within the plant.

Elevated temperatures can affect not only the fertility and development of the fruit part of the plant but also the activities of the pollinators. Many crops depend on the activity of pollinating insects or beneficial insects that prey on plant pests.

Information concerning all the necessary elements for a successful crop remains scarce. Will pollinators be around when the temperature is right for the plant to bloom? Will the temperature support the development of the fruit after the flower is fertilized? In order to plan responses, these types of questions must be answered, and existing knowledge must be used to the extent possible.

An increase in the average minimum winter temperature may be good for some plants, but it is also good for overwintering pests. Corn earworm and corn borers, for example, are devastating pests that have thus far been unable to become established in the Corn Belt of the upper Midwest US because their pupae cannot survive more than about 4 days of temperatures below 14°F. Warmer winters could end that standoff and either prevent or limit corn crops or require pesticide applications. Both practices would change the balance of profitability. Given current reliance on corn for biofuel, and the slim margin of energy production from corn,[e] either these pests will have to be cheaply neutralized or alternative crops would have to replace corn in a warming Midwest.

Invasive plant species can also take advantage of milder climatic conditions. The carrizo cane, for example, is known to be more widely distributed in warmer winter conditions.[33] As its distribution expands into the major river basins of Mexico and the United States, more water is lost in catchments due to the cane's high level of water use.

SOLUTION

Adaptation Potential	★★★
Operational Impacts	★★★
Financial Impacts	★★★
Feasibility and Timing	★★★

Agricultural production under changing climatic conditions can be assured through new plant materials. This plant material should be able to withstand higher temperatures, longer growing seasons, and warmer winters along with increased insect, pest, and disease pressure. Stakeholders can and should support policies and programs for identification and rapid response to pests and diseases, new or newly empowered, which are able to thrive in the altered climatic conditions.

---

[e] Future biofuels may have better margins of energy production. See more about corn-based fuels in the "Security" chapter.

ARS and the National Institute for Food and Agriculture (NIFA), formerly the Cooperative State Research Education and Extension Service, are in a strong position to lead this effort to identify new plant material. Their efforts will entail identification and introduction of alien species—and perhaps genetically modified plant materials. Diversifying plantings in agriculture and forestry will allow land managers to adapt to changing climatic conditions

Our US farm system has continued to progress into a commodity-based approach where a single crop (or, at most, a very few varieties of very few crops) is produced. Diversifying the genetic composition of crop varieties will reduce the risk of catastrophic crop failures due to the quick emergence of a new pest or disease whose occurrence is facilitated by climatic change.

EXAMPLE

For centuries, Andean natives in South America have avoided a potato blight and famine similar to the one that famously decimated Ireland in the mid-19th century. Today, they're working to ensure their continued viability by building a better potato and changing how they farm them.

The Andean natives have long recognized the need for genetic diversity in their potato crops. They grow in excess of 3,000 different kinds of potatoes. Their fields feature many different varieties grown together: a true effort in crop diversification. The Andeans grow so many varieties of potatoes because each one is adapted to a specific microclimate and is resistant to certain diseases and vagaries of climate.

The varieties of potatoes go beyond your Yukon Gold and Russets.

This practice can be contrasted with the Irish, who notably grew a single variety. When the potato famine hit Ireland, many people starved, and a mass exodus occurred. While the Andeans have, to this point, successfully weathered the ups and downs of farming over the centuries through their diversified approach, climate change has already emerged as a threat. It's anticipated that 16 to 22 percent of all wild potato species are threatened with extinction by 2055 as a result of climate change and habitat destruction.[34]

Today, researchers at the International Potato Center in Lima are developing a cross between wild, drought-tolerant varieties and modern potatoes. The US Agency for International Development is also taking on the challenge of supporting the development of new potatoes. The agency is financially supporting the work of researchers from Virginia Tech's College of Agriculture and Life Sciences, who are working with potato farmers in Ecuador and Bolivia to improve their production through land management practices. The work focuses on fighting erosion, changing pesticide-spraying routines, and rotating crops.[35]

Prepare for Increased Salinity in Water Sources

Expect increased salinity

Recommendation: *Support engineering and infrastructure projects that push back seawater or protect crops while embracing new crops tolerant of increased salt in the water and air.*

CHALLENGE

Many types of land-use adaptation will be required in response to climate change.

As sea level rises, salt water will intrude on both surface waters and into ground water. The actual effects of increased salinity will be felt much farther inland than the projections of inundated and flooded lands. This salinity means potable water will be unpleasant to drink, perhaps even unsafe for both humans and livestock. It will kill agricultural crops or at least require changes to protect agriculture and will alter coastal habitat by killing fringing marshes and other ecosystems.

Saltwater incursion occurs because salt water is denser than fresh water. It is a well-known phenomenon in estuaries that ocean water rides far up the estuary underneath a freshwater lens, which floats on top. This salt "wedge" can be pushed farther by tides or sea level rise. It can descend downstream with falling tides, heavy rains, and increased river discharge. Sea level rise isn't the only concern, because changes in the amount of rainfall can have virtually the same effect. Without the force of rainwater flowing down and out of the estuary, seawater will travel farther up the waterway.

In western regions, salinity can also be expected to become an issue as reservoir and river levels drop due to decreased rainfall. The Colorado River's levels are already at record lows, and evaporation and withdrawal has resulted in higher water salinities.

If sea level rise pushes the salt wedge farther up a river system, the consequences to natural habitats, agriculture, and drinking water can be lethal.

A more saline environment is not a favorable one for many estuarine creatures, impacting their ability to properly mature. For example, oysters mature faster in fresher waters and are known to be more susceptible to diseases in waters with higher salinity. Lack of fresh or brackish waters in nursery areas will profoundly affect many commercial species and their food items.

Many agricultural operations pump irrigation water directly from surface waters. Their intake is generally located below mid-depth. Salt wedge encroachment can cause the nearly overnight loss of a crop. This has been observed on the intake for agricultural irrigation in the northern neck of Virginia, where drought allowed the salt wedge to overtake the irrigation intake. The City of Richmond has recently upgraded conductivity sensors in its agricultural pumping stations due to repeated incursions of salt water.[36]

Schooner Bayou Control Structure, where there is a clear distinction between fresh water (foreground) and salt water (background).

Community drinking water intakes also are vulnerable to saltwater incursion. The bottom of the Mississippi River is at about the same elevation at its outlet as it is just south of Natchez, MS.[37] Freshwater outflow from the river keeps the salt wedge well downstream of any municipal water intakes—at least for now. A change in freshwater flow (resulting from diminished rainfall or a change in the elevation of the salt wedge because of sea level rise) could lead to municipal water intakes being in salt water. Several large cities in the United States, including New Orleans, Baton Rouge, Philadelphia, and Richmond, use surface water for drinking and are vulnerable to saltwater inflow.

A salt wedge doesn't impact just surface water; it also interferes with groundwater quality. Many coastal communities retrieve their drinking water from ground water due to saltwater intrusion into surface waters. When salt water intrudes into ground water, drinking and agricultural water can be affected.

Natural habitats will also be affected by increases in surface water and groundwater salt levels. Plants at the seaward edge of a marsh are highly adapted to salinity, whereas interior marsh plants are more sensitive. As sea level rises, storm events and exceptionally high tides push salt water in over the edge of the marsh to the sensitive plants. These plants then die from the exposure, leaving the marsh surface bare and susceptible to erosion.

In short, the marshes die from the inside out. As sea level rises, this will increasingly occur, reducing available marshes. The result will be the loss of marsh habitat, shallow water nursery areas, and the energy-dissipating power of a marsh in a storm. In recent years, several researchers have noticed dramatic plant problems in salt marshes (such as the lower Mississippi Delta and Jamaica Bay, NY) and suggest that these problems are the result of slight increases in salinity.

Adaptation Potential ★★
Operational Impacts ★★
Financial Impacts ★
Feasibility and Timing ★

SOLUTION

Adaptation activities can take many forms when dealing with the salinity effects of sea level rise or salinity increases due to evaporation. Engineering approaches, for example, have for years existed to wall out or push back sea water. These are large infrastructure projects that require the support of local land managers, who will also have other approaches to consider.

Abandoning certain areas for agriculture, for example, may be required, but changing the planted crops to halophytes—salt-tolerant plants with tremendous potential as a biofuel source—is also an option. Installing weirs in irrigation ditches can help guard crops from weather-related increases in salinity of surface waters. Another option is to install warning systems that respond to increased salinity in the intake water by shutting down irrigation systems.

EXAMPLES

Engineering can effectively prevent saltwater intrusion and has done so in a wide variety of places. For example, locks and dams in Washington State have sumps where salt water is pumped back downstream during the locking process. Venice, Italy, installed movable baffles to prevent storm surge and exceptionally high tides from flooding the streets of Venice. They lie on the seafloor and are raised when higher tides are expected. Areas in Louisiana, such as Catfish Point and Schooner Bayou in the Mermantou River, have installed baffles in the waterway to keep the salt wedge from passing upstream.

The City of Los Angeles extracts a third of its water from ground water and, as early as the 1920s, began injection pumping to prevent saltwater intrusion into drinking water aquifers. The act of pumping out the ground water causes a cone of salt water to rise toward the well intake. Pumping in fresh water maintains a gradient pressure that prevents the salt water from reaching the well.

Another approach altogether is to change crops to those that can handle increased salinity. The Halophyte Biotechnology Center at the University of Delaware has worked for decades to identify forage, conservation, and agricultural plants able to prosper with increased water salinity.

Some knowledge exists of plants that are especially susceptible to salt damage such as lettuce, wheat, and soy. Some traditional crops are well suited to slightly elevated salinities. For example, barley could replace wheat crops while lettuces could be replaced with tomatoes or even asparagus.

For higher salinities, the halophyte development site has identified plants such as marsh mallow (*Kosteletzkya pentacarpos*). Marsh mallow produces seed oil suitable for biodiesel and other oil seed crop replacement and has biomass useful for ethanol production. The added advantage of this perennial crop is its utility for habitat even under difficult agricultural growing conditions.

Final Thoughts

Given how political the issue of government control over private lands is, land use will be one of the most challenging arenas for mitigation and adaptation. National "cures" are unlikely to succeed, so local and individual approaches will be the most effective, at least for now.

Additional promotion on the national stage could greatly enhance the results. For example, allowing federal managers to change the management approach to forests and rangeland would have a strong impact on all levels. Supplementary federal enabling would focus certain USDA resource agencies such as ARS, NIFA, the Animal and Plant Health Inspection Service, and the Forest Service on providing technical advancements for identification of new, resilient crops and trees and pest treatment.

Some of our recommendations call for large federal civil works construction projects such as dams, flood protection, and salinity safeguards such as pumps, sumps, and barricades. However, the largest federal role is in providing a unified set of information resources so that flooding and inundation areas can be mapped, infrastructure can be scientifically and fairly allocated, and local governments can take the kinds of zoning and planning measures that only they are empowered to enact.

---

[1] Anthony W. King and Others (eds.), *The First State of the Carbon Cycle Report (SOCCR): The North American Carbon Budget and Implications for the Global Carbon Cycle.* Report by the US Climate Change Science Program and the Subcommittee on Global Change Research, National Oceanic and Atmospheric Administration, National Climatic Data Center, Asheville, NC, November 2007, p. 31.

[2] R.A. Houghton, J.L. Hackler, and K.T. Lawrence, "Changes in terrestrial carbon storage in the United States. 2. The role of fire and fire manage-

ment," *Global Ecology and Biogeography*, Vol. 9, No.2 (March 2000): pp.l 145–170.

[3] See Note 1, p. 34.

[4] USGS National Atlas of the United States of America, 'Federal Lands and Indian Reservations," *NationalAtlas.com* accessed on October 10, 2011, http://www.nationalatlas.gov/printable/fedlands.html.

[5] Patrick M. Condon, P. Duncan Cavens, and Nicole Miller, "*Urban Planning Tools for Climate Change Mitigation*." (*Policy Focus Report)* Lincoln Institute of Land Policy*,* August 2009.

[6] City of Vancouver. *Greenhouse Gas Emission Reduction Official Development Plan. (Adopted by By-law No. 10041),* May 18, 2010.

[7] US Congress, National Historic Preservation Act of 1966. Public Law 89-665 *(*1966).

[8] US Congress, The National Park Service Organic Act of 1916 16 USC. 1 2 3, and 4, as set forth herein, consists of the Act of Aug. 25 1916 (39 Stat. 535) (1916).

[9] US Congress, Organic Act of 1897" *(26* Stat. *1095)* (1897).

[10] Pekka E. Kauppi, and others,"Returning forests analyzed with the forest identity." *Proceedings of the National Academy of Sciences,* Vol.103 No.46, (2006): p.17574.

[11] Christine Wiedinmyer and Matthew D. Hurteau, "Prescribed Fire As a Means of Reducing Forest Carbon Emissions in the Western United States." *Environmental Science Technology* (2010): pp. 1926–1932.

[12] Mark E. Harmon, William K. Ferrell, anf Jerry F. Franklin, "Effects on Carbon Storage of Conversion of Old-Growth Forests to Young Forests." *Science*, Vol. 247, No. 9, (1990) pp. 699–701.

[13] Ellen Morris Bishop, "Monitoring the Forest's Breath," *The Columbian*, July 22, 1998.

[14] Henning Steinfeld, and others, *Livestock's Long Shadow:Environmental Issues and Options,* Food and Agriculture Organization of the United Nations, Rome, 2006.

[15] "Prescribed Burns May Help Reduce US Carbon Footprint," *Science Daily,* March 17, 2010.

[16] Chicago Climate Exchange, *Fact Sheet*, December 2011.

[17] US Environmental Protection Agency, *Report on US Methane Emissions 1990-2020: Inventories, Projections, and Opportunities for Reductions,* EPA 430-R-99-013, September 1999.

[18] John H. Martin, Jr. *An Evaluation of a Covered Anaerobic Lagoon for Flushed Dairy Cattle Manure Stabilization and Biogas Production*, Eastern Research Group, June 17, 2008.

[19] S. S. Jones, and T. D. Murray, *Life and Death in Plants: Studies on Perennial Wheat as a Sustainable Alternative Cropping System,* Washington State University, 2008.

[20]US Department of Energy's Database of State Incentives for Renewables and Efficiencies, accessed on October 10, 2011, http://www.dsireusa.org/incentives/incentive.cfm?Incentive_Code=CA25R&re=1&ee=1.

[21] Nathanial Gronewold, "CDM Critics Demand Investigation of Suspect Offsets," *The New York Times*, June 14, 2010.
[22] Michael Gillenwater, "What is Additionality? Part 2: A framework for a more precise definition and standardized approaches," GHG Management Institute, Discussion Paper No. 002, February 2011.
[23] The Gold Standard Project Registry, http://goldstandard.apx.com/.
[24] S. J. Williams, G. W. Stone, and A. E. Burruss, "A perspective on the Louisiana wetland loss and coastal erosion problem, Introduction," *Journal of Coastal Research,* Vol. 13, No.3, 1997: pp. 593–594.
[25] Robin Kundis Craig, "Stationarity is Dead—Long Live Transformation: Five Principles for Climate Change Adaptation Law," *Harvard Environmental Law Review,*Vol.34, No.1, 2010: pp. 11–73.
[26] Patty Glick, Bruce Stein, and Naomi Edelson (eds.), *Scanning the Conservation Horizon: A Guide to Climate Change Vulnerability Assessment.* Washington, DC: National Wildlife Federation, 2011.
[27] U.S Army Corps of Engineers, "Water Resource Policies and Authorities Incorporating Sea-Level Change Considerations In Civil Works Programs," Engineering Circular EC-1165-2-211, 2009.
[28] Department: Water Affairs, *Working for Water*, Republic of South Africa, accessed on June 17 2011, http://www.dwaf.gov.za/wfw/.
[29] Andrea Thompson, "Surprising New Arctic Inhabitants: Trees" *LiveScience.com*, March 9, 2007.
[30] Jesse A. Logan, Jacques Régnière, and James A. Powell, "Assessing the impacts of global warming on forest pest dynamics." *Frontiers in Ecology and the Environment*, April 2003: Vol. 1 Issue 3, pp. 130–137.
[31] Francis J. Reilly Jr., *Smart Yard Care: Big Rewards from Small Investments in Stewardship.* Blacksburg: Virginia Cooperative Extension/Virginia Tech, 2009.
[32] Firewise Communities/USA. *Firewise Communities,* 2010, accessed June 13, 2011, http://www.firewise.org/.
[33] Georgianne W. Moore, David A. Watts, and John A. Goolsby, "Ecophysiological Responses of Giant Reed (*Arundo donax*) to Herbivory," *Invasive Plant Science and Management,* Vol. 3, Issue 4, 2010: pp. 521–529.
[34] Eliza Barclay, "Warming to Spur Potato Famine in the Andes?" *National Geographic News,* October 1, 2008. http://news.nationalgeographic.com/news/2008/10/081001-peru-potato-climate.html.
[35] "A Virginia Tech Researcher uses Undergrads to Help Improve Potato Farming in the Andes" accessed on June 15 2011, http://www.vt.edu/spotlight/innovation/2011-08-01-ecuador/potato.html.
[36] John Irving, John. *Salt intrusion into the Fraser River*Staff Report—City of Richmond, July 5, 2010.
[37] US Army Engineers New Orleans District, Hydraulics and Hydrologic Branch. *A Saltwater Wedge Affects the Mississippi River.* October 2008, accessed April 30, 2011, http://www.mvn.usace.army.mil/eng/saltwater/salt_leadin.htm.

Structure

The levee had been the community eyesore for years. It needed to be mowed regularly, was a magnet for delinquent behavior, and by blocking views to the pristine waters behind it, hurt property values for nearby homeowners. Funding its upkeep fell on the back-burner when more attractive projects emerged. When the levee was constructed in the 1950s, it was capable of withstanding a 100-year flood—and the water had never risen to such levels. Year after year, after quick consideration, the town deferred major maintenance. Now, one of the town councilmen watched from high ground as the river continued its devastating rise. The town was underwater, the result of weeks of rainfall there and upriver. Recent increases in precipitation and severe storm events meant it would be a long time before the river would recede. The extent of the damage and high cost to restore what was once there made rebuilding unlikely. An entire town had been swept off the face of the earth.

If You Build It ...

What is a structure? We live in them. We drive along them. We work and play in and around them. They provide us with access to basic services, including water and electricity. The oldest human-made structures are more than 75,000 years old, and today's most stunning structures are more than 2,000 feet tall. But structures can also be temporary, and they don't have to be large. Mankind has the ability to develop infrastructure in every setting to suit every need and use.

Accordingly, the global footprint of infrastructure continues to grow as both developing and mature nations evolve. Infrastructure's impact on our environment is undeniable and increasingly significant. The construction, operation, and maintenance of our infrastructure have released greenhouse gas (GHG) emissions into the atmosphere. This influences climate change and will continue to do so as we replace aging infrastructure while building the next generation. We are now forced to mitigate and adapt in response to this byproduct of our own evolution.

The nature of infrastructure influences the adaptation actions you can take.

Our society's physical infrastructure determines the actions functional stakeholders can take to proactively deal with these dynamic impacts. This is a critical juncture for the managers who design and control infrastructure systems at multiple societal and functional levels; they have the authority to make strategic decisions and implement changes. The decisions they make today, and over the next 10–15 years, will ultimately impact the safety, health, and welfare of the public.

The scope of infrastructure is broad and crosscutting. It touches and affects multiple levels of society as well as the environment, and it enables us to function and evolve in the pursuit of both individual and societal goals and ambitions.

Adaptation or Mitigation: A Question of Audience

Mitigation and adaptation strategies for physical infrastructure (as related to climate change outcomes) vary depending on the decision maker. For example, mitigation strategies and tactics can be effectively applied at the individual or microeconomic level. Adaptation strategies and tactics, on the other hand, are more relevant at a societal or macroeconomic level.

Mitigation actions can be bottom-up and technology driven. Or they can be top-down and policy driven. A building's central plant equipment can be retrofitted to reduce GHG emissions, or it can simply be replaced. Mitigation can be driven by industrial competition that spurs the design and manufacture of cost-effective green equipment. These are individual, granular actions that can effectively mitigate GHG emissions over time. At the policy level, mandates to reduce emissions to a previous year's level may drive broader decision making through incentives and penalties.

But no amount of mitigation, no matter how much it reduces GHG emissions, is likely to prevent some form of climate change. This brings us to adaptation, which requires a new model—one that considers a range of possible future climate conditions and associated impacts, some of which will be well outside the realm of experience.[1]

Stakeholders need to identify proactive measures of adaptation. They need to strategically address the likely impacts of climate change on the most vulnerable and critical infrastructure. Doing so will set the foundation for reducing risks to human life, as well as our ecosystems and economy.[2]

Adaptation strategies are more systematic in nature, and a macro view is often most appropriate. For example, multiple stakeholders must be involved in protecting coastal facilities from sea level rise or storm surges to cover the entire affected area; rarely would such an effort boil down to just individual facilities. Public officials, landowners, businesses, communities, and others may be affected, so each has a vested interest in adaptation actions.

This type of coordination requires action on the part of higher management levels that have authority over town, city, or even regional decisions. These leaders are on the frontlines of broader infrastructure management and can make the most significant difference in developing and implementing adaptation strategies that could impact thousands of people.

A Different Set of Variables

The biggest difference between infrastructure and, say, agriculture or public health, is its longevity. Structures, for the most part, are made to last a long time. Roads built as part of the Interstate Highway System in the 1950s are still in operation today. In downtown Chicago, you'll find a Mies van der Rohe high-rise across the street from an early 20th century building still in use for luxury senior living.

With infrastructure, it's hard to push the reset button. Roads, schools, and bridges have and can be maintained. But razing an existing structure can be costly and, sometimes, detrimental to our cultural history. Stakeholders must find adaptation and mitigation solutions that take this into account.

Types of Infrastructure

Sometimes it's best to start with a dictionary to understand where you're going. *Webster's* lists the following primary definitions for infrastructure:

1. The underlying foundation or basic framework (as of a system or organization)
2. The permanent installations required for military purposes
3. The system of public works for a country, state, or region

4. The resources (as personnel, buildings, or equipment) required for an activity.[3]

Infrastructure is typically divided into hard and soft infrastructure. Hard infrastructure refers to large physical networks necessary for the functioning of a modern industrial society. Soft infrastructure, on the other hand, refers to the institutions required to maintain the economic, health, cultural, and social standards of a society; these include financial, education, and healthcare systems.

In the case of climate change, hard, or *physical*, infrastructure is more relevant. The technical structures that support a society, such as roads, water supplies, sewers, and power grids, when viewed functionally, are what *facilitate* things. These elements are the keys to the production of goods and services, to day-to-day living, and to recreation (in the case of our national parks).

For example, with respect to the production of goods and services, our roads enable the transport of raw materials to a factory along with the distribution of finished products to markets. In some contexts, the term may also include basic social services such as schools and hospitals. In a military context, the term refers to the buildings and permanent installations necessary for the support, redeployment, and operation of military forces (national security is discussed in the "Security" chapter).

Infrastructure is not simple. The supporting hierarchies and branches can be very intricate and complex, typically much more so than meets the eye. Regardless, properly designed and implemented infrastructure is so efficient that it is barely noticed as it does what it was designed to do. But when it fails or becomes inadequate, people definitely notice.

Table 5-1 shows how infrastructure elements are naturally interconnected and require integrated skill sets and teams to build. For example, a campus of buildings often shares utilities, communications, roads, and grounds.

Table 5-1. Infrastructure Subcomponents Are Broad and Varied

Infrastructure Type	Description
Transportation	• Roads — urban, rural, and highway networks, bridges, tunnels, lighting and traffic signals, toll plazas, and rest areas • Rail — terminal facilities, level crossings, and signaling and communications systems • Ports — seaports: piers, docks, and lighthouses; airports: air navigational systems • Mass transit — commuter rail systems, subways, tramways, trolleys, and bus transportation
Facilities and other assets	• Facilities — commercial or institutional buildings, hotels, resorts, schools, office complexes, sports arenas, convention centers, campuses, residential housing, etc. • Structures — towers, canopies, fencing, signage, storage containers, etc.
Water systems	• Drinking water systems — engineered hydrologic and hydraulic components, watershed, water purification facilities, uncovered ground-level aquaducts, covered tunnels, underground water pipes, reservoirs, water tanks, and towers • Wastewater systems — sewer and septic systems, pumps, lift stations, and wastewater treatment plants • Stormwater systems — drainage systems, dry ponds, retention ponds, underground detention pipes, and litter traps • Waterways — inland rivers and lakes with supporting locks, canals, and dams • Water control and flood protection systems — tide gates, dikes, culverts, storm barriers, dams, and levees
Parks and recreation	• Parks — visitor centers, campgrounds, cabins, boat launches, piers, docks, trail bridges, internal roads and parking, and other utility structures • Beaches — shoreline protection, public utilities, and boardwalks
Health and safety	• Health — hospitals, treatment centers, urgent and special care, research laboratories, disease control, and pandemic response systems • Safety — antiterrorism and force protection, hazardous material abatement and disposal, decontamination systems, police, fire and rescue, and monitoring and control systems
Electrical power generation	• Power grid — generation, transmission, and distribution — generation plants, interconnected transmission lines, high-voltage transmission towers, underground transmission and distribution lines, substations, and transformers • Nuclear power — power plants, hazardous waste disposal sites, and equipment

The practice of designing, constructing, and operating buildings is usually a collective effort of different groups of professionals and trades. Depending on the size, complexity, and purpose of a particular building project, a project team may include the following:

- *Real estate developers*, who secure funding for a project
- *Financial institutions or other investors*, who actually provide funding

- *Local planning and code authorities*, who enforce zoning ordinances, building codes, and other regulations such as fire and life safety codes
- *Construction managers*, who coordinate the effort of different groups of project participants
- *Licensed architects and engineers*, who provide building designs and prepare necessary construction documents
- *Construction contractors*, who install building systems, such as HVAC, electrical, plumbing, fire protection, security, and telecommunications
- *Facility managers*, who are responsible for a building's operations.

The total value of nondefense public infrastructure is $8.2 trillion.[4] Figure 5-1 shows the areas that make up the bulk of our national, nondefense public infrastructure.

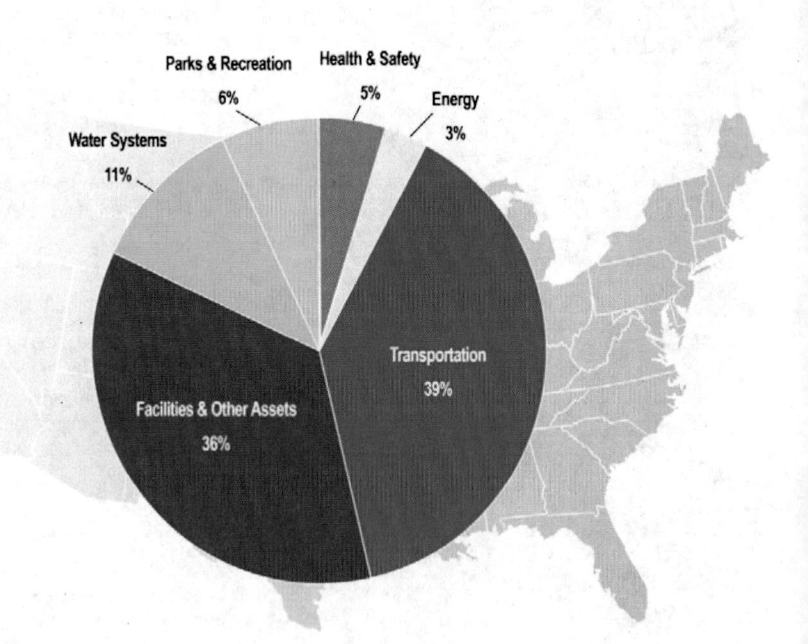

Figure 5-1. Major Infrastructure Components by Percentage of Total Value

Independent assessments have labeled America's infrastructure as being in dire need of repair or replacement. The American Society of Civil Engineers (ASCE), which takes a keen interest in the state of public infrastructure on behalf of its membership, periodically issues its *Report Card for America's Infrastructure*,[5] which grades 15 categories of infrastructure on the basis of capacity, condition, funding, future need, operations and maintenance, public safety, and resilience. ASCE's 2009 report (the most recent) gave the nation a cumulative grade of "D," with the highest grade reaching only a "C+." The required investment across the board over the next 5 years, according to the report, is $2.2 trillion—not exactly what the public sector wants to hear when it's being asked to do more with less (Table 5-2).

Table 5-2. Estimated 5-Year Infrastructure Investment Needs (billions of dollars)

Category	5-year need ($ billions)	Report Card Grade 2009
Aviation	87	D
Dams	12.5	D
Drinking water & wastewater	255	D-
Energy	75	D+
Hazardous waste & solid waste	77	D/C+
Inland waterways	50	D-
Levees	50	D-
Public parks & recreation	85	C-
Rail	63	C-
Roads & bridges	930	C/D-
Schools	160	D
Transit	265	D

The urgency of the situation speaks volumes. For example, more than 85,000 dams in the United States average 51 years of age. Also, more than 85 percent of the nation's estimated 100,000 miles of levees are locally owned and maintained, and their reliability is unknown. With the continuing increase of development behind these levees, the risk to public health and safety from failure has also risen.[5]

Aging, inadequate, or obsolete infrastructure exacerbates the risks of climate change impacts.[6] When we address infrastructure needs, we must consider mitigation and adaptation or risk compounding the problem.

However, due to the complexity of constructing and operating infrastructure, and all the variables that go into it, any actions require proactively engaging stakeholders. They'll be expected to take part in the planning and execution of these plans, which may require significant manpower and financial resources, while supporting administration, contracts, acquisition, logistics, and project management.

Again, such an investment may be difficult to stomach at a time when budgetary priorities are elsewhere; the case for why it will pay dividends down the road, however, is compelling. In fact, the bigger issue, and one that may represent an institutional battle, is not determining *what* must be done, but presenting a case for *why* it deserves attention and public resources.

Mitigation

Success in mitigating GHG emissions from infrastructure will be driven not only by policy, but by dollars. It's also a matter of reinforcing the belief that, like any good investment, some up-front pain in the wallet needs to be weathered to get to the value down the line.

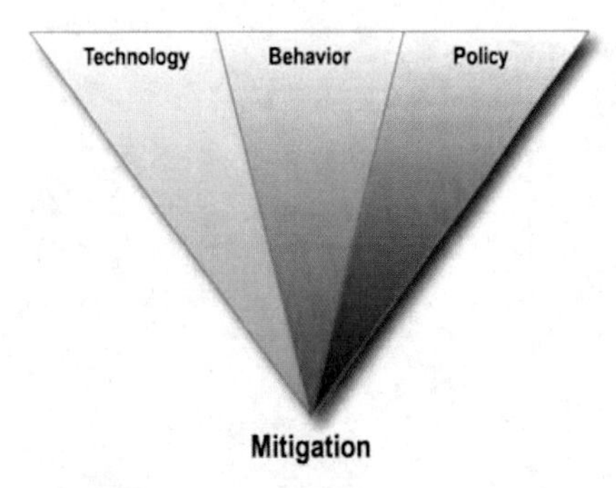

Mitigation strategies must consider available technology, altering human behavior, and new policies.

Organizations that embrace these changes aren't on an island. Organizations in the public and private sectors are evolving to view environmental responsibility as good business. It's increasingly being viewed as a valued investment rather than an avoidable cost.

For example, the Department of Defense (DoD) has targeted a 34 percent GHG reduction, far greater than that expected through President Obama's executive order. Because much of this reduction is to come through facilities and infrastructure, DoD sees it as a long-term investment. Even private-sector companies (Walmart, for example), beholden to their shareholders, have determined that environmental value can result in greater profits.[7]

The emissions from fixed assets, primarily buildings, make up 40 percent of total GHG emissions. Organizations seeking significant business value are finding that reducing GHG emissions in their buildings can yield this value through energy efficiency and other areas of opportunity.

Opportunities for mitigation strategies have three layers: bottom-up, or technology driven; top-down, or policy driven; and cultural behavior modification.

From a policy standpoint, developing and standardizing green building technologies at a national level is a key precursor to any action. It will enable numerous downstream activities, including planning, budgeting, financing, and implementation.

The stronger the policy guidance is, the faster green building technologies are likely to reach a wide audience; this can be spurred by government incentives.[8] For example, in the private sector, tax incentives would encourage the commercial real-estate sector to increase its investment in green real estate. Meanwhile, in the public sector, the adoption of government leasing requirements that give preference to green buildings would spur mitigation actions. As early as 2000, the General Services Administration (GSA) developed green lease policies, becoming a leader in this area.

Green building is the practice of creating structures and using processes that are environmentally responsible and resource-efficient throughout a building's life cycle.

Policy requires accepted metrics and databases that place a tangible value on investment in green real estate to encourage a greater range of lending and investment products for the green building sector.[8] These metrics and databases must be transparent and reliable, clearing the path for decision making on mitigation strategies. Today, the standard bearer is Leadership in Energy and Environmental Design, an internationally recognized green building certification system and provider of metrics. Green building strategies can result in increases in occupant performance measures of 6 to 26 percent.[9]

From a technology standpoint, greener equipment and approaches can be applied across the entire infrastructure life cycle. The greater the incentives are for industry to compete through innovation, the greater the resulting mitigation advancements we can expect. With those could come a reduction in the cost of adopting this new technology.

Finally, behavioral changes that affect mitigation efforts can be applied, not only in the workplace, but in all aspects of work and life. A cultural shift is happening, albeit gradually, but the signs are unmistakable. Each year, more people and organizations engage in recycling, carpooling, using mass transit, teleworking, and purchasing hybrid vehicles.

The surrounding environment generally dictates actions, making it difficult to change behavior if the situations and settings aren't right. These changes must come first, after which behavioral changes will follow.[10] The key is to make desired behaviors easier by designing a support system that enhances the convenience, ease of use, frequency, and proximity of the desired changes.

We recommend the following mitigation activities:

- Develop a GHG assessment and mitigation program.
- Integrate solutions across organizational boundaries.

- Incentivize suppliers to mitigate through procurement guidelines.
- Apply mitigation strategies throughout the infrastructure life cycle.
- Encourage and incentivize behavioral changes.

Develop a GHG Assessment and Mitigation Program

Measure, report, reduce

Recommendation: *Use an integrated organizational asset management program, with clearly stated goals and baselines, as the basis for GHG assessment and mitigation activities.*

CHALLENGE

Many organizations do not have the required baseline information on the GHG emissions, let alone the overall sustainability, of their infrastructure. This usually means they also don't have the management tools for making informed decisions. Some may opt to conduct GHG audits, but that is not enough because an audit only provides a snapshot in time. Meanwhile, GHG mitigation is a quickly evolving knowledge area and part of the budding sustainability growth industry. An organization-wide program, therefore, is required to reap the full benefits of GHG mitigation and overall sustainability.

SOLUTION

Mitigation Potential	★★★
Operational Impacts	★
Financial Impacts	★★★
Feasibility and Timing	★★

First, a GHG assessment and mitigation program works best when it is part of an integrated organizational asset management program (we'll discuss integrated solutions more later). Second, the most effective programs have clearly articulated goals, are systematic in their approach, and are auditable and repeatable.

Here are the key elements of such a program:

- *Strong leadership.* An organizational champion with decision-making authority is a prerequisite to motivate and organize a successful GHG assessment and mitigation team.
- *Clear goals.* Identifying specific GHG emissions reduction goals and objectives to be achieved over 3 to 5 years provides a road map for the mitigation team.
- *Effective communication.* Internal and external marketing of the program, as well as celebration of achievements, keeps the effort on the minds of leadership. Regular communication and feedback with team members creates a strong community and enables the rapid exchange of ideas and lessons learned. Similarly, periodic communication of progress and benefits

to top management increases the program's credibility and helps underscore its importance.

- *An automated approach.* The advances of web-based tools, mobile technology, and cloud computing have made many cost-effective tools and delivery processes for leveraging technology available. Spending time and resources up-front to invest in a technology solution typically pays dividends, but not until the program is established. Automated tools help ease use and improve consistency, quality, data-mining capability, reporting, and archiving.
- *Ongoing support.* Programs fail when the information they provide is not used or becomes outdated. A successful GHG mitigation program is designed to be continuously updated with fresh assessment cycles that enable the organization to build on its achievements and establish a cycle of continuous improvement.

EXAMPLE

Through its September 2010 *Strategic Sustainability Performance Plan*, the US Department of Energy (DOE) established a systematic, auditable, and repeatable sustainability performance program.

DOE established strong leadership through a governance model that included a Senior Sustainability Steering Committee, a Sustainability Integration Team, and working groups. For long-term implementation and oversight, it will establish a Sustainability Performance Office to coordinate corporate oversight, foster behavioral change, evaluate performance, facilitate information management, and report progress toward sustainability goals.

DOE established clear GHG emission reduction goals for fiscal year (FY) 2020 (with FY 2008 as a baseline), which include

- reducing Scope 1 and Scope 2 GHG by 28 percent, and
- reducing Scope 3 GHG by 13 percent.

DOE set objectives to reduce Scope 1 and Scope 2 GHG emissions, including the following:

- Prioritize investment in efficiency measures and infrastructure improvements on the basis of carbon intensity.
- Reduce the use of petroleum-based fuels.
- Deploy operations and maintenance best practices.
- Install asset-level metering.

- Assess and upgrade the real property portfolio to meet the high-performance sustainable building guiding principles (GPs).

To reduce its Scope 3 emissions, DOE will

- expand the use of teleconferences, videoconferences, and web-based meetings to reduce employee air travel;
- reduce transmission and distribution losses through on-site power generation; and
- reduce waste generation by increasing sustainable purchasing and recycling.

One of DOE's first priorities is to complete building assessments for its enduring buildings. It is also focusing on buildings closest to complying with the GPs. Headquarters then tracks GP assessment and compliance status using the sustainability section of DOE's real property database. DOE uses a host of technology tools for assessing and managing its varied asset portfolio. Finally, it ensures ongoing support through internal and contracted expertise.

DOE used its governance structure to effectively communicate this plan to its team and the organization.

Other segments of the federal government are developing GHG assessment and mitigation programs, including DoD and the military. Military organizations, in particular, excel in leadership, clear GHG reduction goals, and communication. They benefit from their well-defined chain of command, which comes with accountability, particularly at the installation level.

Integrate Solutions across Organizational Boundaries

Recommendation: *Employ total asset management (TAM) to align processes and resources to meet GHG mitigation goals.*

CHALLENGE

As organizations expand over time, they also can become overly complex and decentralized. Organizational hierarchies, divisions, and business units are necessary to some degree, but can also result in localized or short-term benefits at the expense of greater organizational efficiencies. Too often, these divisions stand in the way of an organization's realizing its full potential.

Sometimes, the rigidity of these organizational structures and cultures make it difficult to change them. The phrase "turning the battleship" comes to mind in describing the speed of enacting change as an organization grows.

Historically, the phrase "buckets of money" has been used to identify funding that these organizational units receive each year. For example, in one bucket is funding for facility maintenance and repair, while the money for facility modernization is in another, keeping them separate. They are both separate from energy management, which is separate from safety and health, which is separate from security, and so on. Each task has its own bucket, and no one ever seems to consider that they are all drawn from the same well. Only very recently have organizations begun taking a more integrated approach to infrastructure management.

SOLUTION

Mitigation Potential	★★
Operational Impacts	★★
Financial Impacts	★★★
Feasibility and Timing	★★★

Organizations should employ TAM, a strategic approach to physical asset planning and management. An organization develops a 10-year asset plan that details and aligns its priorities and strategies with its mission. Such a plan carefully considers the limits of available resources.[11]

The idea is to look across the entire infrastructure management organization and align processes and resources with a common set of goals and objectives.

Stakeholders develop business rules, desired outcomes, and key measurement indicators and commit to following them. The key is to start early because TAM is much easier to implement in the early stages of organizational development (but it can also be employed in mature organizations over time).

The GHG mitigation benefits of this approach are achieved by significantly reducing duplicate or overlapping processes and the expended energy and resources that go along with them. Committing to a TAM approach requires an organization to start on a path of consolidating and integrating legacy systems, recommitting to a common organizational "language."

There will be some individual winners and losers, but ultimately the organization will benefit.

EXAMPLE

The US Intelligence Community (IC), in response to the 9/11 attacks, began employing an integrated effort based on the TAM system.

The attacks revealed that the IC agencies had been duplicating efforts, operating closed off from one another, and simply not sharing information. Individually, they gave the protection of information a higher priority than using it collectively to solve issues of national security.

Today, more information is being shared and coordinated than ever before, a cultural shift that has filtered down to infrastructure and asset management. One particular defense organization is using the TAM model to help coordinate a major business transformation effort, part of which includes the management of infrastructure. Its approach will reduce duplication, consolidate legacy systems, and standardize and integrate management systems across the entire organization. The approach is driven by a desire to improve mission efficiency and cost, but it will also significantly reduce the agency's carbon footprint.

Secure building at Bolling Air Force Base designed for low emissions.

The integration of security and energy efficiency (also discussed in the "Security" chapter) has trickled down to increase attention to federal facility security, safety, and force protection issues since 9/11. At first glance, being energy efficient, green, or sustainable would seem to take a back seat to force protection and security design attributes.

But the connection between the two comes into focus when you view security and safety measures in a total project context, including their impacts on occupants and the environment, regardless of the necessary level of protection. Today's security designs are based on a multihazard approach, where planners assess the likelihood and impact of all hazards on a facility, including criminal, terrorist, accidental, and natural threats.

Joint Base Anacostia-Bolling, a 905-acre military installation in Washington, DC, is home to the Defense Intelligence Analysis Center. Built in 2003 with both security and sustainability in mind, the facility is a hub for the exchange of intelligence. Its design includes an earthen shelf that provides ongoing energy, and GHG and cost savings by reducing the facility's heating and cooling demand. The same shelf is able to stop a truck traveling at up to 40 miles per hour, offering superior protection to the center's occupants.[12]

Incentivize Suppliers to Mitigate through Procurement Guidelines

Recommendation: *Influence suppliers through procurement guidelines that support the organization's mitigation efforts.*

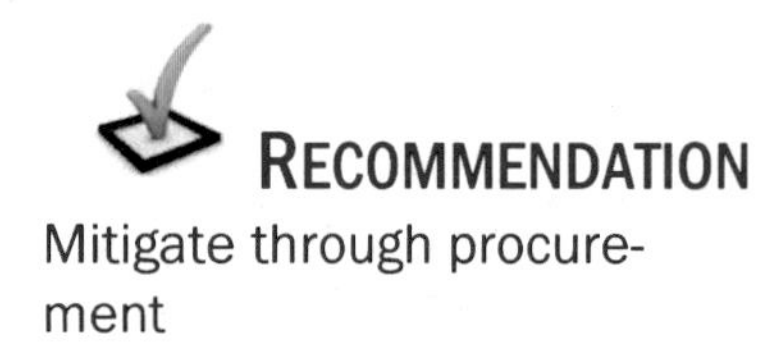

Mitigate through procurement

CHALLENGE

Procuring products and services (and the related solicitation and contracting practices) is a process filled with inefficiencies, inconsistencies, and exceptions. If this process were a road, it might be the Cross Bronx Expressway in New York City, which Empire State drivers will tell you is perpetually strewn with potholes and bumpy patches, orange cones, new traffic patterns, and pockets of traffic.

The procurement process will likely remain unpredictable, but embedding mitigation and sustainability guidelines in its language will establish a new threshold for entering the competition. The challenge here is to not only write mitigation policies and incentives into individual statements of work, but also into the organization's general terms and conditions.

SOLUTION

Mitigation Potential	★★
Operational Impacts	★★★
Financial Impacts	★★
Feasibility and Timing	★★★

The relatively simple fix is to implement procurement guidelines that are strong and clear enough to influence suppliers. For example, an organization can establish that it will only hire suppliers that, in general, have documented programs and records in GHG mitigation. As organizations build this type of requirement into their procurement guidelines, suppliers will have no choice but to comply. They will be driven to compete through their mitigation practices as well.

Green procurement guidelines influence each segment of the infrastructure life cycle (which comes up in the next recommendation):

- *Program.* When hiring a real-estate firm to select a site, stipulate that it account for the cost-benefit tradeoffs of GHG mitigation metrics within each alternative.
- *Design.* When hiring a consulting engineer to design the infrastructure, stipulate that it use building information modeling (BIM) tools where integrated information can be shared by all contractors throughout the project's life cycle. Also, require that it apply the most current design standards and offer additional green alternatives.
- *Build.* When hiring both construction contractors and equipment manufacturers, mandate that their supply chains include

the use of documented GHG mitigation practices. Award performance-based contracts and reward innovations.

- *Operate.* When hiring operations and maintenance contractors, stipulate that, for the work they perform for you, they use only products and practices that are part of a GHG mitigation program. Look for proposals that demonstrate ways to cost-effectively increase the useful life of infrastructure or its components while minimizing GHG emissions.

A crew laying asphalt.

The private sector is preparing to provide potential clients with this information because it's good business, meaning that stakeholders can and should insist on its inclusion in all infrastructure procurements.

EXAMPLE

Changes to the procurement process are being embraced by private-sector firms looking for work, and the necessary support structures to help organizations meet these new requirements are being developed.

State and local government have a number of options when paving roads and highways, for example. Their choices include warm-mix asphalt, a percentage of recycled asphalt, and perpetual pavement. Some are more GHG friendly than others. States have begun engaging in alternative bidding, where they look at initial cost, construction time, and the future costs of a number of options.

To support the needs of these states, the Asphalt Pavement Alliance has issued guidance to its members for participating in alternative bidding processes. Under alternative bidding, private firms present road and highway agencies with a host of paving options that weigh cost and construction time, future costs, life cycle with discount rates, performance periods, rehabilitation plans, salvage value, and analysis periods.[13]

This guidance targets asphalt and Portland cement concrete pavement, but it could easily be expanded to include other mitigation-friendly alternatives.

The Los Angeles Community College District is in the midst of a massive $6 billion modernization program. To support their Sustainable Building Program, it has required the use of BIM, and any building team hoping to win a contract for design-build for new buildings or retrofits must agree to a rigorous BIM program.[14] The stringent process sets expectations of compliance with strict rules on workflow, information sharing, and early design collaboration.

Apply Mitigation Strategies throughout the Infrastructure Life Cycle

RECOMMENDATION

Mitigate cradle to grave

Recommendation: *Apply BIM to mitigation activities to generate and manage building data during the infrastructure life cycle to inform future decisions.*

CHALLENGE

Too many decisions are made and actions taken in isolation, without a broader view of their effect on future phases of a life cycle. Understanding how one phase links to the next can benefit the cycle. An infrastructure portfolio typically has projects in each phase of the life cycle (Figure 5-2).

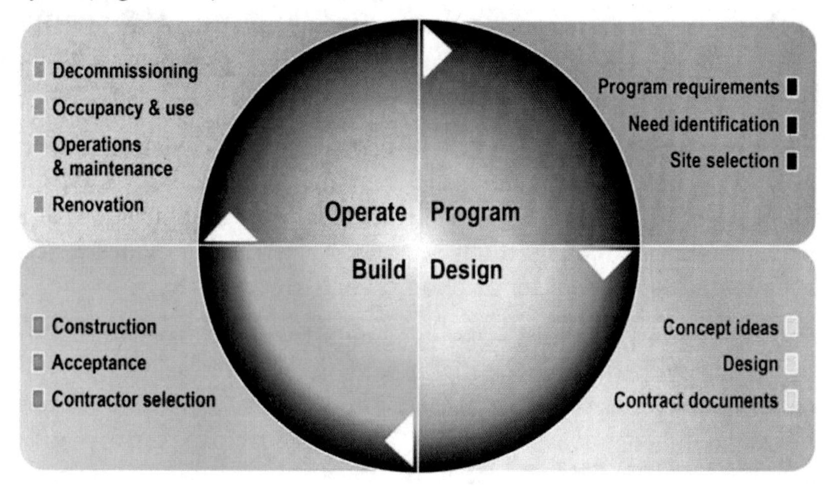

Figure 5-2. Incentivized Mitigation Must Be Applied across All Phases of the Infrastructure Life Cycle

SOLUTION

Mitigation Potential	★★★
Operational Impacts	★
Financial Impacts	★★★
Feasibility and Timing	★★

BIM is an object-oriented technology where each object has numerous linked attributes tracked for the entire life cycle of the structure. An object, for example, can be an HVAC unit, where its attributes are data elements such as operating specifications, warranty information, energy usage, component lists, and programmed operating parameters. Using BIM technology can foster interoperability among the many stakeholders involved in the infrastructure life cycle.

Through BIM and other tools, these attributes can be viewed in functional layers so that consulting engineers can match the manufacturer's specifications to the required cooling load for the building being designed. The commissioning agent can view the sequence of operations and commissioning checklist. The facility manager can view normal operating parameters, run-time history, and adjust set points. The service contractor can look at the unit's repair history, spare parts inventory, and preventive maintenance schedule.

The facilities chief can compare the energy usage of this unit—at the end of its useful life—with newer, more efficient technologies in making a replacement decision.

BIM can certainly be used to facilitate the design and construction of net-zero buildings, which have zero net energy consumption and zero carbon emissions. The intent of these buildings is to reduce energy demand, generate on-site power, cycle water, and reduce or eliminate waste.

EXAMPLE

GSA has strategically and incrementally adopted the use of BIM since 2003. It led a pilot program of more than 30 projects to address programming, design, and construction challenges.

The program, managed by the GSA Office of the Chief Architect, establishes policy that requires the incremental adoption of BIM for all major projects. The program is leading BIM pilot applications and incentives for current and future capital projects and provides expert support and assessment for ongoing capital projects.

Part of the process involves assessing industry readiness and technology maturity, partnering with BIM vendors and professional organizations, and formulating a GSA BIM toolkit that includes GSA solicitation and contractual language. GSA has developed a BIM guide series covering individual functional topics.

Encourage and Incentivize Behavioral Changes

Recommendation: *Support changes in the organizational culture in line with mitigation activities through incentives and, sometimes, disincentives.*

CHALLENGE

The greatest overall impact on mitigation efforts is often found in the end users of infrastructure—a facility's occupants, drivers on roads, and consumers of power and water. Unfortunately, actions having immediate returns are often given priority over those with long-term benefits.

Measuring the impact of certain behaviors is more difficult, making it harder to convince people of their benefits.

SOLUTION

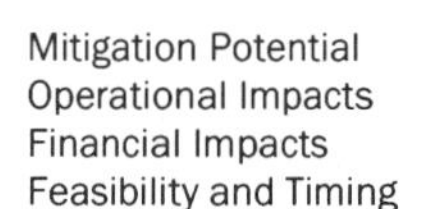

Mitigation Potential	★★
Operational Impacts	★★★
Financial Impacts	★★
Feasibility and Timing	★★★

Components of behavior change strategies include changes in habits and practices, consumer education, targeted information programs, incentives and rebates, technology promotion, and lifestyle changes.[15] Behavior change enablers include financial, technological, cultural, and informational elements.

Most behavior-based campaigns are not effective without incentives. Sometimes disincentives are necessary. By linking on-the-job energy savings behavior to rewards, the pool of stakeholders is instantly expanded. For example, a business unit may set a target of energy usage for the year, and part of the organization's performance review is whether it reached that target. A portion of money from energy saved might be given back to users (to do so, facilitators would need to ensure that funds can be transferred from the utility account to one that directly benefits users).[16]

Finally, as mentioned earlier, the key to achieving behavioral changes is to design an environment in which they are as easy and inviting as possible.

Developing behavior change programs around smart phones can help. The proliferation of smart phone technology and acceptance over the past several years has been remarkably rapid. If you have a smart phone, it travels with you, and push technology allows you to be alerted and notified to a number of messages. In particular, touch-screen technology has increased the speed and convenience with which users can do just about anything on a smart phone.

Smart tablets enable energy-efficient construction practices.

For example, how many people reading this chapter have used their smart phone to check into a flight and download a boarding pass QR code? This behavioral change from the old standard check-in process eliminates printing, paper, the use of check-in kiosks, and waiting in line for attendants. In short, this behavioral change has succeeded because it was designed to be easy to implement, mitigating GHG emissions by adopting a paperless system, while also saving time for the user.

Industry-wide changes in construction, even small ones, can make a big difference in mitigating GHG emissions. First, provide proactive training, communication, and incentive programs on energy-saving techniques, while identifying desired behaviors and targets. Again, smart phone technology can be used to manage and track deliveries, minimize wait times, and monitor vehicle idling time.

EXAMPLE

A number of smart phone apps, some free, others available at a moderate cost, are giving people the opportunity to change their habits or support new practices in an easy, convenient, and portable interface. The array of apps is stunning, and a number of them support participation in GHG mitigation activities.

One app, called GreenFuel, allows users to search for alternative fuel stations in their area, even providing information for compressed natural gas, biodiesel, electric, and hydrogen.[17] It comes in a full version for a modest price, or a more limited "lite" version, which only gives the user results for compressed natural gas within a 15-mile radius. Another similarly priced app, called Eco Maps, plots out the most efficient route for trips around town, offering the user convenience in the form of saving time, money, gas, and the environment.[18]

Adaptation

Mitigation strategies are most effective when coordinated with one another, but they can still be effective in isolation, since they are all geared toward a common threat—GHG emissions. Adaptation strategies, on the other hand, must be coordinated with respect to a specific threat among many.

Climate change poses an array of impacts and threats to infrastructure. Each requires an adaptation strategy.

On the basis of the trend of higher summer and winter temperatures, certain types of events will become more intense or more frequent. Two of the most likely are flooding and extended heat waves.

Increased Temperatures

The average global land surface temperature for August 2011 was the second warmest, behind 1998, since record-keeping began in 1880.

The temperature was 1.51°F above the 20th century average. This is similar to the July 2011 temperature anomaly and continues a streak of 142 consecutive months (since November 2000) that the monthly global land temperature has been above the long-term average.[19] Increased temperatures have a number of common effects on infrastructure:

- Asphalt rutting
- Rail buckling
- Heat or lack of ventilation on urban transit systems
- Low water levels on inland waterways
- Thermal expansion of bridges
- Overheating of diesel engines
- Enhanced wildfire risk
- Upper temperature limits for workers stemming from poor air quality
- Extended heat waves, especially in places where cooling systems are underdesigned or undersized for extended heat events
- Shifts in the freeze-thaw cycle that impact transportation through inland waterways.

Increased Precipitation

Average precipitation has increased in the United States and worldwide. Since 1901, it has increased at an average rate of more than 6 percent per century in the contiguous states. The percentage of heavy precipitation (intense single-day events) has also increased. Eight of the top 10 years for extreme single-day precipitation events have occurred since 1990.[20] The impacts of precipitation on infrastructure include the following:

- Concrete deterioration
- Altered runoff patterns
- Flash flooding.

Sea Level Rise and Storm Surges

Worldwide, sea level has risen at a rate of 0.6 inches per decade since 1870. The rate of increase has accelerated in recent years to more than an inch per decade.[20] Rising sea level, combined with increasingly intense precipitation, amplifies the risk of damaging storm surges. Sea level rise and storm surges impact infrastructure as follows:

- Erosion of coastal highways
- Flooding of tunnels
- Increased risk to levee systems, meaning greater potential for widespread flooding of surrounding areas
- Higher tides at harbor facilities and ports
- Low-level aviation infrastructure risk
- Less bridge clearance
- Washouts of coastal highways and rail tracks.

The key to a successful adaptation plan is to anticipate future challenges. Following this, actions must be initiated that will ultimately reduce vulnerability and increase the resilience of infrastructure. Our recommended adaptation strategies meet both requirements:

- Conduct a risk assessment and analysis
- Develop a proactive adaptation plan
- Establish and communicate early warning systems and evacuation plans
- Develop flood control systems
- Improve infrastructure resilience through protection and hardening.

Conduct a Risk Assessment and Analysis

Assess climate risks

Recommendation: *Use risk as a tool for plotting a course of action for infrastructure adaption, and employ the appropriate assessments and analyses to acquire meaningful data.*

CHALLENGE

Each organization or community has a different risk profile for climate events based on their unique circumstances. Vast differences include mission, location, number of people, degree of integration with other organizations and communities, and so on. To identify the most effective adaptation measures, each particular community or organization must understand the degree and type of risk posed.

SOLUTION

Adaptation Potential	★★★
Operational Impacts	★★★
Financial Impacts	★★★
Feasibility and Timing	★★★

Risk management shows stakeholders the areas on which to focus. According to the ASCE's *Guiding Principles for the Nation's Critical Infrastructure*,[6] risk management is the application of a systematic process for identifying, analyzing, planning, monitoring, and responding to risk so that critical infrastructure will meet service expectations. Managers and stakeholders should participate in this process for their infrastructure.

In this case, risk is defined as shown below:

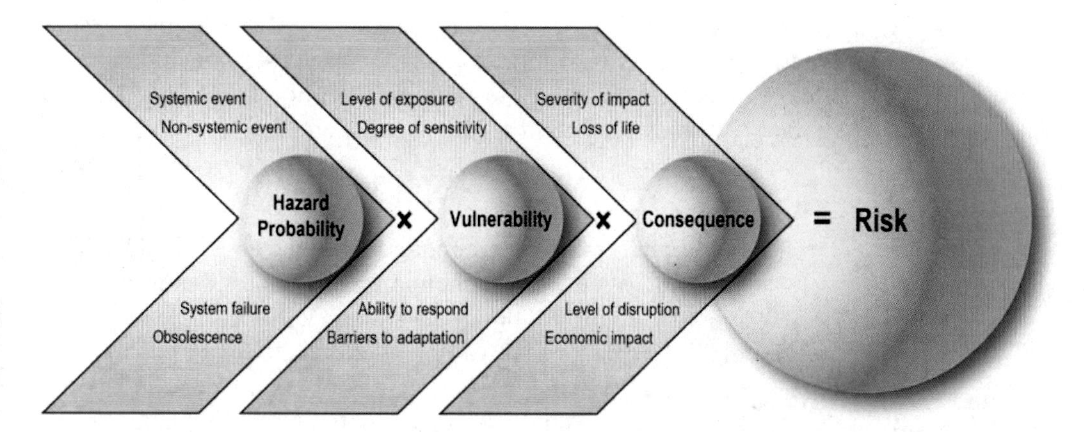

Probability of Hazard

This first component is the probability that a hazard will occur and that critical infrastructure will not perform to required levels. The probability of a hazard will certainly vary by geographic location. A simple probability scale is sufficient when used in conjunction with the other components of the equation: (1) remote, (2) unlikely, (3) likely, (4) highly likely, and (5) near certainty.

Risk is a big part of this book! The risk framework discussed here can easily be applied to every functional area in this book.

Vulnerability

Vulnerability is the degree to which a system is exposed and unable to cope with adverse effects. A number of variables must be identified and documented to properly assess vulnerability, including the following:

- *The level of exposure to a hazard*, assessing the geographic location and related physical elements
- *The degree of sensitivity to a hazard*, linked to the age of a certain infrastructure, its location, its capacity, its quality, and so forth (even a relatively mild event could have serious consequences in some cases)
- *The ability to respond*, or the resources available to respond to a climate change event, which considers financial, human capital, and physical equipment resources, among others
- *Any barriers to adaptation*, physical, financial, social, regulatory, or political, each or a combination of which may prevent or delay adaptation measures.

Consequences

Consequences are the range of possible impacts of an event, such as loss of life, economic impact, environmental damage, or cultural loss.[4] The Department of Homeland Security's National Critical Infrastructure Prioritization Program uses consequence-based criteria to identify national critical infrastructure. These criteria include loss of life, economic costs, ability to reroute, length of detour, and time to rebuild if damaged. The consequence scale arranges outcomes as follows:

1. Minimal or no impact.
2. Additional resources are required, but the ability to adjust exists—no loss of life, economic impact under $1 million, geographic footprint limited to a single facility complex or a single facet of a larger area (a power outage, for example), time to repair may take up to 6 months or can be done in as little as a few days. Operations can continue with minimal disruption.
3. Minor impact—no loss of life, economic impact up to $10 million, geographic footprint impacts an isolated section of a city or area, rebuilding may take 6 months to a year, continuation is possible with some workarounds or temporary solutions.
4. Major impact—potential for some loss of life, economic impact in the tens or hundreds of millions of dollars, geographic footprint impacts a portion or portions of a major city or region, rebuilding may take 1 to 3 years, but continuation is possible with workarounds and temporary relocation.
5. Cannot continue—large-scale loss of life, economic impact in the billions, geographic footprint impacts a major city and surrounding areas, rebuilding process may take 5 to 10 years or more.

Consequences must also include the increasing cost and stringency of obtaining insurance because a greater insurance burden increases overall financial risk to organizations. Private industry may move away from standard insurance practices, putting greater burden on state and local governments to provide some measure of recourse to those uninsured or uninsurable. This is a case where today's investment to adapt to risk could save millions or billions of dollars down the line.

Risk Matrix

The variables of this formula remain a challenge because predicting the timing of most events is difficult or impossible. In turn, the opportunity for response is usually very limited. However, we know that major events will continue to happen, and we can begin to establish probable long-term frequencies of events and ranges of possible outcomes based on historical data. Extending these data beyond what has occurred in the past provides insight into what may be a worst-case scenario.

Figure 5-3, a risk analysis matrix for climate change outcomes, shows this relationship in three dimensions:

- Horizontal axis, probability of hazard
- Vertical axis, level of consequence
- Size of bubble, degree of vulnerability (the larger the bubble is, the greater the vulnerability).

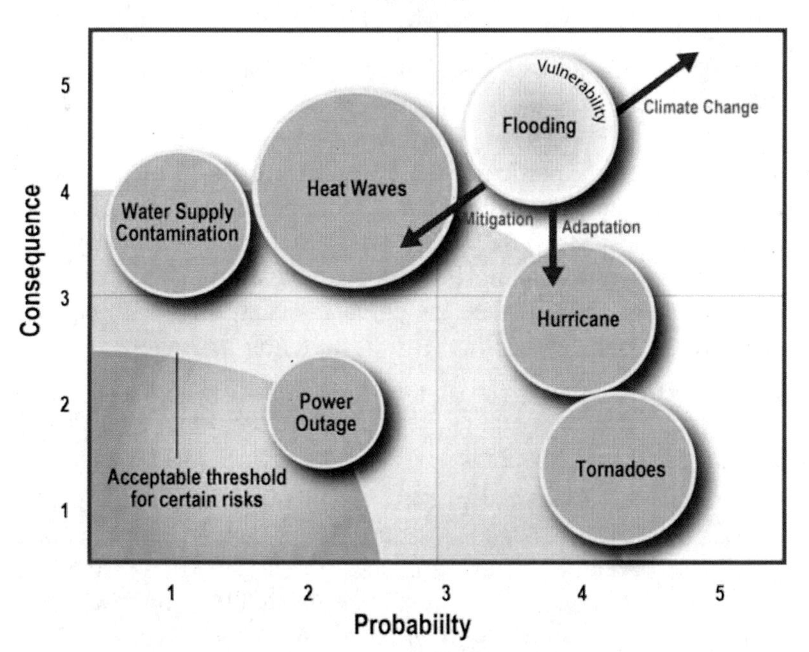

Figure 5-3. Three-Dimensional Representation of Risk Analysis

The curving lines underneath the circles represent acceptable levels of risk. The thought is that by identifying these risks and taking adaptive action, an organization can remain within those thresholds.

This matrix will vary on the basis of individual circumstances, hence the need for individual risk assessments. An infrastructure manager in

Colorado, for example, will not have the same flood or heat wave probability as a counterpart in Mississippi.

From their probability, degree of infrastructure vulnerability, and consequence of impact, flood events and heat waves have the potential to be most devastating.

Floods can be caused by multiple sources; they'll endanger lives, and they have potentially staggering economic impacts and lengthy recoveries. Heat waves, on the other hand, will increase in frequency over the coming decades and pose a threat to lives, particularly those of the elderly (as discussed in the "Health" chapter). Focusing adaptation strategies on these two types of events has the greatest potential to save lives and reduce economic impact.

It is also useful to attempt to predict the direction of movement of the circles on the basis of changes in risk variables and to speculate on how implementing adaptation strategies will influence that direction.

EXAMPLE

Several states and municipalities have begun pilot assessments of risk to their road infrastructure. One is the Washington State Department of Transportation (WSDOT). The state received a federal grant worth $189,500 supporting participation in a pilot project to better understand how extreme weather brought by climate change could damage transportation infrastructure.[21] The program follows a step-by-step process set up by the Federal Highway Administration to examine the risks to and potential impacts of climate change on Washington's major bridges, highways, and other transportation structures.

Washington's statewide effort included studying WSDOT-owned and -managed facilities that were potentially at risk to sea level rise inundation, river and stream channel migration, melt effects, extreme temperature effects, drought threats to wetland creation, mitigation sites, roadside vegetation, soil moisture/flux, invasive species, worker health, wildfire, precipitation changes, threats to slope stability, stormwater management, erosion control, landslides, and "road survivability." It also considered wildfire safety and emergency response.[22]

The outcomes of the Washington program (and other pilot programs) will be used at the federal level to improve the step-by-step process and make it available to other state transportation departments and organizations.

Develop a Proactive Adaptation Plan

Recommendation: *Develop and implement a strategic, proactive plan consisting of adaptation activities that take into account partners and stakeholders and coordinates their efforts.*

RECOMMENDATION

Plan to adapt

CHALLENGE

According to the Council on Environmental Quality (CEQ),

> *Adaptation measures should focus on helping the most vulnerable people and places reduce their exposure and sensitivity to climate change and improve their capacity to predict, prepare for, and avoid adverse impacts.*

This broad, sweeping statement accurately reflects the nature and challenge of adaptation planning, which requires many moving and interconnected pieces as well as multiple groups of stakeholders.

The goal of adaptation is not to be free of risk—that's impossible—but to manage risk well.[23] Proactive leadership is required to ensure the understanding, planning, and implementation of adaptation strategies. This effort also needs the support of a strong, effectively communicated public awareness campaign.

Actions will be required by government, businesses, and individuals to increase resilience. Actions will be needed to reduce the vulnerability of established systems to unavoidable climate impacts.[24] This investment has much at stake and may require substantial resources. As a result, these measures must be developed carefully, with thoughtful planning and a reliance on up-to-date science and analytical tools. Creativity must be emphasized to ensure the development of practical, cost-effective actions and technologies.

Because so much shared infrastructure must be protected, the most effective adaptation strategies will involve the coordinated efforts of multiple stakeholders.

SOLUTION

A strategic infrastructure adaptation plan that accounts for climate change impacts requires a few basic considerations. Organizations should evaluate each step with respect to their unique set of climate risks (Table 5-3).

Adaptation Potential	★★
Operational Impacts	★★
Financial Impacts	★★★
Feasibility and Timing	★★

Table 5-3. Adaptation Actions Appropriate for Various Stakeholders

Stakeholder	Adaptation Action
Public sector	• Promote greater public awareness • Initiate discussion among multiple levels of government, industry, and community • Develop meaningful policy guidance, typically through a climate change adaptation task force • Communicate to the broader public
Businesses	• Innovate through competition • Strive for win-win scenarios where adaptation measures can be implemented that are also economically advantageous • Provide input regarding policy development
Communities and individuals	• Interact with local governments • Provide checks and balances on government and business initiatives • Give feedback as customers and end users

Identify and Coordinate with Stakeholder Groups

Stakeholders include the public sector (federal and local government), businesses, communities, and individuals. These groups must work together to implement adaptation strategies.

For example, a county official in charge of housing developments realizes that an increase in heat waves elevates the risk of degradation and strain on building materials. Countering this risk will require input from the construction industry to develop forward-looking design standards capable of withstanding extended heat waves. While developing this adaptation strategy, which involves upgrading design standards, input should be gathered from surrounding businesses, communities, and individuals. The goal should be to gauge sensitivity to the higher prices that may result from the increased construction cost brought on by these new standards.

Plan Strategically

Consider both long- and short-term strategies to maximize benefits and satisfy as many stakeholders as possible.

Long-term strategic planning involves community and stakeholder support and buy-in. Long-term actions may involve large infrastructure investment and pursuit of wholesale changes at macrosocietal levels. They potentially involve diversifying or shifting supply and demand. These actions will cross-pollinate with many functional areas, spurring public debates and decisions on land use, development, and population growth projections, to name a few.

Short-term planning, on the other hand, involves more tactical decisions. Certain decisions must happen at a national level, but much can be achieved at a local or regional level.[2] Local and regional governments have jurisdiction over planning decisions, zoning, and building codes.

For example, a building owner's property is located in a low-lying area. His greatest risk is from basement flooding and water damage due to increasingly frequent and intense precipitation. A short-term solution may be to improve the capacity to respond to flooding through better pumping equipment. A longer-term solution would be to reengineer and renovate the surrounding drainage system, build protective barriers, or ultimately to divest and relocate.

Include Climate Risk in Policy Development, Including Insurance

Adaptation measures may be viewed as insurance policies against climate events, where worst-case scenarios are considered and preparations made. Adaptation measures may include energy security through insurance against long-term price increases, volatility, and disruptions. Some benefits will not be in the form of positive economic value, but rather in minimizing losses. These might not be revealed, however, for several decades.

For example, as insurance companies begin pricing climate change impacts into their products,[25,a] facility managers will want to economize by finding means to minimize potential damage to their organizations' insured assets.[26] The Securities and Exchange Commission now requires US-registered corporations to report their climate change-related risks,[27] including the effects of legislation, international agreements, and regulation, plus potential physical impacts.

Stakeholders should review existing policies to make sure they don't inadvertently increase vulnerability to climate change.[20] For example, construction of levees may initially be a solution but may increase

---

[a] On March 17, 2009, the National Association of Insurance Commissioners adopted a mandatory requirement that insurance companies disclose to regulators the financial risks they face from climate change as well as actions companies are taking to respond to those risks. All insurance companies with annual premiums of $500 million or more are required to complete an Insurer Climate Risk Disclosure Survey every year, with an initial reporting deadline of May 1, 2010. The surveys must be submitted in the state where the insurance company is domiciled.

The increased probability of floods may force some owners to relocate.

long–term vulnerability if the perception of safety in the area behind the levee results in increased development (a form of "maladaptation").[19]

Be Current on Adaptation Science and Service Capabilities

Remaining current on climate knowledge and adaptation best practices can help facility managers and public officials make the best decisions when developing a proactive adaptation plan. Science facilitates these decisions and technology offers many powerful tools.

One of the goals of the CEQ is to improve the integration of science into decision making. A key recommendation in this area is to develop an online data and information clearinghouse for adaptation. Tools such as an interactive climate impact map may be useful for decision makers.

Adaptation policies must also be flexible enough for unpredictable circumstances.[28]

The Climate Adaptation Knowledge Exchange, or "CAKE," is one source for staying current on adaptation science and service capabilities. Its goal is to build a shared knowledge base, sourced by authoritative materials, and establish an innovative community of practice.

Develop Metrics and Prioritize Strategies for Best Value

Develop metrics and feedback mechanisms to measure whether adaptation actions are achieving desired outcomes.[b] Adapting to a project as it unfolds will allow revisions as outcomes are evaluated. Establishing metrics gives stakeholders the ability to track and highlight progress while identifying weak spots. For example, metrics may include establishing required design standards to withstand a 100- or 1,000-year event or monitoring the percentage of successful early warning system tests. Other metrics may come from risk analysis.

Once the risks have been analyzed, acceptable levels defined, and appropriate adaptation strategies identified, the next step is to prioritize them with an eye on maximizing their value and benefit. This should take into account adaptation strategies and consider public health and safety, economic benefits, capacity for adaptation, and environmental impacts that will yield a "best value" result.

---

[b] Please refer to the second and third recommendations in the mitigation section of the Security chapter for examples of energy and GHG metrics and their application to risk.

For example, a facility manager has conducted a risk analysis and determined that his highest risk element, extended heat waves, carries a probability score of "4," has a consequence score of "3," and brings with it "high vulnerability."

The manager then works proactively to upgrade obsolete cooling equipment, increases the facility's ability to respond through an early warning system, establishes a communications plan, and promotes a policy giving facility occupants the ability and option to telework. These actions have now reduced the facility's risk profile metrics.

Some of the prioritization criteria to consider in evaluating adaptation strategies include the following:

- *Public health and safety*, where levels of risk must be communicated and decisions made with public well-being in mind.
- *Economic benefits*, where tradeoffs between budget line items will be much easier to make if a tangible payback can be substantiated. Adaptation strategies will only follow the trajectory of other GHG reduction efforts if the energy and environment industry can show how it can be monetized. Another economic tradeoff to consider is the cost of *not* investing in infrastructure adaptation. These costs would include greater and more frequent emergency repairs and more frequent maintenance of existing systems that may be at, or over, capacity.
- *Capacity for adaptation*, which contains three critical elements: social equality, income, and development of health systems. Development of adaptive capacity involves some leadership and participation nationally from the executive branch. This must be combined with local policy development, strategic investment, and increased flexibility in infrastructure, work processes, and workforce. For example, high-density populations in high-risk areas do not have the same capacity to adapt as low-density populations. The economic and physical constraints vary by community and demographic.
- *Environmental benefits*, which involve adaptation strategies that also provide mitigation benefits. Instead of repair and replacement, improvements and modernization can also yield significant GHG mitigation benefits. Another facet to consider are strategies that enhance existing efforts—it is much easier to modify or expand that which is already

happening than it is to start something new from scratch, which requires funding and staffing.

EXAMPLE

The Commonwealth of Pennsylvania has developed several proactive plans that not only measure risk, but also offer forward-looking actions that can begin the process today for adapting to climate change.

The *Pennsylvania Climate Adaptation Planning Report: Risks and Practical Recommendations*, issued by the Pennsylvania Department of Environmental Protection in January 2011, is the first statewide effort to identify practical strategies for addressing climate change impacts.[29]

The report includes recommendations for climate change adaptation of four sector-specific working groups: infrastructure, public health and safety, natural resources, and tourism and outdoor recreation. For the transport sector, the report recommends reviewing research for materials that have the potential to withstand higher temperatures to prevent buckling of roadways and bridges. It also recommends performing more intense inspections of transportation infrastructure after high-impact events in areas subject to erosion. It also suggests adopting green infrastructure, promoting walkable communities, and integrating adaptation and mitigation strategies as part of government agency planning and operations.

In 2009, the Pennsylvania State University's Environment and Natural Resources Institute wrote a report, *Pennsylvania Climate Impact Assessment: Report to the Department of Environmental Protection*,[30] but it did not include infrastructure. However, this report provides a solid foundation for future work by describing the expected impacts of climate change on Pennsylvania. It discusses temperature and precipitation impacts; examines the implications for water resources, forests, wildlife, aquatic ecosystems and fisheries, and agriculture; and looks at what these impacts mean to energy, human health, tourism and outdoor recreation, insurance, and economic risk.

Establish and Communicate Early Warning Systems and Evacuation Plans

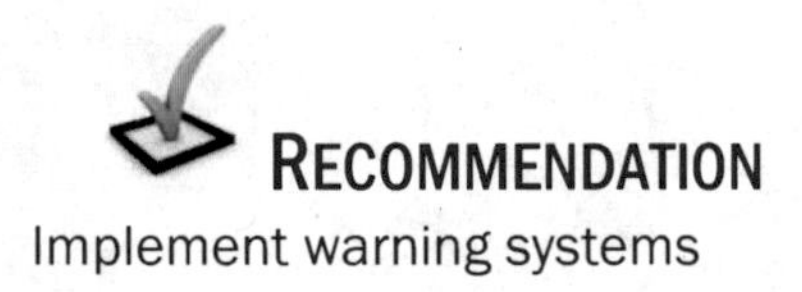

Recommendation: *Prepare preventive safety measures, including preparedness and response capability, and communicate them to stakeholders and the public.*

CHALLENGE

Extreme weather events are the most urgent threat to infrastructure: they are more frequent and, at times, allow little time to react. Although high-intensity weather events are more frequent, the loss of life has decreased due to advanced early warning systems and evacuation plans.

The sooner that early warning systems are implemented for climate change events, the lower the longer-term costs will be.

SOLUTION

Adaptation Potential	★★★
Operational Impacts	★★★
Financial Impacts	★★
Feasibility and Timing	★★★

Natural disasters will happen regardless of climate change, but with our current research on and scientific knowledge of climate change, we can predict which types of events will happen more frequently and begin preparing for them. Preparedness and response capability, on the other hand, requires strong government support and leadership.

Infrastructure can be used to help save lives. Infrastructure systems that furnish sufficient protection and response capacity for events include wireless communication, air-conditioned shelters, and health systems.[2] Information technology infrastructure has a broader reach and greater interconnection than ever before. Tornado warnings, for example, can be quickly disseminated from National Oceanic and Atmospheric Administration satellites, to the National Weather Service, to a local county government, and directly to someone's smart phone. Today, many local governments have opt-in programs for emergency notifications through e-mail or text messaging.

Similarly, public facility infrastructure can be used for emergency shelter during severe weather events. This could be a school providing shelter and emergency supplies during a hurricane or a shopping mall operating for extended hours during a heat wave to provide available cool spaces.

EXAMPLES

New York City officials, fearing a significant storm surge from Hurricane Irene in late August 2011, declared for the first time in city history a mandatory evacuation of residents living in low-lying coastal areas.

New York City's hurricane evacuation plan designates three separate zones: A, B, and C. The city's lowest-lying region, most susceptible to hurricanes and storm surges, is Zone A, while Zones B and C are areas that will be flooded by higher-category hurricanes. In advance of Irene, without knowing for certain how intense the storm would be, city officials issued evacuation orders to people in Zone A and parts of Zone B. They expected the storm to diminish in intensity but possibly remain a Category 1 when it reached Manhattan.

Nearly 270,000 Zone A residents from neighborhoods in Manhattan, Brooklyn, Queens, Staten Island, and Coney Island faced threats of severe flooding and were ordered to leave their homes. All subway, bus, and ferry system services were suspended, another first for the city.[31]

New York Mayor Michael Bloomberg regularly briefed the public, city firefighters and social workers assisted with the evacuation, and emergency shelters were established throughout the city.

Hurricane Irene did not impact these urban areas as severely as feared, but the evacuation was viewed as a success because the nation's largest city had in place an evacuation plan based on up-to-the-minute models of flood behavior.[32] These models indicated that the storm might have the power to flood large parts of Manhattan, which *did* happen to cities and towns from New Jersey all the way to Vermont.

Develop Flood Control Systems

Develop flood controls

Recommendation: *Establish flood control systems that make sense for the locality by considering all options and evaluating them for their impact in relation to the required investment.*

CHALLENGE

Flooding is the common denominator of multiple types of events. Many factors cause floods: heavy rainfall, severe winds over water, unusually high tides, tsunamis, or failed dams, levees, retention ponds, or other structures. Periodic floods occur on many rivers, forming a surrounding region known as a flood plain.

An intense summer thunderstorm can bring with it flash flooding, and floods can result from a hurricane-spurred storm surge, extended spring precipitation in a flood plain, or coastal impacts from a gradual sea level rise. The force of water is very powerful and can cause extreme damage.

About 40 percent of the world's population lives within 60 miles of the coast, well within reach of coastal storms.[14] Over 36 million people, roughly 12 percent of the US population, live along the Atlantic and Gulf coasts from North Carolina to Texas. This is the target zone for Atlantic hurricanes and the threat of flooding from storm surges.

Also, 100 million people live less than 1 meter above mean sea level.[14] For example, a 1-meter rise in sea level would cover significant areas of New York City around the Hudson River's Upper Bay—an area that includes several major international airports.

Combined with storm surges, this sea level rise will not only impact coastal areas, but also have extremely damaging effects on sanitation systems, power plants, and the subway system. More gradual effects would include the salination of freshwater systems that supply residents with water.

Flooding damages property, causes erosion, and endangers the lives of humans and other species. Prolonged high floods may delay traffic in areas that lack elevated roadways. Structural damage can occur in bridge abutments, banklines, sewer lines, and other structures within floodways. Waterway navigation and hydroelectric power are often impaired. Financial losses due to floods are typically in millions of dollars each year.

Storm surge is simply water pushed toward the shore by the force of the winds swirling around a storm and is a function of storm strength, location, tides, elevation, and speed. An advancing surge from a hurricane can combine with normal tides to increase the average water level by 15 feet or more.

In addition, wind-driven waves are added to storm tide and can pound away at structures. Water weighs approximately 1,700 pounds per cubic yard; extended pounding by frequent waves can demolish any structure not specifically designed to withstand such forces. Because much of the United States' densely populated Atlantic and Gulf coastlines lie less than 10 feet above mean sea level, the danger from storm tides is tremendous. Wave and current action associated with the tide can also cause extensive damage.

The current situation in Venice, Italy, offers insight into what challenges lie ahead for low-lying coastal cities. Can you imagine canals and transportation by gondola in parts of Manhattan?

SOLUTION

Adaptation Potential	★★
Operational Impacts	★
Financial Impacts	★
Feasibility and Timing	★★

Developing flood control systems doesn't have to be expensive. Sometimes, the appropriate solution for a local community includes planting vegetation to retain extra water, terracing hillsides to slow downhill flow, and the construction of floodways—human-made channels designed to divert floodwater. However, other techniques include the construction of levees, dikes, dams, reservoirs, or retention ponds to hold extra water during times of flooding. The previously recommended assessments determine what suits each case best.

Coastal flooding has been addressed in Europe and the Americas with coastal defenses such as seawalls, beach nourishment, and barrier islands.

Tide gates are used in conjunction with dikes and culverts. They can be placed at the mouth of streams or small rivers. Tide gates close during incoming tides to prevent tidal waters from moving upland and open during outgoing tides to allow waters to drain out via the culvert. The opening and closing of the gates is driven by a difference in water level on either side of the gate.[33]

EXAMPLE

Europe is at the forefront of flood control technology. Many countries across Europe are at or below sea level, so the problems of floods and rising sea levels are ever increasing. The largest and most elaborate flood defenses, referred to as Delta Works, are in the Netherlands. The Oosterscheldekering (Eastern Scheldt storm surge barrier) is its crowning achievement. Here, visitors see a plaque that says, "Hier gaan over het tij, de wind, de maan en wij" ("Here the tide is ruled by the wind, the moon and us [the Dutch]").

These works were built in response to the North Sea flood of 1953, in the southwestern part of the Netherlands, and are said to protect the country against a 10,000-year flood. Humans manually operate the dam, and an electronic security failsafe system serves as a backup. Extensive policies guide its operation. One Dutch law sets the conditions for when the dam is allowed to close, requiring water levels at least 3 meters above regular sea level before the doors are completely shut. Workers close each sluice gate monthly for testing purposes, and emergency procedures are tested regularly. Once a test is passed, the shutters are again immediately opened. This ensures a minimal amount of impact on tidal movements and the local marine ecosystem.[34]

The full dam has been closed just 24 times since 1986 as a result of water levels exceeding or being predicted to exceed the 3-meter mark.

Italy is the site of storm surge barriers constructed at key points in major waterways, albeit at an extremely high cost. In Venice, for example, an artificial island is being built to bridge the 800-meter Lido inlet. Also, sunken barriers in the Venetian lagoon are raised during times of high tide to control flooding.

Countries such as the Netherlands and Italy could prove important models for other countries around the world. These sorts of gigantic projects represent massive investments but could be the key to combating rising sea levels, an increase in the frequency and severity of some natural disasters, and even increased durations of dry or rainy seasons.[35]

Coastal storm surge barrier at the Delta works in the Netherlands.

Improve Infrastructure Resilience through Protection and Hardening

RECOMMENDATION

Protect and harden

Recommendation: *Use solidified structures and modified design and construction practices to improve the resilience of existing infrastructure.*

CHALLENGE

Much of our nation's infrastructure was built several decades ago and is now in need of upgrades, repair, or replacement. The combination of age-related deficiencies and functional obsolescence is challenging. Add to this increasing vulnerability due to climate change, and you are looking at a challenge on the cusp of magnifying exponentially. But that's not all. Any attempts to address these challenges must also take into account the continued effect of climate change and projected future conditions.

The New Orleans flood control system, as it is being rebuilt and hardened, is an example of a project that will have to incorporate more stringent design considerations, improved construction practices and quality standards, and early warning and evacuation systems.

SOLUTION

Adaptation Potential	★★
Operational Impacts	★★
Financial Impacts	★★
Feasibility and Timing	★★

Improving resilience is the equally important counterpart of reducing vulnerability. Solidifying structures and modifying design and construction practices will reduce the risk inherent to the infrastructure for which the vulnerability cannot be reduced.

Improving infrastructure resilience should be thought of in terms of increasing its protection and hardening. It's almost the opposite of a

White roofs reduce thermal load and save energy.

retrofit—using new and emerging technologies and practices to enhance current infrastructure. This is a means for doing more with less, allowing what is already standing to better cope with hazards. Improved building codes and design standards, the development of new design concepts, and evolving construction practices are all devices for achieving this on a local level.

If, for example, more frequent and prolonged heat waves occur over the next decades, then significant, proactive infrastructure adaptation measures must be taken to minimize impact. Among the adaptation measures to consider are energy-efficient cooling equipment; improved design standards of residential buildings to increase insulation effectiveness; transformation of historical structures into modern, functioning facilities; increased green spaces within urban areas; painting buildings white to reduce the heat island effect; incorporation of renewable energy sources (solar, for example) to generate the most when the sun is shining and the cooling load is the greatest; and the expansion the power grid capacity to handle the increased electrical load.

EXAMPLE

Since 2006, utility companies in Florida, a perpetual target of hurricanes and tropical storms, have partnered with the University of Florida's Public Utility Research Center to research hardening the electric infrastructure to better withstand and recover from tropical storms.

The project stems from an order issued by the Florida Public Service Commission instructing the state's investor-owned electric utilities to participate in collaborative research for developing storm resiliencies in the state's electric utility infrastructure. They were also asked to explore technologies that reduce storm restoration costs and outages to customers.

The initiative's current projects focus on the feasibility of undergrounding existing electric distribution facilities through case analyses of existing Florida underground projects. They're also gathering data and analysis of hurricane winds in Florida and exploring the possible expansion of a hurricane simulator that can be used to test hardening approaches. Finally, they're investigating effective approaches for vegetation management.

The state's goal in insisting that utilities undertake this effort is to increase their coordination with local governments and to implement effective storm-hardening plans while developing a culture of storm preparedness.

Final Thoughts

The challenge of working with infrastructure is that there are substantial constraints facing the stakeholders who actually have the ability to enact change and make decisions. These managers must contend with laws and regulations and oversight and enforcement bodies. Consider how no activity in many cases can begin until the proper permits have been obtained.

In other words, infrastructure planning must consider this process as much as—or more than—the threats, vulnerabilities, and impacts that climate change or other threats pose. Doing so may involve the employment of experts, use of the media, education of policymakers, and formation of political coalitions or other mechanisms to get things done.

The inclusion of all these groups demonstrates the diversity of the stakeholders that play a role in implementing mitigation and adaptation strategies for infrastructure management. Success will require the dedicated teamwork and ingenuity of all, and each must do its part to coordinate to achieve meaningful, far-reaching results. Efforts, from policy development to technological innovation, must come at the issue from both sides.

GHG mitigation, as part of an overall energy-efficiency and sustainability plan, can and must mean improved economic value to organizations. To feed this budding growth industry and virtuous cycle, solutions must not only achieve their green targets but spur economic growth.

Adaptation success will require abundant creativity, innovation, and persistence to envision solutions that may not be needed for several decades but which must be started now. The 100-year-event infrastructure risk thresholds of the past are shrinking dramatically, and adaptation solutions must evolve to reduce those risks.

Greater day-to-day awareness of climate change impacts will drive individual behaviors as well as those of organizations and communities. These behaviors will help ensure that our infrastructure is protected and used most effectively.

Finally, as more data become available regarding the results of implementing infrastructure mitigation and adaptation strategies, the growing costs and risks of inaction will become increasingly clear. The time to take forward-looking steps to mitigate and adapt infrastructure to climate change impacts is now.

Changing infrastructure is a daunting task in the current environment, where doing more with less is a necessity. But if stakeholders can't communicate the enormous cost of failing to invest now, efforts to mitigate and adapt to climate change will never gather the necessary momentum. This is where creativity and innovation are vital.

---

[1] National Research Council, Division on Earth and Life Sciences, Board on Atmospheric Sciences and Climate, *Adapting to the Impacts of Climate Change* (Washington, DC: National Academies Press, 2010).

[2] Teri Cruce, *Adaptation Planning—What US States and Localities are Doing* (Arlington, VA: Pew Center on Global Climate Change, August 2009).

[3] Merriam-Webster Dictionary, definition of "infrastructure," *m-w.com*.

[4] US Bureau of Economic Analysis—note: facilities, schools, and other assets have been combined into "Facilities and Other Assets."

[5] American Society of Civil Engineers (ASCE), *Report Card for America's Infrastructure* (Reston, VA: ASCE, March 25, 2009).

[6] ASCE, *Guiding Principles for the Nation's Critical Infrastructure* (Reston, VA: ASCE, 2009).

[7] Chris Corps, *Green Building in North America Background Paper 2c: Toward Sustainable Financing and Strong Markets for Green Building: Valuing Sustainability* (Montreal: Commission for Environmental Cooperation [CEC], 2007).

[8] Leanne Tobias, *Green Building in North America Background Paper 2c: Toward Sustainable Financing and Strong Markets for Green Building: Green Building Finance* (Montreal: CEC, 2007).

[9] US Green Building Council (USGBC), *Making the Business Case for High Performance Green* (Washington, DC: 2003), p. 6.

[10] Peter Bregman, "The Easiest Way to Change Behavior," *Harvard Business Review Blog*,March 11, 2009, http://blogs.hbr.org/bregman/2009/03/the-easiest-way-to.html.

[11] New South Wales Government, *Total Asset Management (TAM) Introduction*, http://www.treasury.nsw.gov.au/tam/tam-intro.

[12] Richard Paradis and Bambi Tran, "Balancing Security/Safety and Sustainability Objectives," *Whole Building Design Guide* (Washington, DC: National Institute of Building Sciences, June 9, 2010).

[13] Asphalt Pavement Alliance, *Keys to a Successful Alternate Bidding Process*, IM-50 DEC 2010 (Lanham, MD: 2010).

[14] Jeff Yoders, "LACCD's $6 billion BIM connection," *Building Design+Construction*, http://www.bdcnetwork.com/laccd%E2%80%99s-6-billion-bim-connection.

[15] Karen Erhardt-Martinez, John Laitner, and Kenneth Keating, *Pursuing Energy Efficient Behavior in a Regulatory Environment* (Berkley, CA: California Institute of Energy and Environment, August 2009).

[16] Andrea McMakin, Regina Lundgren, and Elizabeth Malone, *Revised Handbook for Promoting Behavior-Based Energy Efficiency in Military*

Housing (Richland, WA: Pacific Northwest National Laboratory, March 2000).

[17] Energistic Software, "GreenFuel," http://www.greenfuelapp.com/.

[18] Casey Lewis, "Eco Maps for the iPhone," http://ecoroutemaps.com/.

[19] US Department of Commerce, National Oceanic and Atmospheric Administration, National Climatic Data Center, "State of the Climate: Global Analysis, September 2011," http://www.ncdc.noaa.gov/sotc/global/.

[20] US Environmental Protection Agency (EPA), *Climate Change Indicators in the United States,* EPA 430-R-10-007 (Washington, DC: US EPA, April 2010).

[21] Victoria Tobin, Noel Brady, and Nancy Boyd, "Washington receives funding for study of impacts of climate change on transportation," October 4, 2010, http://www.wsdot.wa.gov/News/2010/10/04_FHWAmoneyforrisk assessment.htm.

[22] Butch Wlaschin, "Climate Change Adaptation Strategies for Infrastructure Managers," (presentation, Green Streets and Highways Conference, Denver, CO, November 17, 2010).

[23] California Climate Change Center, California Energy Commission, *The Future is Now–An Update on Climate Change Science, Impacts, and Response Options for California*, CEC-500-2008-077 (Sacramento, CA: September 2008),

[24] The White House Council on Environmental Quality, *Progress Report of the Interagency Climate Change Adaptation Task Force: Recommended Actions in Support of a National Climate Change Adaptation Strategy*, October 5, 2010.

[25] National Association of Insurance Commissioners (NAIC), "Insurance Regulators Adopt Climate Change Risk Disclosure," *NAIC News Release*, March 17, 2009.

[26] Report prepared by Ernst & Young and Oxford Analytica, *Strategic business risk 2008 —Insurance*, EYG no. EG0015, http://www.aaiard.com/11_2008/2008_Strategic_Business_Risk_-_Insurance.2.pdf.

[27] Securities and Exchange Commission (SEC), "Section 501-15: Climate change related disclosures," *Codification of Financial Reporting Policies*, (amended February 2010). For an informative discussion of the SEC regulation, see SEC, *Commission Guidance Regarding Disclosure Related to Climate Change*, Release Nos. 33-9106; 34-61469; FR-82.

[28] California Natural Resources Agency, *2009 California Climate Adaptation Strategy —A Report to the Governor of the State of California in Response to Executive Order S-13-2008* (Sacramento, CA: 2009).

[29] Pennsylvania Department of Environmental Protection (PDEP), *Pennsylvania Climate Adaptation Planning Report: Risks and Practical Recommendations*, 7000-RE-DEP4303 (Harrisburg, PA: January 2011).

[30] James Shortle and others, *Pennsylvania Climate Impact Assessment: Report to the Department of Environmental Protection*, 700-BK-DEP4252 (Harrisburg, PA: PDEP, June 29, 2009).

[31] The Huffington Post, "Mandatory Evacuation Ordered for Zone A residents in New York As Hurricane Irene Approaches," *www.huffingtonpost.com*, August 26, 2011, http://www.huffingtonpost.com/2011/08/26/mandatory-evacuation-irene-hurricane-new-york-city_n_938251.html.

[32] Annalee Newitz and Robert Gonzalez, "Why the hurricane Irene evacuations in New York City were a success," *io9.com*, August 30, 2011, http://io9.com/5835596/why-the-hurricane-irene-evacuations-in-new-york-city-were-a-success.

[33] Guillermo Giannico and Jon Souder, "The Effects of Tide Gates on Estuarine Habitats and Migratory Fish," *Rep. No. ORESU G-04-02* (Corvallis: Oregon State University, 2004).

[34] Delta Works Online, http://www.deltawerken.com/.

[35] Colin Woodard, "Netherlands Battens Its Ramparts Against Warming Climate," *Christian Science Monitor*, reprinted in *National Geographic News*, September 4, 2001, http://news.nationalgeographic.com/news/2001/08/0829_wiredutch.html.

Vehicles

The refineries had been operating at peak capacity, but the airline fuel processed was just sitting in storage tanks. Houston, recovering from its third major hurricane landfall in 18 months, was now a difficult city from which to move goods by truck. A few years ago, city planners had discussed expanding Houston's access to major railways, but the cost was deemed too high. Now, with all the damage the city had sustained from repeated 25-foot storm surges, a new railway system was an unaffordable luxury. The airlines were now ignoring the city that was the nation's energy hub, opting instead to get their fuel from the Port of New Orleans. Houston was slowly losing its clout in the energy industry.

Climate Change in the Driver's Seat

America has a love affair with cars. We didn't invent them, but we certainly perfected their production for the marketplace. We've built roads that honor these vehicles, and for many people, a vehicle represents the same sort of freedom and self-expression that a horse did for the cowboys of yore.

Our businesses and government, our way of life, are tied to our vehicles. The US transportation system—4 million miles of roads and streets used by more than 255 million registered passenger vehicles[1]—is vital to how our government operates, our nation's economy, and its general well-being. The potential impacts of climate change on transportation cannot be ignored.

In the "Structure" chapter, we talk about the potential effects of climate change on transportation infrastructure. Infrastructure has a substantial impact on vehicles and fleets.

Increases in temperature may degrade pavement integrity and deform rail tracks. Higher temperatures can lead to the subsidence of permafrost, limiting the use of roads, rail lines, and airports in areas not equipped to handle this sort of disruption. Sea level rise could flood or destroy roads, bridges, airports, and rail lines in coastal areas. More frequent and severe precipitation events may temporarily disrupt some transportation infrastructure, increase road washout, and affect the structural integrity of roads, bridges, and tunnels.

If you have a vested interest in fleet management, you probably aren't directly responsible for transportation infrastructure. You will, however, be forced to adapt to climate change impacts on the transportation system. Early action requires evaluating the risk to fleet facilities from climate change, ranking them by vulnerability, and making plans to protect high-risk assets. If infrastructure becomes less reliable, different transportation modes may need to be adopted, and employees may become concerned with the realities of their commute; your responsibilities will likely expand to provide flexibility when these concerns arise.

However, the largest potential impact of climate change for the fleet manager will be dealing with a new national focus: the reduction of transport-related greenhouse gas (GHG) emissions. The US transportation system is a major source of GHG emissions, the second largest behind electric power. Between 1990 and 2009, tailpipe GHG emissions from transport grew by 22 percent, from 25.2 to 27.1 percent of total US GHG emissions (Table 6-1). These increases can be attributed to higher travel demand at a time when fuel economy improved only slightly.

Table 6-1. US Tailpipe GHG Emissions from Transportation Activities by Teragrams of Carbon Dioxide Equivalents (Tg CO_2 eq.)

Sector/Gas	1990	2009
Transportation	1,548.3	1,816.9
CO_2	*1,500.9*	*1,732.7*
CH_4	*4.5*	*1.6*
N_2O	*42.9*	*22.4*
Others	—	*60.2*
All other sectors	4,633.5	4,816.3
Total	**6,181.8**	**6,633.2**

Inevitably, controlling vehicle GHG emissions will be a growing focus of the federal government. The role of GHG emissions on potential climate change is already having a policy impact on fleets and vehicles. This impact has resulted in US regulation of vehicle GHG emissions, which could potentially reduce transportation use and shift travel to more energy-efficient modes; current policy also promotes the use of lower-carbon alternatives to petroleum fuels. Furthermore, climate change may lead to additional challenges for vehicle emission mitigation, such as heat-related effects on ground-level ozone.

The regulatory process is already under way. In 2009, the Environmental Protection Agency (EPA) issued an endangerment finding, determining that the current and projected concentration of GHG threatens public health and welfare. In this finding, it included the contribution of new vehicles. This procedural move was a critical one, as it enabled EPA to launch the first US-wide vehicle (and other) GHG emission standards under the Clean Air Act. These standards were established jointly with the Corporate Average Fuel Economy (CAFE) standards.

The CAFE standards, around since the mid-1970s, were meant to improve average fuel economy in vehicles. They were a direct response to the Arab Oil Embargo of 1973. They've been considered effective tools: a 2002 study found that gas consumption would be 14 percent higher if not for their role in improving fuel economy.[2]

The new standards require vehicle manufacturers to reduce their fleetwide average GHG emissions. Between model years (MYs) 2012 and 2016, auto manufacturers must reduce their fleets' average GHG emissions from 295 to 250 grams of CO_2 per mile.

And for federal fleets, Congress and the president have already started imposing vehicle GHG emission-reduction requirements. The president, as previously discussed, issued Executive Order (EO) 13514, "Federal Leadership in Environmental, Energy, and Economic Performance," while Congress passed the Energy Independence and Security Act (EISA) of 2007, Section 141 of which prohibits federal agencies from acquiring light-duty motor vehicles and medium-duty passenger vehicles that are not low-GHG-emitting vehicles.

Let's skip ahead a decade to the years surrounding 2025 and look at the potential landscape for vehicles and fleets. By then (in the scenario with which we're playing), the federal government will have issued comprehensive regulations limiting vehicle GHG emissions through standards. It will have moved to further enhance fuel efficiency. Manufacturers will alter their product offerings to account for these restrictions, and the result will be a new generation of passenger cars, medium-duty passenger vehicles, and trucks.

Vehicles will be lighter (using materials such as aluminum and carbon fiber) with more efficient engines, hybrid electric drive trains, and improved drag resistance. They will be considered "clean." Internal combustion engines will all be equipped to use alternative fuels, and that's what they will use. Electric vehicles will hum along our nation's roadways.

This vision also involves increasingly restrictive constraints on fleet GHG emissions. President Obama's EO targets a 28 percent reduction in federal agency GHG emissions from 2008 through 2020; this is only an initial salvo. The longer-term targets are likely reductions of 50 percent or more.

All fleet managers know their primary responsibility is to maintain cost-effective, reliable, and safe transportation services that meet mission needs. This is true in the public and private sectors (especially for those who do business with the government). Over the next few decades, the concerted effort to limit vehicle GHG emissions, without breaking budgets, will encourage fleet managers in both sectors to learn from each other and develop partnerships to meet these common goals.

Mitigation of Fleet GHG Emissions

Fleet managers will have to rethink their transportation needs to achieve their mitigation goals.

First, they must ensure their fleets are the right size. Fleets must be sized for their missions, which helps align the demand for fleet resources with an organization's transport needs.

Second, they will have to identify the transportation options that offer realistic low-carbon emissions and determine how low they can afford to go while still meeting the organization's mission. Once that's done, fleets will be able to shift modes appropriately. Fleet managers' jobs will expand from being vehicle administrators to travel coordinators, responsible for managing organizational use of air, rail, and mass transport.

Third, they must acquire more fuel-efficient vehicles and improve driver operational behaviors with respect to fuel use. Both actions will lower carbon emissions.

Enhanced CAFE standards can be used to ensure new vehicles reach the market with greater fuel efficiency than today. However, aggressive GHG emission targets will force fleet managers to get creative in order to discover the lowest-GHG-emitting vehicle (or mode) for ensuring their mission is supported.

Doing so means descending into a world of vehicle class alphabet soup, where abbreviations rule and distinguish exactly how a vehicle operates. First stop, hybrid electric vehicles (HEVs), which are growing in market presence today and which rely on a combination of fuel and battery. Because of their current availability, they will fill a large portion of any fleet's fuel efficiency improvement needs.

However, shifting the composition of a fleet from gasoline and diesel vehicles (in which HEVs are technically included) to alternative, lower-carbon vehicle technologies means going one step further.

Three classes are expected to be cost-competitive mainstream offerings by 2020: electric vehicles (EVs), pure battery electric vehicles (BEVs), and plug-in hybrid electric vehicles (PHEVs). These alternatives are expected to make up large parts of most fleets if they can be deployed effectively. Hydrogen, an effective carrier of energy from low-carbon sources, may also emerge as a commercially viable technology. EISA renewable fuel mandates are expected to provide fleets with widespread access to low-carbon cellulosic E85;[a] this means the option of E85 flex-fuel vehicles (FFVs) is on the table as well. Access will continue to expand as other niche fuels, such as compressed natural gas (CNG), provide fleet managers with access to other, new low-carbon vehicle technologies.

Low-GHG-Emitting Vehicles

Here's a quick reference of abbreviations used for vehicle classes in this chapter:

- HEV—hybrid electric vehicle
- EV—electric vehicle
- BEV—pure battery electric vehicle
- PHEV—plug-in hybrid electric vehicle
- FFV—flex-fuel vehicle.

Behavior is also an area of mitigation opportunity. Fleet managers must implement guidelines and oversight to ensure that drivers operate the organization's vehicles in a GHG-friendly way—observing speed limits, avoiding excessive idling, and regularly maintaining vehicles.

At the end of the day, many of these decisions will rest on an organization's bottom line. How the cost will impact operations depends on fleet and vehicle use characteristics, but many of these technologies may save operating costs (in fuel and maintenance, for example), paying for the up-front investment (which may look to the uninformed as an increased purchase cost). That said, other technologies

[a] E85 is an ethanol fuel blend of up to 85 percent ethanol fuel, with the rest made up of gasoline.

may only be justified because they provide a favorable cost per kilogram of CO_2 reduced.

A Primer on GHG Emissions from Transportation Activities

Federal fleets now only measure tailpipe GHG emissions (direct fuel combustion) not life-cycle emissions.

Two types of emissions are associated with vehicle transportation: *direct* (or "tailpipe) and *indirect*. They differ enough to impact how you develop your mitigation strategy and proposed activities. The federal fleet and CAFE programs currently *do not include* life-cycle (fuel or vehicle cycle) emissions in calculating GHG emission performance or targets.

DIRECT, OR TAILPIPE, GHG EMISSIONS

Tailpipe GHG emissions are the primary source of vehicle GHG emissions and result from the combustion of fuel. This source is also referred to as "tank-to-wheel" GHG emissions. This is where CO_2 enters the picture, as it represents the overwhelming majority of GHG emissions from fuel combustion; the rest consist of small quantities of nitrous oxide and methane. Tailpipe GHG emissions are the only ones from fuel combustion attributed exclusively to the transportation sector by EPA; other related GHG emissions linked to vehicle and fuel use are allocated to other end-use sectors (such as petroleum systems in the energy sector and iron and steel production in the industrial processes sector).

GHG emissions from the combustion of fuel depend on four things:

1. Volume of fuel combusted
2. Fuel energy density
3. Fuel carbon content
4. Fraction of carbon oxidized to GHG.

GHG tailpipe emissions (like most other GHG emissions) are usually calculated indirectly. These calculations are based on the fuel quantity, type of fuel combusted, and efficiency of the vehicle combustion process; each is determined under laboratory conditions.

CO_2 emissions from biofuel combustion, referred to as biogenic emissions, do not add carbon to the active global carbon cycle (they are "carbon neutral").

The process for reporting GHG emissions for fossil (petroleum) fuels differs from that for biofuels (referred to as biogenic GHG emissions). That's because, when the source is renewable biomass feedstocks, combusting biofuels generate GHG emissions that are offset by photosynthesis: the process for growing the biomass absorbs CO_2 from the atmosphere. As a result, biogenic emissions are considered carbon neutral. They do not add new carbon to the active global carbon cycle. Fossil fuels, on the other hand, consist of carbon captured millions of years ago, which is added as new carbon to the atmosphere when combusted.

INDIRECT GHG EMISSIONS

Indirect vehicle GHG emissions include those associated with the fuel cycle, where the extraction, production, and distribution of fuel take place (referred to as "well-to-tank" emissions). Add these to tail-pipe emissions, and you get what's known as "well-to-wheels" emissions (Figure 6-1).

Figure 6-1. Components of Direct and Indirect Life-Cycle GHG Emissions

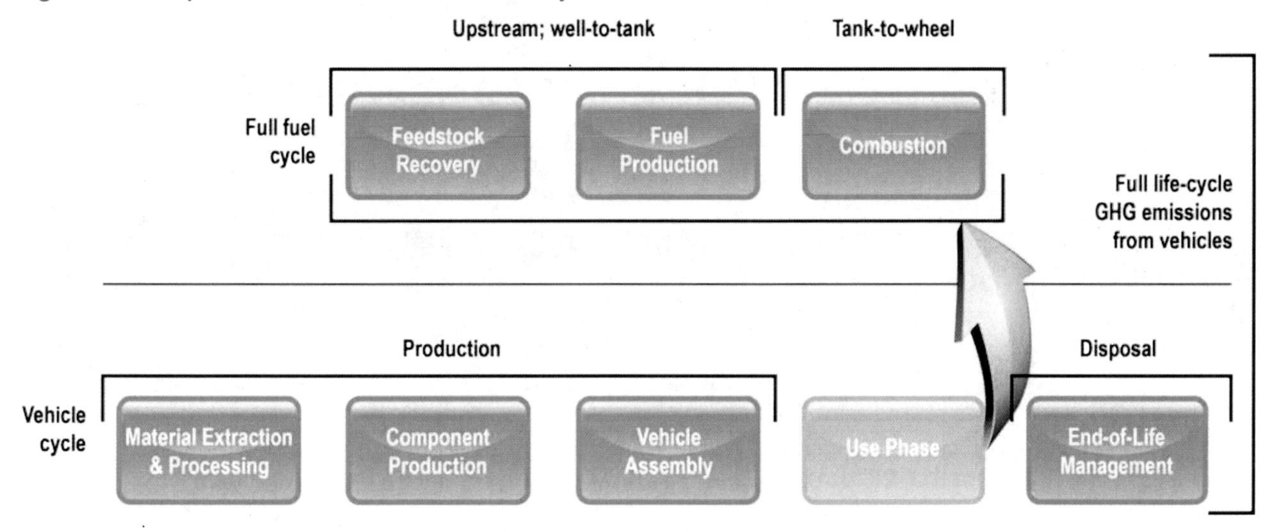

For example, in the Clean Air Act, EISA defines renewable fuels on the basis of full fuel life-cycle GHG emissions. It states that the combined amount of GHG emissions (direct and indirect) must be combined with "all stages of fuel and feedstock production and distribution" and adjusted for the relative potential for global warming.[3]

Other estimates of life-cycle GHG emissions from vehicles also add indirect emissions from the manufacture, distribution, and disposal of vehicles (referred to as the vehicle cycle).

Indirect vehicle and fuel cycle emissions are a large component of vehicle-related GHG emissions, often accounting for up to a third of total vehicle life-cycle emissions. Through two separate models, the Greenhouse Gases, Regulated Emissions, and Energy Use in Transportation (GREET) Model and the Life-Cycle Emissions Model

(LEM),[b] EPA has estimated that GHG emissions from fuel and vehicle cycles add 38 to 50 percent to tailpipe emissions for light-duty gasoline vehicles; for heavy-duty diesel trucks, the results are lower, an additional 21 to 45 percent (Figure 6-2).[4]

Figure 6-2. GHG Emissions from Fuel and Vehicle Cycles Relative to Tailpipe Emissions

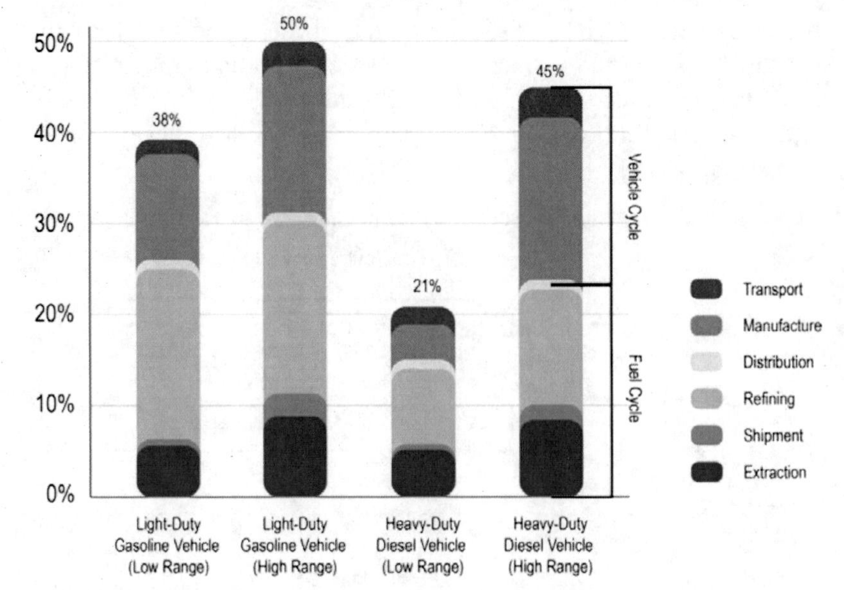

GHG emissions from the fuel cycle are typically higher than those from the vehicle cycle; the implication is that selecting fuels with lower production carbon intensity will have a greater benefit than focusing on the process used to manufacture the vehicle.

Trends in GHG Emissions from US Vehicles

Table 6-2 compares US tailpipe GHG emissions in 1990 and 2009 by vehicle type. It reveals some interesting trends in American vehicle usage. In 2009, passenger cars were the largest source of vehicle tailpipe GHG emissions (40.3 percent), followed by light-duty trucks, which include sport utility vehicles (SUVs), pickup trucks, and minivans. A combination of medium- and heavy-duty trucks and buses doesn't even represent a quarter of emissions totals.

[b] The GREET Model was developed by Argonne National Laboratory. The LEM Model was developed at the University of California at Irvine.

Table 6-2. GHG Emissions from Vehicle Fuel Combustion (Tg CO_2 eq.)

Vehicle Type	1990	2009
Passenger cars	657.4	627.4
Light-duty trucks	336.6	551.0
Medium & heavy-duty trucks	231.1	365.6
Buses	8.4	11.2
Total	**1,233.5**	**1,555.2**

Over those 19 years, GHG tailpipe emissions from vehicles have increased by more than 26 percent. During this time, our vehicle engines grew more efficient, so what happened? Why would GHG emissions increase?

The first reason is the rise of the SUV. Over the last 30 years, families, and even individuals, migrated away from passenger cars in favor of light-duty trucks, primarily SUVs. These vehicles are notoriously less fuel efficient. So, although passenger car emissions decreased by 4.6 percent over the last two decades, the use of these larger, less-efficient vehicles completely negated any gains. During a period in which passenger car emissions decreased, GHG emissions from SUVs, minivans, and trucks *increased by nearly 64 percent.* This was the result of one thing: consumer preferences. Drivers saw lower gas prices and opted for size. Automakers responded by adopting new technology to produce vehicles with greater acceleration, improved safety, bigger cabin size, and other appealing features. Improved fuel economy remained on the backburner.

Second, people simply drove more. Vehicle travel demand in the United States, measured in vehicle miles traveled (VMT), increased by 39 percent between 1990 and 2009.[5] Population growth (24.5 percent) tells part of the story, but the real smoking gun is the strong US economy, which spurred an increase in urban sprawl (new housing created new suburbs, with longer commutes to urban hubs) and more disposable income (resulting in the ability and preference to own personal vehicles). Add to the equation lower fuel prices for much of this time frame, and you have a potent mix. The result was increased vehicle trips, freight demand, and per capita VMT.

Then things changed: 2008 ushered in a faltering economy and spikes in fuel prices. Improvements in fuel economy, after 19 years of being exclusive to passenger vehicles, began to reach the rest of the consumer market. As a result, transportation-sector GHG emissions have fallen, VMT growth slowed, and fuel economy improved.

Recommendations to Mitigate GHG Emissions from Transportation Activities

GHG emissions are the product of three elements of vehicle transportation:

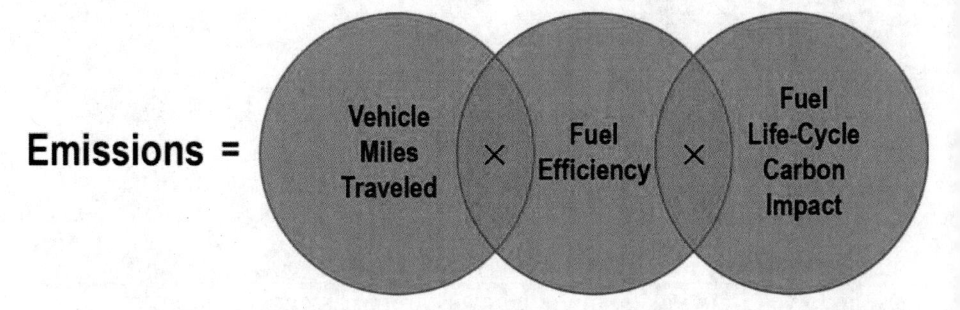

Simple math holds true here. Fleet managers can reduce GHG emissions by slashing one or more of the components of this equation. In fact, this equation provides fleet managers with four guiding principles for reducing GHG emissions:

1. *Conservation.* Reduce travel demand or VMT.
2. *Efficiency.* Reduce the amount of fuel burned per mile (vehicle fuel efficiency).
3. *Substitution.* Replace petroleum use with lower-carbon alternative fuels.
4. *Mode shift.* Shift to lower-carbon transportation modes.

These guiding principles are the framework fleet managers must use to develop a strategic GHG emission-reduction plan. They provide fleet managers with the flexibility to tailor their plans to match the organization's fleet profile and meet its mission.

In this vein, fleet managers should become tailors. Each fleet location deserves its own evaluation before determining a GHG emission-reduction strategy. This involves considering site-specific characteristics, including transportation needs and characteristics, alternative fuel availability, fleet size, and fleet vehicle composition. Fleet managers should also employ best practices, including right-sizing their fleet to the mission need.

Fleet managers supporting the federal government have several resources at their disposal, including EO 13514 guidance and the *Comprehensive Federal Fleet Management Handbook.* Prepared by the Department of Energy (DOE), these publications offer comprehensive guidance for evaluating and implementing the principles contained in the EO.[6,7]

The good news for fleet managers is that CAFE standards, expected to continue their current trend of increasing, will ensure that normal turnover of fleet vehicles will reduce fleet GHG emissions. Currently, the National Highway Traffic Safety Administration (NHTSA) and EPA issue standards for passenger cars and light-duty trucks only. The current expectations of manufacturers are that they'll increase combined fuel efficiency by 25 percent,[c] from MY 2011 to 2016, and from MY 2012 to 2016, reduce combined GHG emissions by 15 percent.[d]

This means the price of vehicles will increase slightly. NHTSA estimates that technologies needed to meet these standards will increase the average purchase price of a light-duty vehicle by $434 in MY 2012 and $926 in MY 2016. Many purchasers will see benefits that outweigh the costs, so long as they keep the vehicle for a reasonable amount of time. NHTSA projects that the standards enable the "average car buyer of a 2016 model year vehicle to enjoy a net savings of $3,000 over the lifetime of the vehicle, as up-front technology costs are offset by lower fuel costs."[8]

---

[c] CAFE fuel economy standards increase from 27.3 mpg in MY 2011 to 34.1 mpg in MY 2016.

[d] CAFE GHG emission standards reduce from 295g CO_2 per mile in MY 2012 to 250g per mile in MY 2016.

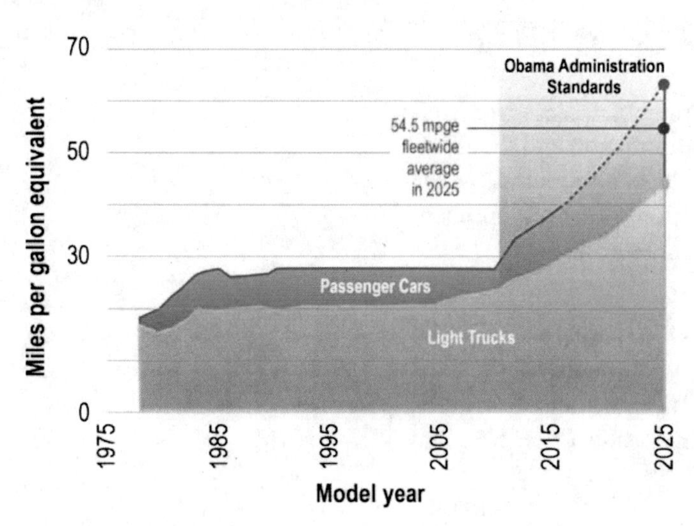

Light-duty vehicle fuel economy standards, 1978–2025.

Fleet managers can expect similar increases in average light-duty fuel efficiency requirements past MY 2016.[e] In July 2011, President Obama proposed increasing new GHG and fuel efficiency standards for passenger cars and light-duty trucks through 2025. The proposal would be the largest ever mandatory increase in fuel economy, requiring manufacturers to reduce their GHG emissions to 163 grams per mile (a 45 percent reduction over 13 years) and increase their average fleet-wide fuel economy to 54.5 miles per gallon (mpg) by MY 2025 (an 81 percent increase over the same period).

EPA and NHTSA recently established a new program designed to improve the fuel efficiency of medium- and heavy-duty vehicles for MY 2014 through 2018 compared with MY 2010. For diesel vehicles, these standards would provide an average increase in fuel efficiency of 15 percent and an average reduction in GHG emissions of 17 percent; in gasoline vehicles, a 10 percent increase in fuel efficiency and a 12 percent reduction in GHG emissions. Similar reductions are expected for semi trucks and vocational vehicles.

Evaluating Mitigation Solutions

Fleet managers provide transportation services that support an organizational mission; this is their core responsibility. They measure their success on the basis of cost-effectiveness, service consistency, vehicle utilization, and safety. These responsibilities and related metrics, however, will grow to include GHG emission reductions.

Driven by the scale of GHG emissions from transportation activities, the president's EO has already established "an integrated strategy toward sustainability in the federal government, [making] reduction of GHG emissions a priority for federal agencies." Federal fleets will

[e] The average fuel economy of light-duty vehicles will likely be less than the CAFE standards due to credits for dedicated alternative fuel vehicles and dual-fuel vehicles, and banked credits from exceeding CAFE standards in previous years.

help reach this vision by establishing reduction of GHG emissions as a core goal of federal fleet management.[9]

Therefore, it's not a matter of whether to reduce fleet GHG emissions but how to do so. How attractive are the available options? To figure this out, fleet managers must weigh the impacts of providing transportation services against the costs of implementation. A few solutions, such as focusing on demand to optimize transportation services, may actually only have a limited impact on services—which is what you want—even as they provide cost savings and emission reductions (a win-win situation!).

Not every solution is a clear winner, but still, each has its benefits. For example, adopting medium-duty HEVs as part of a fleet may provide net operating savings that pay back initial investment within a few years. Other GHG solutions, such as the use of ethanol in E85 FFVs, may not provide net cost savings but may offer an attractive cost per CO_2 equivalent (CO_2e) reduction. If a carbon market evolves, these credits may provide revenue sources that offset some if not all of the costs.

The four criteria we use to compare fleet GHG emission-reduction mitigation solutions remain the same as outlined earlier in this book:

1. GHG reduction potential
2. Operational impacts
3. Financial impacts
4. Feasibility and timing.

Unique in this case is the context and comparison. Fleet managers should be aware that sometimes even the best solutions only reduce GHG emissions by half. The reality of transportation activities is that they use energy, which generally emits GHGs. The only way for a fleet to be GHG emission free is by eliminating transportation use entirely or by using solely renewable energy.

We offer five key mitigation recommendations, four of which center on these four guiding principles for mitigating fleet GHG emissions from transportation activities. The fifth recommendation involves using vehicle capabilities to assist in mitigating emissions from the electricity sector. These recommendations are as follows:

- Reduce transportation demand.
- Improve fuel efficiency.
- Use lower-emission fuels.
- Optimize transportation modes.
- Store renewable energy.

Reduce transportation demand

Reduce Transportation Demand

Recommendation: *Work to reduce transportation demand by focusing on trips and routing.*

CHALLENGE

Reducing transportation GHG emissions requires focusing on lowering the "transport demand" variable in the emissions equation. The fewer miles traveled, the fewer tailpipe and fuel cycle GHG emitted. This, unfortunately, doesn't always mesh with an organization's mission. People, products, and services need to get from point A to point B.

That said, transportation is an inefficient business. In 2011, the Pew Center on Global Climate Change (now the Center for Climate and Energy Solutions) quipped, "The greatest oil reservoir in the world is the empty seats in American cars."[10]

One option for reducing transportation demand is removing excess fleet vehicles. This is tempting as it can allow fleets to completely eliminate all life-cycle GHG emissions associated with those transportation services. However, this mitigation solution has a relatively low ceiling on the scale of potential emission savings. Fleet managers will still be in the business of providing transportation services. We estimate that right-sizing even the most oversized fleets may offer only a 25 percent VMT reduction opportunity.

SOLUTION

Mitigation Potential	★
Operational Impacts	★★
Financial Impacts	★★★
Feasibility and Timing	★★

Conservation is the most effective way to reduce fleet GHG emissions, and it boils down to reducing vehicle use to only the amount needed to accomplish the agency mission. Reductions in fleet VMT bring with them a full reduction in tailpipe and fuel cycle GHG emissions, along with limited impacts on fleet services if implemented correctly.

Transportation demand can be reduced in three ways: consolidate or make fewer trips, eliminate trips altogether, or improve routing.

Decreasing or consolidating trips is buoyed by the efficiency such an option offers. Fleet managers can use several methods to decrease the number of trips and the associated fuel consumption, including trip consolidation, more efficient scheduling, and car pooling.

When evaluating opportunities for eliminating trips, fleet managers and operators should also identify individual trips that involve the same or similar routes and the scheduling of these trips. The fleet manager and operators should then determine opportunities for

- consolidating trips on multiple days or at different times to nearby destinations into a single day or time, or
- staff members with similar destinations and schedules to carpool by sharing fleet vehicles and combining their individual trips into a single vehicle trip.

Let's consider how the Navy might use trip consolidation. Its current Pearl Harbor fleet operations on the island of Oahu include many trips that share similar routes. For example, individuals heading to the east, toward Fort Shafter and Marine Corps Base Hawaii, or to the north, toward Schofield Barracks or Wheeler Air Base, could forego separate trips and instead consolidate these routes into a single trip.

For the same Pearl Harbor fleet, the fleet manager can also help establish carpools to share trips to Fort Shafter, Marine Corps Base Hawaii, Schofield Barracks, or Wheeler Air Base. Why have four different drivers each take 10-mile trips across the island when they can reduce their VMT by 75 percent if they just share a single vehicle?

Another option is to eliminate trips altogether. Each trip that a fleet manager eliminates provides an equivalent reduction in VMT and associated fuel use. Fleet managers and operators can use two primary methods to eliminate trips:

1. *Remove unnecessary routes.* Evaluate all the individual trips that fleet vehicles make each day and determine how many are unnecessary and can be eliminated.
2. *Use video- and web-conferencing tools for meetings.* Eliminate the need for personnel to travel for meetings, thereby reducing VMT.

Finally, optimizing the routes traveled can be an effective strategy to reduce trip VMT.

Time and money are easily wasted by inefficient driving habits. Something as minor as eliminating left-hand turns has been used effectively by private delivery companies to significantly reduce travel requirements.[11] Fleets are encouraged to explore both internal and external options to track and manage vehicle usage through scheduling and standardized routing. Numerous private companies offer software and consulting services to assist fleets with route and scheduling assets.

EXAMPLE

In responding to a 20 percent decline in mail volume over the past 3 years, the US Postal Service has undertaken a variety of projects and initiatives designed to reduce overall transportation demand. It has reduced the number of processing and distribution centers. Similarly, it has reduced the number of trips required to carry mail transport containers by 143,780 trips per year—simply by realigning service centers for mail transport equipment. Highway contract transport routes have further been streamlined through trip reduction and the elimination of underutilized routes, which decreased VMT by 4.5 million miles in FY 2010 alone. Also, nearly 89,000 of the Postal Service's mail delivery routes are either "fleet of feet" or "park and loop" routes, where mail carriers deliver personal mail on foot.

Its goal is a 20 percent reduction in overall petroleum use by 2015 as well as a 20 percent reduction in contractual transportation petroleum use by 2020.[12]

The Postal Service is also investigating the hardest cut of all by eliminating a day of delivery service (trimming mail delivery to 5 days per week), something that would require congressional approval. This has the potential to reduce annual petroleum use by 20–25 million gasoline gallon equivalents. This initiative alone would reduce GHG emissions anywhere between 315,000 to 500,000 metric tons per year.

Improve Fuel Efficiency

Improve fuel efficiency

Recommendation: *Improve fleet fuel efficiency by lowering the fuel intensity per mile through acquisition and operational changes.*

CHALLENGE

Improving fleet fuel efficiency is the second most effective way to reduce transportation GHG emissions (after conservation). Such an effort requires a focus on lowering the fuel intensity per transportation unit (mile) variable in the emissions equation. This, however, delves into promoting culture shifts and behavioral changes—asking people to change the way they drive and how much attention they pay to their vehicle's performance.

It also requires wide changes in an organization's acquisition process and operational tactics to truly increase the overall fuel efficiency of vehicles (and subsequently reduce fuel use). Because of the investment required to improve fuel efficiency, achieving short-term gains may be difficult.

Idling, a terribly inefficient use of fuel, is engrained in driver habits, whether because of where the driver lives or because of the kind of work they do. Climate change is a hurdle to changing this behavior. Colder areas have remote starters for the purpose of getting a car warmed up before a trip. And vehicle air conditioners don't work when the engine isn't running, making hotter areas a difficult place to earn wide fuel efficient habit adoption as well. These are all things to consider when developing a communications plan to educate drivers on new idling restrictions. Law enforcement personnel are among those who spend a great deal of time in vehicles that need access to auxiliary systems and that must be ready to move at a moment's notice.

SOLUTION

Mitigation Potential	★★★
Operational Impacts	★★
Financial Impacts	★★
Feasibility and Timing	★★

Fuel efficiency can be achieved through four key actions characterized by impacts on acquisition or operations:

- *Acquiring vehicles that are more fuel efficient*, including HEVs (acquisition)
- *Maintaining vehicles* to improve fuel efficiency or replacing inefficient vehicles that have exceeded their useful life (operations)
- *Driving more efficiently*, for example, observing the speed limit and avoiding aggressive driving (operations)
- *Avoiding excessive idling* (operations).

The *acquisition* piece of this puzzle, while simple in presentation ("get new vehicles"), is actually multitiered. It involves considering new technology and changing standards. In a nutshell, though, it's about optimizing the fleet by swapping out the elements that aren't performing to the highest standard.

Fleet managers will greatly reduce GHG emissions through normal replacement of vehicles in the fleet. This shouldn't represent a big change to business; leasing arrangements as well as normal wear and tear result in regular vehicle replacements.

Typically, organizations should focus on replacing their least-efficient vehicles first. This includes older or oversized light-duty vehicles as well as medium- or heavy-duty vehicles. These vehicles use more fuel per mile, so even a small percentage in improvements to fuel efficiency can provide large absolute GHG reductions.

For example, the reduction in fuel use by increasing the fuel economy of a 5 mpg vehicle by an additional 5 mpg is 10 times more than replacing a 20 mpg vehicle with a 25 mpg vehicle (assuming the same mileage).

What's encouraging from a GHG mitigation standpoint is that increased CAFE requirements and technological improvements will ensure that these new vehicles will be far more fuel efficient. As a result, they'll save significant GHG emissions over the next few decades. The CAFE standards will, however, impact the price of a vehicle (as previously mentioned, the technology required to meet future increases in the CAFE standards is expected to tack several hundred dollars onto the price tag).

The challenge remains the short term, as large improvements in fleet fuel efficiency are difficult to accomplish quickly. Any potential benefit will depend on the rate of fleet vehicle turnover. For example, the overall turnover rate for the federal fleet is currently slightly more than 10 years, making fuel efficiency improvements a solution cast in a medium to long time frame.

Of course, fleet managers don't need to rely simply on the increases in the CAFE standards and can go a step further by shifting to hybrid and plug-in hybrid technologies. The benefit is fuel savings (along with GHG reductions). The US Energy Information Administration estimates that a focus on acquiring compact HEVs rather than similar gasoline vehicles could provide an additional 38 percent fuel savings (and equivalent GHG reductions) between 2010 and 2030. Similarly, PHEVs could raise those additional fuel and GHG savings to 50 percent or more depending on the source of electricity generation.

Hybrid electric drive trains offer a substantial boost in fuel efficiency (30 percent or more). They also come at a cost. For example, you'll pay an extra $3,400 for the hybrid version of the 2011 Ford Fusion, but you'll get fuel efficiency 50 percent greater than the standard version.[13] As a result, payback is shortened through higher vehicle use and fuel prices. At an average usage of 12,000 miles per year and 2011 gas prices of $4.00 per gallon, fleet managers can recover the Fusion's hybrid price premium in about 5 years.

Currently, limited medium- and heavy-duty hybrid electric drive vehicles are available. Because these vehicles use far more fuel than light-duty vehicles, fleet managers can achieve large potential absolute fuel savings. Again, this is good business; the favorable economic prospects of integrating hybrid electric drive trains in heavier classes will lead to more widespread availability over the next decade.

Once more, there's an investment up front. Available medium- and heavy-duty HEVs cost from $20,000 to more than $100,000, while offering a 10 to 40 percent fuel efficiency improvement. However, like the Ford Fusion example, in high-use applications the fuel savings will pay for these costs early in the vehicle's lifetime. For example, Azure Dynamics, which produces a hybrid version of the Ford E-450 commercial stripped or cutaway chassis, estimates a 39 percent fuel savings and an estimated payback of the incremental cost within 4 years, based on a high-usage drive cycle.[14]

Most of the inherent fuel efficiency improvements of new vehicles will come at relatively low sticker cost premiums, providing potentially large operational cost savings over the vehicle's life. Other technologies designed to improve vehicle fuel efficiency, such as hybrid electric and plug-in hybrid electric drive trains will increase the purchase price of vehicles. However, fuel savings will save costs over the vehicle's operating lifetime, their magnitude depending on the price of fuel and the use of the vehicle. For HEVs, cost savings are achievable today. PHEVs are not cost-effective today, but over the next 10 years, incremental purchase prices are expected to decline to the point where net cost savings for specific fleet operations are possible.

Current vehicles should not be replaced with more fuel-efficient models without a planned strategy. One of the core considerations of fleet management is to ensure that the fleet is right-sized to the organization's mission. Fleet managers can accomplish this goal by employing a vehicle allocation methodology (VAM), which helps affirm that vehicle fleets are not more costly than necessary, correctly sized in terms of numbers, and of the appropriate type for accomplishing agency missions.

The process already considers whether mission requirements can be met with smaller, more fuel-efficient vehicles. Fleet managers are encouraged to specifically consider the fuel efficiency ramifications of their decisions. They're allowed to consider how they relate to fleet vehicle mix and to factor in vehicle fuel efficiency as they implement the VAM process.

Right-sizing vehicles to missions provides the opportunity to not only attain improvements in fuel efficiency (which are a characteristic of newer vehicles), but also to achieving additional efficiency gains by shifting to smaller vehicle size classes. For example, a 5-year difference in models can be a huge gain for fuel efficiency. While replacing a 2006 six-cylinder Ford Focus with a 2011 version increases a fleet's fuel efficiency by almost 10 percent. Furthermore, shifting to a smaller Ford Focus will offer an additional 28 percent improvement in fuel efficiency.[15]

The other opportunities for increasing fuel efficiency are found on the *operational* side.

Operational strategies to reduce fuel consumption involve changing the behaviors of operators and other staff members. Granted, the percentage of GHG reductions from each of these strategies is generally low, but they also require fewer up-front costs, can be implemented widely across the fleet, and, ultimately, can be very cost-effective. Still, something as simple as creating a pool from the cost savings and using those funds for a party or prize might be enough motivation to spur this along.

The most obvious strategy is ensuring regular vehicle maintenance and following best practices, a strategy similar to preventive medicine (see the "Health" chapter). In this case, the body is the vehicle. Regular checkups and healthy maintenance practices include what US General Services Administration (GSA) Federal Management Regulation (FMR) B-19 calls "preventative maintenance programs and driver inspections." They are the equivalent of the annual physical, only with usage checkpoints guiding the process instead of a calendar.

The Fueleconomy.gov website, which is maintained by EPA and DOE's Office of Energy Efficiency and Renewable Energy, provides the following preventive maintenance tips designed to improve fuel efficiency:

- Keep vehicle engines properly tuned.
- Keep tires inflated to the recommended tire pressure.
- Check and replace air filters regularly.
- Use the recommended grade of motor oil for your vehicle to increase fuel efficiency.

Fleet managers can also improve fleet fuel efficiency by selecting tires that have reduced rolling resistance. Tire manufacturers are continually striving to reduce rolling resistance by improving tire materials, structure, and tread design. A 10 percent reduction in rolling resistance will typically improve fuel efficiency by roughly 2 percent. During the 2008 presidential race, both major candidates stressed the importance of keeping tires fully inflated as a simple way for Americans to conserve fuel and be more efficient in their driving.

The idea of efficient driving can help fleet vehicle drivers improve fuel efficiency and reduce petroleum consumption and GHG. For the federal fleet, GSA recommends that fleet managers "develop and implement a communication plan to ensure that strategies for improving fleet fuel efficiencies are disseminated agency-wide." This involves creating awareness in all drivers that driving more efficiently has the same impact on fuel. The communication plan should include the following messages:

- Drive at speeds that conserve fuel (less than 60 mph).
- Use cruise control, when appropriate, on the highway to maintain a constant speed.
- Drive safely and responsibly, avoiding fast starts and hard braking.
- Remove excess weight, such as unnecessary items in the trunk.

One overlooked area is driving without moving—when a car is just idling. Idling vehicles typically burn from ¼ to 1 gallon of fuel per hour.[16] The result is the worst in efficiency; you're not going anywhere, yet you're still polluting the air, wasting fuel, and causing excess engine wear.

Reducing idle time is a challenging strategy to implement and monitor, but it is effective at conserving fuel, reducing engine wear, and saving money while reducing emissions and noise. More than 30 states have laws establishing a maximum idling time, which means there is no choice but to adopt measures that limit unnecessary vehicle idling.

Typically, the following actions can help reduce unnecessary vehicle idling:

- Turn off your engine when you are parked or stopped (except in traffic) for more than a minute.
- Avoid remote vehicle starters, which encourage unnecessary idling.
- Avoid drive-throughs: walk inside instead.

Truckers can use web-based tools to find electric recharging sites.

Fortunately, a variety of technologies are available to reduce idling, while addressing some of the reasons people idle their vehicles.

For light- and medium-duty vehicles, three types of technologies are available for warming a vehicle without idling: coolant heaters, air heaters, and energy recovery systems. Coolant heaters keep the engine warm, air heaters blow hot air into the vehicle interior, and energy recovery systems use engine heat to keep the car's cooling system and heater operating.

Some technology works to make idling less of a necessity. For heavy-duty vehicles, onboard equipment such as automatic engine stop-start controls and auxiliary power units can be used anywhere. Truck stop electrification enables trucks to hook up to stations that provide power and other amenities.

Many HEVs will soon have engines that shut off temporarily when the vehicle is idling. Fleets can reduce the idling of law enforcement vehicles by using battery systems rather than the engine to support auxiliary power demands for onboard systems.

Idaho National Laboratory won the 2010 Lean, Clean and Green Presidential Award, for improving fuel efficiency, increasing the use of bio-based alternative fuels, and reducing GHG emissions.

EXAMPLE

In 2008, as part of a DOE initiative to reduce its energy intensity by 30 percent, the National Renewable Energy Laboratory (NREL) and LMI studied opportunities to reduce petroleum use and increase alternative fuel use at the Idaho National Laboratory (INL), which had, at the time, DOE's largest fleet. One of the biggest opportunities was to improve the fuel efficiency of INL's aging bus fleet, which accounted for more than 60 percent of that organization's fuel use.

These model year 1980 through 1990 diesel coach buses, used to transport employees to the remote INL site, have a fuel economy of roughly two-thirds and capacity of 80 percent that of newer buses. LMI estimated that replacing each old bus with a new bus would increase average fuel economy from 4.7 to 7 mpg and save more than 2,500 gallons of fuel per year. The actual opportunity may be even greater; we recorded one bus using 57 gallons to travel 209 miles (or 3.7 mpg). Acquiring new buses is also cost-effective, with favorable leasing terms coupled with fuel savings and lower maintenance costs.

Since this study, INL has been aggressively implementing our recommendation; full replacement of older buses will reduce petroleum use by more than 231,000 gallons and reduce GHG emissions by 1,725 metric tons.

Use Lower-Emission Fuels

Recommendation: *Maximize displacement of petroleum with lower-carbon alternative fuels, and, in turn, lower the fuel carbon impact variable in the emissions equation.*

CHALLENGE

Currently, strategies to shift to lower-carbon fuels typically increase fleet costs. With the exception of CNG and electricity, all alternative fuels cost more on an energy-equivalent basis than gasoline or diesel.[17] Fleet managers must also plan for up-front capital costs when switching to alternative fuels. Vehicle acquisition costs may be higher because the use of alternative fuels typically requires some power train modification.

Although alternative fuels can significantly reduce use of conventional fossil fuels, the extent of GHG emissions reduction varies by fuel type. A number of options are available—fuels that qualify for the "alternative" label include a blend of 85 percent ethanol and gasoline, biodiesel, CNG, liquefied natural gas, liquefied petroleum gas, and electricity—so each must be weighed in any fleet strategy as all fuels do not offer the same benefits.

Capital costs for other alternative fuel strategies may be high. Alternatives such as electricity require a completely different vehicle drive train. In many cases, fleets may also incur costs to purchase and install refueling infrastructure dedicated to alternative fuels.

SOLUTION

Mitigation Potential	★★★
Operational Impacts	★★
Financial Impacts	★
Feasibility and Timing	★★

Among alternative fuels, biofuels offer the greatest potential for GHG emissions reduction. For GHG emission purposes, the federal fleet excludes biogenic GHG emissions in measuring compliance with agency GHG emissions goals. For example, the use of E85 instead of gasoline will reduce federal fleet vehicle GHG emissions by 85 percent because it counts only emissions from the 15 percent that is gasoline when looking at targets.

Still, other alternative fuels are effective at reducing GHG emissions. Using natural gas instead of diesel, for example, reduces tailpipe GHG emissions by 28 percent. Use of electricity in EVs completely eliminates tailpipe GHG emissions, though agencies must count the emissions used to generate electricity when they're non-renewably sourced.

For the primary alternative fuels used in the federal fleet today, B20 (containing 20 percent biodiesel and 80 percent standard diesel) and E85, modification costs are low to none. B20 can be used in most current diesel engines without the need for any mechanical changes, and E85 is used in E85 FFVs, which are already on the market and similar in price to conventional vehicles. Also, FFVs are already a large and growing component of the federal fleet—they made up more than 31 percent of light-duty vehicles in 2010.

Then there are the first-generation alternative fuels, including corn-based E85 and soy-based 20 percent biodiesel blends. Over the full fuel life cycle, these fuels reduce GHG emissions between 14 and 21 percent compared with their conventional fuel counterparts.

In 10 years or so, the fuel life-cycle GHG emission reduction—as well as the cost-effectiveness—of lower-carbon alternative fuels is expected to increase dramatically. Because of the EISA Renewable Fuel Standard 2 requirements, two other options are expected to become available on a wide scale: cost-competitive cellulosic ethanol (fuel from biomass, which provides more than a 90 percent reduction in life-cycle GHG emissions) and lower-cost electric vehicles (which provide an average 47 percent reduction in life-cycle GHG emissions). Down the road, but not for another 20 to 40 years, hydrogen may also emerge as a viable fuel technology. Hydrogen offers potential emission reductions of 50 percent in 2030 and 80 percent in 2050.[18]

In 2007, EPA estimated GHG emissions for the entire fuel life cycle for a range of alternative fuels. Figure 6-3 compares the estimated GHG emissions of these alternatives with those of petroleum fuel.

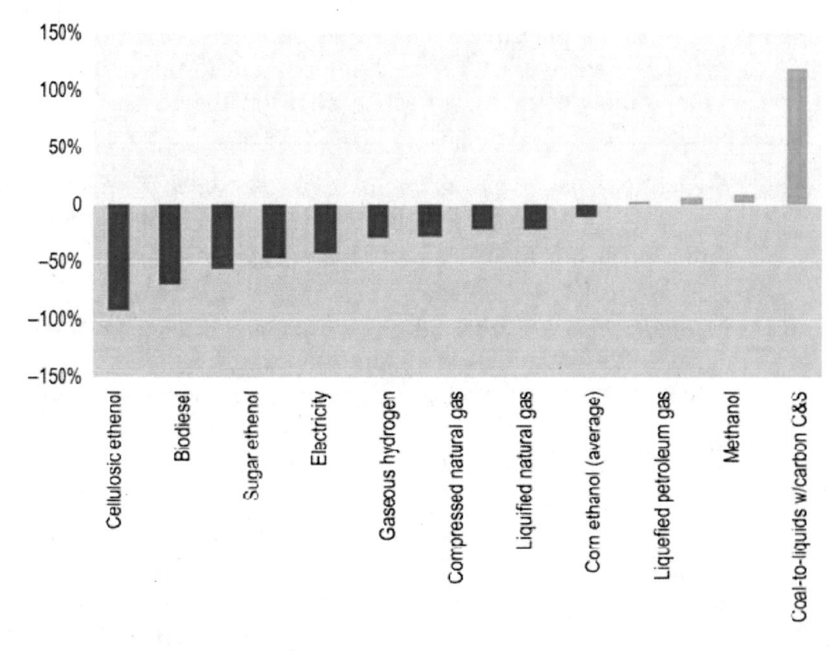

Figure 6-3. Comparison of Life-Cycle GHG Emissions for Alternative Fuels Relative to Conventional Fuels

Note: C&S equals capture and storage.

Strategies that incorporate lower-carbon alternative fuel typically make the most sense for fleets located where alternative fuel is currently available; the same can be said for high-use locations where alternative fuel infrastructure can be installed (except for electricity). A great resource for locating both public and private alternative fuel pumps is the Alternative Fuels and Advanced Vehicles Data Center (formerly the Alternative Fuels Data Center), which offers tools such as FleetAtlas,[19] TransAtlas,[20] and the Alternative Fueling Station Locator.[21]

Fleets at lesser-use facilities without access to these alternative fuels should focus on other GHG emission-reduction strategies. These include the acquisition of EVs and, if favorable to the site and mission, facilitating development of infrastructure to support alternative fuel at local commercial or other fleet stations.

In the next 10 to 20 years, BEVs and PHEVs will become cost-effective for specific fleet applications, including high annual use within a range of less than 100 miles. When this economic shift occurs, fleet managers should change from other more costly lower-carbon fuel options to EVs. They should then focus on using zero-emission renewable electricity in these vehicles rather than coal-based electricity from the grid.

The EO 13514 *Comprehensive Federal Fleet Management Handbook* offers a six-step framework for identifying optimal lower-carbon alternative fuel strategies at each fleet location on the basis of fleet characteristics:

1. *Evaluate alternative refueling options at or near the fleet location.* Is alternative fuel available where vehicles currently refuel or at other nearby locations accessible by fleet vehicles?
2. *Estimate potential alternative fuel use at the fleet location.* What is the potential maximum number of alternative fuel vehicles (AFVs) or biodiesel-capable diesel vehicles, assuming accelerated replacement? What is the theoretical maximum use of each alternative fuel at the fleet location if fuel availability were not an issue?
3. *Decide on optimal alternative fuel strategies (except electric) at each fleet location.* What are the optimal alternative fuel strategies on the basis of fleet characteristics (potential annual fuel use and local availability of alternative fuel use)?
4. *Decide on optimal EV strategy at each fleet location.* At locations that have or will have access to alternative fuel, prioritize acquisition of vehicles capable of using that alternative fuel. Then, determine the availability of EVs to replace conventionally fueled vehicles that are not candidates to be replaced with AFVs or biodiesel-capable diesel vehicles. At locations without access to dedicated EV-charging infrastructure, fleet managers will need to install the charging infrastructure required to support the efficient operation of fleet EVs.
5. *Evaluate the cost implications of potential strategies.* Implementation of some of these lower-carbon alternative fuel strategies may strain fleet budgets. Fleet managers will need to determine whether potential strategies can be implemented at competitive life-cycle costs. For example, many E85 FFVs are available at no incremental cost, and at high-use locations, infrastructure costs can be allocated over a large number of vehicles. Fleets also can work with DOE's Clean Cities Program, which can help in developing infrastructure at or near their fleet location.
6. *Establish vehicle acquisition strategies to support alternative fuel strategies.* Success in implementing alternative fuel strategies depends not only on fuel availability but also on vehicles that can use the fuels and drivers who can properly and consistently refuel vehicles with alternative fuel. Fleet

managers should maximize the number of vehicles capable of using alternative fuels at locations that have existing alternative fuel infrastructure or that are candidates for new infrastructure.[22]

One critical consideration is the economics involved in such a decision. Organizations don't have the luxury of drawing resources from a bottomless pit. Effective investments make sense strategically *and* financially.

EXAMPLE

In 2007, Congress asked the Department of Defense (DoD) to identify measures to increase the use of biofuels to displace petroleum. DoD responded with five recommendations that centered on increasing the use of alternative fuel where available and promoting or installing new alternative fuel pumps in locations near where DoD uses a lot of fuel.[23]

This tactical vehicle is testing the use of biodiesel.

The low-hanging fruit for DoD is to change fleet behavior to better use E85 and B20 pumps when close by. Just using the biofuel pump instead of the petroleum-based fuel pump when refueling at biofuel stations could increase DoD's E85 use by 2.1 million gasoline gallon equivalents, increase B20 use by 4.0 million gasoline gallon equivalents, and reduce tailpipe nonbiogenic GHG emissions by more than 23,000 metric tons of CO_2e. Driving the extra mile to refuel at the nearby E85 and B20 station rather than the conventional fuel station could avoid an additional 31,500 metric tons of GHG emissions. Both of these recommendations are very cost-effective ways to reduce GHG emissions, requiring no capital investment and only slight to moderate increases in operating costs.

Full implementation of the DoD recommendations, including installing new E85 or B20 pumps near high-use fleet locations, offers the potential to increase DoD E85 use by 25.5 million gasoline gallon equivalents, increase B20 use by 10.8 million gasoline gallon equivalents, and eliminate more than 208,000 metric tons of tailpipe nonbiogenic GHG emissions.[24]

Optimize Transportation Modes

Optimize transportation modes

Recommendation: *Determine the optimal transportation modes for supporting the organization's mission and work to adopt less carbon-intensive alternatives while supporting the necessary investments to make adoption a reality.*

CHALLENGE

Any effort to shift fleet use from vehicles to mass transportation is complicated by the limited availability of and access to these services and possible impacts on the organization's mission.

It's a fleet manager's responsibility to focus on using fleet vehicles to provide required transportation services for the organization. Managing GHG emissions will be an increasingly more important aspect of this job, and fleet managers will need to implement strategies that influence fleet operators. Key to this effort is compelling operators to shift to less carbon-intensive travel modes. This may be a tough sell for those who focus on the fact that the fleetwide GHG emission-reduction impact of this solution is less than 10 percent.

Some modes of transportation make sense for an organization's profile and mission but require a massive investment of time and resources. For example, extending the Washington area's highly used Metrorail to the Northern Virginia suburbs and Dulles International Airport was first discussed in the 1960s;[25] it is just now becoming a reality. Some cities, like Houston, simply lack a centralized downtown. So, no single solution is right for every fleet.

SOLUTION

Mitigation Potential	★
Operational Impacts	★★
Financial Impacts	★★★
Feasibility and Timing	★★

In most cases, mass transportation is far more efficient than individual vehicles in GHG emissions per passenger mile traveled (PMT). For example, the life-cycle GHG emissions per passenger mile from the Bay Area Rapid Transit (BART) system are 37 percent of a passenger car and 23 percent of a pickup truck.

Table 6-3 shows 2008 estimates of life-cycle GHG emissions for different transportation modes that may be available to fleet managers.

Table 6-3. Estimates of Life-Cycle GHG Emissions by Transportation Mode (in g CO_2e per PMT)

Vehicle Type	Life-Cycle Emissions	% Difference to Sedan
On-road vehicles		
Sedan	380	—
SUV	450	+18%
Light trucks	620	+63%
Transit bus		
Bus (off peak)	680	+79%
Bus (peak)	85	–78%
Rail transit		
Bay Area Rapid Transit	140	–63%
Caltrain	160	–68%
Massachusetts Bay Transit	230	–61%

Using public buses during peak times can reduce GHG emissions by an additional 31 percent compared with rail transit. However, during non-peak times, when ridership is lower, buses may actually release more GHG emissions per passenger mile than other vehicles. It's a balancing act, because encouraging employees to use existing bus systems during off-peak hours does not increase GHG emissions of the bus system and can, in fact, reduce federal GHG emissions. Still, fleet managers should avoid operating bus or shuttle transport at sub-peak conditions.

Because of the challenges associated with shifting fleet use from vehicles to mass transportation (including, as mentioned, the limited availability of and access to these services and possible impacts on the organization's mission), these services need to align with organizational transportation needs to make sense. Mass transportation provides many benefits in large and densely populated metropolitan areas, such as Washington, DC, for the federal fleet.

Although the fleetwide GHG emission-reduction impact of this solution is less than 10 percent, implementing these strategies can both save money and reduce travel time—two other key considerations that should be included in any decision.

EXAMPLE

Public transportation is only as appealing as its efficiency, timeliness, and accessibility. Salt Lake City made mass transportation more attractive to riders by improving its bus system. In 2008, the city began significantly improving its bus routes by connecting them to existing transportation systems. This was accomplished by creating the MAX Bus Rapid Transit System.[26] The city created bus lanes along the nearly 11 miles of bus route, and in the most congested areas of the city, the buses travel down designated center lanes in both directions, similar to a light-rail system. Buses are given the right of way at intersections, and track signal detection is used to give buses priority at stoplights. Since its implementation, the MAX line has seen a one-third increase in ridership and a 15 percent reduction in travel time for its average 4,200 riders a day.

The success of the MAX line has led Utah Transit Authority to make plans to connect the bus routes with up to 80 miles of new rail tracks to other cities in the region within the next 20 years.

Store Renewable Energy

Use electric vehicles

Recommendation: *Use vehicle-to-grid (V2G) systems to reduce electricity generation impacts.*

CHALLENGE

As the source of a third of US GHG emission, the generation of electricity is a major focus of climate change mitigation efforts. Electricity generated from coal, petroleum, and natural gas account for all direct emissions, which is astounding considering that these fuels represent only 69 percent of our electricity generation.

Mitigating GHG emissions from electricity generation involves shifting generation from these high GHG-emitting fuels to renewable technologies that avoid GHG emissions.

One barrier to increasing the use of intermittent renewable sources, such as solar and wind, is that the electricity supply (generation and distribution) has to match demand at each moment in time. If winds gust at night when demand is low, energy collected from wind systems may be lost. Similarly, if clouds block the sun when demand is high, solar systems may not be able to meet peak demand needs.

Also, technologies that support these efforts are still in their infancy.

Mitigation Potential	★★
Operational Impacts	★★★
Financial Impacts	★★★
Feasibility and Timing	★★

SOLUTION

V2G power involves the distributed electricity storage of a fleet of EVs, such as pure BEVs and PHEVs, communicating with the power grid to supply electricity when needed. V2G offers utilities a very low-cost and effective system for stabilizing renewable energy generation. Willett Kempton, a V2G pioneer, estimates that the service may provide value to utilities of up to $4,000 a year per vehicle.[27]

In redeploying energy stored during peak load times, when vehicles are often parked, utilities can defer or avoid high GHG-emitting peak-generation capacity, such as natural gas and petroleum, in favor of zero-emission-renewable electricity. Once renewable energy generation expands, utilities may even be able to eliminate GHG emissions from coal-based load generation.

Installing V2G capability for EVs requires three components:

1. Connection to the grid to provide power
2. A real-time control system to communicate with the grid
3. A precision meter on the vehicle.

It costs roughly $500 to get V2G ready for an electric vehicle running on a battery. When the revenue exceeds the cost of providing power stored in the vehicle, V2G systems pay back these capital costs and generate income for the fleet. EVs are well suited for these applications since their batteries are designed to quickly respond to numerous and large power fluctuations.

For fleet managers, the economics of adding V2G capability to their EVs can be very favorable. For example, the projected annual net profit for a fleet of 250 BEVs, after paying back capital costs, ranges from $150,000 to $2.1 million (or $600 to $8,400 per vehicle).[28] The high variability reflects the regional value of V2G services, the capacity of electric connections (electrical plug connections range from 6.6 to 15 kW), and the capacity of the battery.

V2G technology is still in the earliest stages of development. The economic benefits are theoretical and may be far less attractive in practice. Fleet managers should consider this a medium- to long-term solution, as the commercial availability of V2G is likely many years away.

One agency in particular, DoD, may find that the storage capability of electric vehicles also offers a low GHG national security solution (see the "Security" chapter); this could support energy security and emergency response capacity. DoD could investigate the use of EVs as a means for providing mobile electric supply and storage in theater and for emergency responses. This would serve as a solid alternative to GHG-intensive generators, especially if the fleet can provide immediate electricity generation to maintain mission-critical functions such as communications, medical services, and command center operations. DoD can leverage mass-production EVs as low-profile microgrids with low-noise signatures and small GHG footprints.

Perhaps the more promising option is to combine gensets (engine generators) with batteries. This would allow users to cut generator use significantly at night, an appealing option for the military (no tactical commander wants to tie up vehicles if it isn't absolutely necessary).

EXAMPLE

A group of faculty members at the University of Delaware has been studying V2G technology for nearly a decade. These researchers have been working closely with AC Propulsion, a company that specializes in alternating-current drive train systems for electric vehicles. Their work has shown that AC Propulsion's engines are compatible for both vehicle transportation and grid support and has resulted in a prototype known as the eBox. The eBox performs like any other vehicle, capable of going from 0 to 60 mph in 7 seconds, and reaching a top speed of 95 mph. More impressive is the eBox's energy consumption of 220 to 320 watt-hours per mile and its driving range of 100 to 145 miles.[29]

The researchers have determined that, in order for the eBox to be commercialized, there would have to be enough vehicles to support the need to produce power at any one time. With drive time for an average car approaching 1 hour per day, about 100 vehicles with eBox capability would be required to ensure reliable service to the grid.[27]

Fleet Adaptation to Climate Change

The work of fleet managers involves controlling more than just vehicles. They manage infrastructures such as garages, vehicle maintenance facilities, and refueling centers. They are also accountable for the reliability of the transportation services they provide, ensuring that operators can travel as needed.

Climate change impacts fleets in two primary ways. First, some fleet infrastructure will become vulnerable to the impacts of climate change. The impacts in question include sea level rise and storm surges in coastal areas, increased frequency and severity of precipitation events, and thawing of sea ice and permafrost in Alaska.

Second, we can expect reduced reliability and availability of some transportation infrastructures throughout the United States.

The fleet focus on climate change has been primarily on GHG mitigation. Adaptation is equally critical. If climate change impacts accelerate, planning must accelerate as well, including the assessment of fleet infrastructure and operations vulnerabilities, and the creation of adaptation plans.

Adapting proactively is far less costly and much more effective than responding to climate change impacts as they happen. Proactive adaptation planning may reveal opportunities to take action to reduce or eliminate climate change vulnerabilities. For example, a fleet manager may be able to identify and modify existing plans to place a new refueling center in a location that is vulnerable to coastal flooding or to add the ability to collect stormwater to the design of a new fleet parking facility in a region at risk of increased precipitation.[30]

Climate Change Impacts on the US Transportation System

Five primary climate change effects can impact transportation infrastructure:

1. *Increases in temperature.* Increased temperatures and the frequency of high-temperature days can have detrimental effects on roads and bridges. Pavements, especially asphalt, will be subject to deterioration and reduced integrity, which includes softening and rutting. High temperatures may shift more construction to the nighttime hours and lengthen the construction season. Bridges may fail or be closed due to increased maintenance forced by the thermal expansion of bridge expansion joints.
2. *High winds.* High winds can make travel dangerous. Larger vehicles, such as tractor trailers, are already susceptible to rollover, and sustained high winds can increase that risk. Efforts to navigate the roads may be impeded by wind-related damage to trees, limbs, and signs.
3. *More frequent and intense precipitation events.* The increased frequency and intensity of precipitation has the potential to add to flooding that impacts roadways, rail lines, and tunnels in low-lying areas; the same floods could overload drainage systems, causing backups and street flooding, road washout, landslides, or mudslides that damage roads. Soil moisture levels could affect the structural integrity of roads, bridges, tunnels, and standing water on the road base. Some of these impacts may limit travel temporarily, while others may close routes permanently.

Heavy downpours are projected to increase due to climate change. These can temporarily disrupt and damage transportation routes.

4. *Rising sea levels.* Climate change has the potential to increase sea levels, which could have major impacts on transportation infrastructure near coastal areas. These impacts may be amplified by the high population density in these areas; population density in coastal areas is significantly higher than in the rest of the nation (300 people per square mile along the coast compared with just 98 per square mile elsewhere). The primary effects on transportation infrastructure from potential rise in sea level are permanent flooding of some roads and areas that creates a vulnerability to periodic flooding and storm surges; more frequent severe flooding of underground tunnels and low-lying infrastructure; erosion of road base and bridge supports; and bridge scour, a situation in which the sand and rock around bridge abutments or piers wash away.
5. *Greater intensity of coastal storms.* In coastal areas, the combination of increased storm intensity and rising sea levels can amplify the effects of each on transportation infrastructure.[31]

Our recommendations for adapting vehicle fleets to contend with the impacts of climate change are functional with respect to how they fit into an organization's overall strategy:

- Refocus the fleet organization on addressing risks of climate change impacts.
- Identify climatic trends affecting agency transportation operations.
- Conduct vulnerability assessments and risk appraisals of fleet infrastructure.
- Create adaptation action plans for each vulnerability.
- Increase the flexibility to shift between different transportation modes.

Refocus the Fleet Organization on Addressing Risks of Climate Change Impacts

Recommendation: *Begin early preparation for climate change impacts by reconsidering how the fleet is organized—where uncertainties are in relation to cost, risk, scale, and timing—and use these factors to determine a flexible, forward-thinking approach.*

CHALLENGE

Organizationally, tackling the potential impacts of climate change on fleet operations is a difficult problem for fleet managers. The complexities in addressing climate change include uncertainties in the scale and timing of impact and the associated risks and costs to the fleet.

Faced with a lack of urgency to deal with long-term impacts, unknown risks, and likely high costs, organizations tend to avoid investing in an effort to head off potential climate change impacts. However, climate change will have real effects on an organization's ability to provide reliable transportation services. Organizations must reorient their perspective and focus on the risks of climate change to their transportation services.

Initially, organizations may be resistant to adapting to climate change because little direct evidence currently points to the urgency of these impacts on their operations.

SOLUTION

Adaptation Potential	★★
Operational Impacts	★★★
Financial Impacts	★★
Feasibility and Timing	★

Initially, organizations will be resistant to adapting to climate change since there will be little direct evidence of the urgency of these impacts on their operations. However, organizations will need to recognize that climate change is a potentially significant problem for tomorrow and decisions to address these issues should be a priority today. Although the potential effects are uncertain, the benefits of early preparation far outweigh the probability-weighted costs of negative outcomes if the effects occur as predicted.

Initiatives are under way that can serve as guides for any fleet's adaptation efforts.

In October 2009, President Obama established the Interagency Climate Change Adaptation Task Force to recommend how policies and practices already in place in the federal sector can help fuel and reinforce a national adaptation strategy for climate change.[32] In its 2010 progress report, the task force said "climate change preparation and response should be integrated into core policies, planning, practices, and programs whenever possible."[33]

It established a series of goals to ensure that each agency focuses on addressing the risks of climate change impacts on its operations. For fleets, the most critical is the goal to "encourage and mainstream adaptation planning across the federal government." Agencies should implement flexible, forward-thinking plans that identify and address climate change vulnerabilities while ensuring resilience to expected impacts.

The task force recommended three primary actions to reach this goal:

1. *Implement adaptation planning within federal agencies.* Each climate change adaptation plan should focus on the unique impacts on the agency (or organizational) fleet operation. Within each plan, the organization should establish specific goals and metrics to review whether actions are achieving desired outcomes.
2. *Employ a flexible framework for agency adaptation planning.* The task force developed a six-step flexible planning framework to help make certain that plans are tailored to the specific needs of each agency fleet. These steps are (1) set a mandate with clear goals and success; (2) understand how climate is changing; (3) evaluate how these changes impact fleet operations; (4) identify, prioritize, and implement adaptation measures; (5) assess the effectiveness of actions and adjust efforts; and (6) build awareness and enable organization through education, training, and mandates.
3. *Use a phased and coordinated approach to implement agency adaptation.* The initial phase for each fleet is to build awareness internally and demonstrate the need for adapting to climate change impacts on operations. Next, each fleet should develop a high-level plan, setting priorities and goals. Fleets can then expand this plan to identify targeted near-term actions and long-term planning efforts.

EXAMPLE

According to the Pew Center on Global Climate Change, 15 states either have climate adaptation plans in progress (4 states) or completed and in place (11 states).

Eight additional states have climate action plans that recommend state development of an adaptation plan.[34] South Carolina, for example, established the South Carolina Climate, Energy and Commerce Advisory Committee (CECAC). In a July 2008 report, CECAC made several recommendations concerning transportation and land use, including adopting clean-car standards and improving mass-transit, bicycle, and pedestrian networks. It also provided analysis of the state rail system and its potential for expansion.[35]

In addition to such recommendations, CECAC proposes that South Carolina develop a "'Blue Ribbon' Commission on Adaptation to Climate Change to develop a state Climate Change Adaptation Plan."[36] According to the committee, one of the key components of the state adaptation plan should be to provide recommendations for responses to potential climate change impacts to, among other things,

South Carolina's transportation systems and human-made infrastructure. The stated purpose of this recommendation was to minimize risk to the state's inhabitants, natural systems, and resources.[37]

Identify Climatic Trends Affecting Agency Transportation Operations

Recommendation: *Ensure plans involving fleet adaptation to climate change impacts take into account local concerns, with a focus on key data and variables, in order to develop appropriate strategies for specific locations.*

CHALLENGE

Because both climate change impacts and fleet operations vary enor-mously from place to place, climate change will affect fleets differently at the local and regional levels. Fleet managers will need to adapt to these impacts to ensure they can accomplish their missions.

For example, climate change will impact fleet operations in Alaska differently than those in Minnesota. The same is true for the Gulf Coast. No blanket strategy can account for the vast differences in impacts and how they will be felt.

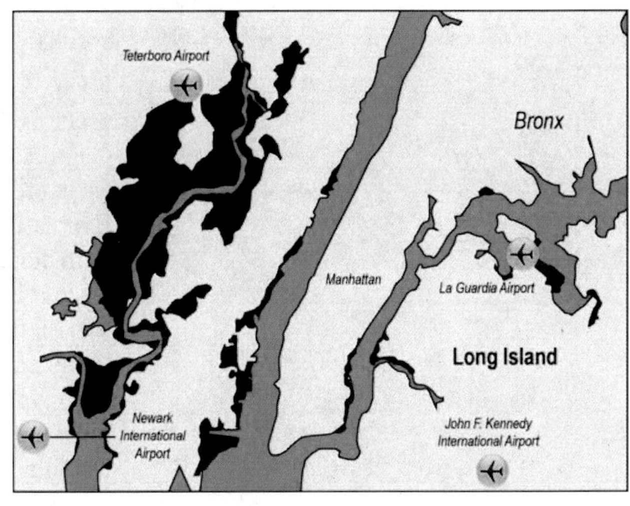

Of the four major airports near NYC, three are below 1 meter elevation. These can be damaged from storm surge long before being inundated by sea level.

Rising temperatures in Alaska may expand the ongoing retreat of permafrost areas (causing increased land subsidence risks to roads and fleet infrastructure) just as rising winter temperatures in Minnesota are actually improving fleet operations (allowing for the increased passage of roads and less use of treatment chemicals). And, down south, sea level rise might make roads and fleet operations in the Gulf Coast more vulnerable to coastal storm flooding as well as to storm surges from hurricanes.[38] The quality and quantity of projections of climate change impacts will vary from region to region. Fleet managers will need to evaluate the usefulness, applicability, and the soundness of sources of potential climate change impacts.

The most critical question is whether operations will be impacted by changes in climate conditions (such as rise in temperature or change in precipitation) or climate impacts (such as rise in sea level and increase in coastal flooding). Is the ability to provide transportation services sensitive to one of these impacts, and how will they affect operations?

SOLUTION

Adaptation Potential	★★
Operational Impacts	★★
Financial Impacts	★★
Feasibility and Timing	★★

Fleets must tailor adaptation plans to the circumstances at their location and to fit their operations. So, the first step to adapt fleet operations is to ask, "How will climate change affect each of my fleet locations, and will it pose a risk for my fleet operations?"

Evaluating the potential impacts of climate change on your fleet locations is not an exact science. Projections differ from one study to another, depending on factors that include the global climate model used, assumptions, and emission scenarios. The following tips will help you understand the potential impacts to your fleet locations:

1. Focus on key climate change variables, such as temperature, precipitation, wind, sea level, and other weather.
2. Gather data on ranges of future impacts, to develop worst-case, probable, and best-case scenarios.
3. Evaluate projected seasonal variations and compare them with current and historical data.
4. Recognize that projections of temperature are typically more certain than those of precipitation.
5. Anticipate that newer climate change models (5 years old or less) typically will be more accurate than older ones.
6. Document the assumptions and time frame of the studies used to project climate change impacts.[30]

Two data sources, when they overlap, can help identify the impacts of climate change on fleets: the inventory of fleet operating areas and projected primary climate changes in those local and regional areas. By using geographic information systems, fleet managers can automate their analysis and simplify future updates. This is also a way to determine whether impacts will be direct (e.g., from changed weather conditions) or indirect (e.g., from sea level rise).

After understanding the potential impacts of climate change on the locations where a fleet operates, a fleet manager should determine the operations that may be affected by these projected climate changes. Of course, some areas will be subject to larger potential changes in the climate than others.

EXAMPLE

The state of Maryland has undertaken "a range of aggressive actions … related to climate change, both on mitigation and adaptation."[39]

Not only has the Maryland Commission on Climate Change published the *Comprehensive Strategy for Reducing Maryland's Vulnerability to Climate Change Phase II: Building societal, economic, and ecological resilience* (January 2011), the Maryland State Highway Administration (SHA) is creating a climate adaptation plan.[40] Although the plan is not yet complete, the Maryland SHA has identified the anticipated climate change impacts that the state may face, such as increased temperature of approximately 2°F, storm surge, increased storm frequency and intensity, increased precipitation in spring months, stronger hurricanes, increased likelihood of storms referred to as "100-year events," and a sea level rise of 3–4 feet.[f]

It has determined that in response to such changes, adaptation measures should be directed toward pavement rutting and buckling, increased precipitation, flooding, power loss, traffic disruptions, sea level rise inundation of coastal areas, more frequent and costly evacuations, scouring of bridge foundations, and failure of bridge decks.[41] The planning process put into place now involves identifying climate changes, assessing risk and vulnerabilities, developing risk-based adaptation strategies, and identifying opportunities for co-benefits, implementation, and monitoring.[39]

Conduct Vulnerability Assessments and Risk Appraisals of Fleet Infrastructure

Recommendation: *Fleet managers should appraise their existing infrastructure for vulnerabilities and determine the areas exposed to climate change impacts and how capably the fleet can adapt.*

CHALLENGE

With an inventory of fleet operations that may be affected by projected climate changes, the next question fleet managers must answer is, "How vulnerable are fleet operations to these climate impacts?" The greater the exposure to climate change is, the larger the potential impact to fleet operations.

[f] The amount of sea level rise with which it's working isn't expected until at least the year 2100.

Fleet managers also face a high degree of uncertainty, from the timing of events to their severity.

Location is critical and must be considered; Alaska is certainly not Manhattan, and climate change will impact both of these areas (and every other area nationwide) differently.

SOLUTION

Adaptation Potential	★★
Operational Impacts	★★
Financial Impacts	★★
Feasibility and Timing	★★

A vulnerability assessment (see the "Security" chapter) helps address the degree to which these fleet operations are susceptible to climate change and may be unable to cope with adverse effects.

Ultimately, creating adaptation plans for your fleet requires understanding of and a response to four primary questions:

1. *To what extent is each fleet operation exposed to climate change impacts?* The extent of exposure has two components: magnitude and timing. Magnitude is determined by the scale of impacted fleet operations and the intensity of those impacts. For example, the magnitude of climate impacts in lower Manhattan may be very high. These fleet locations, which lie at elevations between 6 and 20 feet above sea level, could see increasingly frequent and severe flooding from hurricanes and nor'easters. On the other hand, for fleets in Alaska's more developed areas, the magnitude of impacts may be relatively low. They simply might require flexibility or alternative modes of transport when, for example, roads fail due to retreat of permafrost.
   Fleet managers also must prioritize adaptation plans on the basis of the timing of the potential impact. Some smaller potential fleet impacts in the near term will be given priority over those in the distant future, even those that rank as high potential impact events. In fact, sudden impacts will take precedence over similar, but gradual, impacts. Nonetheless, fleet managers must remain vigilant and plan for longer-term impacts to ensure plans are proactive rather than reactive.
2. *What is the threshold where sensitivity to impacts increases?* Fleet managers should next estimate how severe the impacts must be before they become significant factors that affect fleet operations. Referred to as the "impact threshold," this is the point at which a fleet can no longer perform its mission due to these impacts. In the lower Manhattan example, the impact threshold is the point at which future sea level rise or storm surges flood fleet locations. In the Alaska example, this

point may be the temperature where permafrost melting impacts roads used by fleet vehicles.

3. *How critical is the impacted fleet operation?* A consequence scale can be an important tool in evaluating the criticality of impacts. Some fleet operations and infrastructure are more critical than others in fulfilling an agency's mission. Fleet managers should consider their relative importance when they assess the vulnerability of each fleet operation potentially exposed to climate change impacts.
4. *What is the capacity of the fleet to adapt to the impacts?* Climate change is far less of a concern for fleets that can easily adapt their operations to avoid impacts. Organizations can relocate or substitute some fleet operations, while making others less vulnerable or more resilient. In Alaska, for example, a fleet may be able to adapt to roads degraded by permafrost by using other modes of transportation, different routes, or more resilient vehicles. In lower Manhattan, a fleet may build improved flood barriers and add a stormwater pumping option to their infrastructure.[42,43]

EXAMPLE

Alaska is facing a significant potential increase in its public infrastructure costs as a result of climate change, one that researchers at the Universities of Alaska-Anchorage and Colorado-Boulder estimate could be in the billions.

Road damage due to permafrost melting.

The university analysis encompassed approximately 16,000 elements of public infrastructure in Alaska. From various climatic model scenarios, they estimated the annual replacement cost of the state's infrastructure with and without the influence of climate change, concluding, "climate change could add $3.6–6.1 billion (+10% to +20% above normal wear and tear) to future costs for public infrastructure from now to 2030 and $5.6–7.6 billion (+10% to +12%) from now to 2080."[44] This study also notes that the majority of additional costs between the present and 2030 will result from transportation infrastructure (roads, airport runways, etc.). Granted, large expenses accompany transportation infrastructure in Alaska regardless of circumstances, but the researchers note that many airports and some roads lie in areas that may be greatly impacted by climate change.

The study noticeably adopts the elements of the risk equation, where *risk* is the product of *probability of hazard*, *vulnerability*, and *consequence*.

The researchers used various climatic model scenarios to represent the probability of the hazard (climate change). The indicators of vulnerability include the count of elements of infrastructure and their unique characteristics (including useful life), along with the local permafrost condition, susceptibility to flooding, and proximity to the coast. Consequence is seen in the replacement costs for elements of infrastructure. Combined, these elements gave the researchers the level of risk associated with the impacts of climate change on the cost of infrastructure replacement.

Create Adaptation Action Plans for Each Vulnerability

Create adaptation plans

Recommendation: *Tailor adaptation plans to address each vulnerability in the fleet, including the development of a time frame and metrics for success.*

CHALLENGE

The real challenge for fleet managers is selecting the best adaptation strategies for a particular fleet.

After determining the vulnerability of their infrastructure and classifying facilities that are critical or essential, fleets need to evaluate adaptation options and implement adaptation plans that ensure the ability to continue to provide transportation services as needed. Fleet managers must identify partners that can help carry out strategies that are critical to ensuring mission success.

SOLUTION

Adaptation Potential	★★★
Operational Impacts	★
Financial Impacts	★
Feasibility and Timing	★★

Fleet managers should work with facility or location managers and other key players to identify the impact site adaptation plans will have on the fleet.

- *How can my fleet adapt to climate change impacts?* For each fleet operation exposed to climate change impacts, fleets must develop a list of potential adaptation strategies. Table 6-4 provides some examples for different climate change effects.

Table 6-4. Example of Climate Change Impacts on Fleet Operations

Climate Change	Fleet Operation Impacts	Potential Adaptation
Increase in temperature	Deterioration of pavement Bridge failure Vehicle overheating Increased tire wear Retreat of permafrost	Increased road maintenance Route flexibility and planning Increased vehicle maintenance Alternative modes of transport Expanded or changed time of use
Sea level rise and storm surge	Flooding of low-lying infrastructure Erosion and failure of roads Land subsidence Increased road closures Weather-related delays and traffic disruptions	Reinforcement and protection of infrastructure Relocation of assets Added drainage capacity
More intense storms	Flooding of low-lying infrastructure Increased road closures Reduced structural integrity of roads	Route flexibility and planning Alternative modes of transport
Stronger winds	Reduced utilization of trucks Increased road closures	Added drainage capacity Reinforcement and protection of infrastructure Increased road maintenance Alternative modes of transport Route flexibility and planning

- *What are the best adaptation strategies for my fleet?* Developing a proactive adaptation plan should start with strategies that address the most urgent climate change impacts. These include the impacts with the highest risk, highest likelihood, and greatest impact on fleet operations. However, fleets must also weigh the benefits of adapting their operations against the costs of doing so. This involves analyzing the capital and operating costs that may be necessary for providing both the minimum and preferred level of service given the climate change impact. These costs need to be compared with the value of providing the fleet operation; fleet managers must have an answer to the question, "What is the cost to the organization if the level of service is reduced by climate change impacts?" Using this means for analyzing adaptation options, fleets can establish adaptation investment priorities and generate the funds they need.
- *When should my fleet implement these strategies?* Each climate change impact will have a time frame during which the risk of adverse effects will increase if no action is taken. Each plan should establish a time frame that dictates when adaptation strategies need to be implemented (in the near and long terms). Fleet managers should ensure that their plan aligns with investment priorities and funding.
- *How well is my adaptation plan doing?* Projections of future climate change impacts are sure to vary over time, especially

those for the long term. They may be more or less severe (or sudden) than predicted in the initial adaptation plan. Fleet managers should review these adaptation plans annually, reevaluating the effectiveness of each strategy and revising them on the basis of performance or updated projections.

New York City's fleet is one that may be very susceptible to the impacts of climate change. The city maintains numerous facilities, including depots and refueling centers, in low-lying areas vulnerable to temporary and permanent flooding due to increases in sea levels and intense precipitation events. Operators may find themselves parking their vehicles where standing water could be located or using fuel pumps that could be blocked or located in standing water.

After inventorying the city fleet facilities, the local city manager should use projected climate change impact scenarios to determine those activities most at risk. The city may need to plan to move some of these facilities to less vulnerable areas. Others may need some retrofits or construction to ensure their future availability and reliability.

EXAMPLE

If the infrastructure supporting a fleet is vulnerable, so is the fleet itself. On December 1, 2010, the state of Oregon released *The Oregon Climate Change Adaptation Framework*, which addresses potential climate-related impacts and vulnerabilities specific to Oregon, including vehicle fleet-related infrastructure. It also provides a gap analysis, suggests priority actions, and proposes other necessary actions required for the state to better adapt to climatic impacts. According to Jeff Weber, who serves at the Oregon Department of Land Conservation and Development, "This framework positions Oregon to take effective early steps to avoid some of the most costly potential consequences of climate change."[45]

This framework addresses transportation infrastructure in many different contexts. It notes that increased flooding, landslides, coastal erosion, and wildfires may disrupt transportation infrastructure. Such disruptions, in turn, can impede the delivery of essential goods (such as water and food) and services.[45] Thus, one of the priority actions in the report is to "improve capability to rapidly assess and repair damaged transportation infrastructure, in order to ensure rapid reopening of transportation corridors."[46]

The report lists this priority action under the risk of increased extreme precipitation events and flooding. However, it clearly applies to several other risk areas identified.

The next steps related to this priority involve assessing the state's capacity to repair and reopen transportation infrastructure after flooding. It also looks at developing guidance for the Oregon Department of Transportation (ODOT) and other agencies to help them implement effective procedures after a major transportation route has been closed by extreme weather or flooding and reviewing other states' best practices for repairing and reopening transportation corridors after flooding-related closures.[47] The report identifies other critical actions, including more diligent monitoring of "landslide-prone slopes near transportation infrastructure" [48] and assessing the impact of higher air temperatures on transportation infrastructure (bridge expansion joints, pavement, rail-tracks, etc.).[49]

The appendixes of *The Oregon Climate Change Adaptation Framework* also describe current state actions planned through 2011, a preliminary gap analysis, and needed actions. This information is broken out by state agency. Activities under way benefit transportation, including the removal of "hazard trees" along Highway 101, maintaining wind warning systems at various locations along Oregon's coast, and reopening transportation corridors as soon as possible after extreme weather events and flooding. In addition, the Oregon TripCheck website allows drivers to view road closures and delays on line and in real time.[50]

In terms of gap analysis and needed actions, the framework suggests that the Oregon Department of Geology and Mineral Industries work with various stakeholders, including ODOT and affected communities, to develop "a program to systematically map landslide inventory, landslide susceptibility and mitigation techniques in transportation and lifeline corridors."[51] The report identifies one of ODOT's shortcomings—its data needs. The point emphasized is that better data are necessary, especially in regard to when sea level rise will occur, as well as a the capacity to identify transportation infrastructure vulnerable to coastal erosion.[52] With this report, Oregon has identified specific risks to be addressed, preliminary actions to be taken to combat climate-related risks, and gaps in knowledge or data that must be filled in the future to effectively adapt to climate change.

Increase modal flexibility

Increase the Flexibility to Shift Between Different Transportation Modes

Recommendation: *Develop an adaptation plan that allows fleets to be proactive and flexible in adjusting the modes of transportation that make up the fleet.*

CHALLENGE

Once climate change impacts disrupt a transportation route necessary to meeting an organization's mission, the fleet manager will likely enter a crisis mode to find alternatives.

In many regions, surrounding transportation infrastructure may also be affected by climate change impacts. Some of these impacts, such as those from severe precipitation and storm surges, may lead to periodic temporary closing of roads, rail lines, or airports. Other impacts, such as sea level rise and thawing of permafrost, may permanently close infrastructure.

Fleet managers will need to assess where these issues may occur and develop contingency plans. If surrounding infrastructure becomes less reliable, fleet managers will have to incorporate flexibility in providing transportation services to shift between different transportation modes.

SOLUTION

Adaptation Potential	★★
Operational Impacts	★★
Financial Impacts	★★
Feasibility and Timing	★★

Fleet managers must proactively develop contingency plans for transportation routes vulnerable to climate change impacts. This may mean that they expand their responsibilities from the traditional provision of vehicles to acting as a transportation manager with the flexibility to adopt new transportation modes on the fly.

For impacts that create short-term issues, the fleet manager should develop short-term contingency plans. These may include alternative routing if available, a shift to an alternative travel mode, or finding a substitute for travel altogether.

For example, we expect increased frequency and severity of storms, such as the tornadoes in the Southeast and Midwest in spring 2011, which may limit passage on some roads and bridges. Fleets located in areas vulnerable to tornadoes, such as Oklahoma, should have established plans to find alternative routes in case primary routes are blocked. They may also plan for the availability of off-road-capable vehicles or even web-conferencing facilities so that they can continue to meet mission needs.

Other impacts may permanently affect transportation infrastructure. Although the short-term impacts on the fleet may be the same, the solution will need to be more permanent, cost-effective, and manageable. Fleet managers will have to shift planning from crisis management to a more thorough analysis of the lowest impact along with the most cost-effective alternative transportation modes.

EXAMPLE

Studies of climate change scenarios in Western Europe indicate incidences of low water levels in the Rhine River will increase. Researchers analyzed freight prices for inland shipping on the Rhine, finding an inverse relationship: as water levels decreased, freight prices increased. A very dry summer on the Rhine in 2003 led to an estimated €91 million in losses, much higher than the average "welfare loss" of €28 million in 1986–2004, which was attributed to low water levels in the river.[53] If the climate change scenarios are correct, researchers argue that the increased occurrence of low water levels may make inland shipping less attractive relative to road or rail transport; the result would be a shift from water transport to rail and road transport.[54]

Multiple disruptions in transportation planning occur when gasoline supplies are compromised by climate change–driven disasters.

Final Thoughts

Climate change is real and will likely impact how our government and nation operate, certainly from a vehicle fleet standpoint. Climate changes will affect fleet managers in two ways. First, they will have to reduce fleet GHG emissions. Second, they will need to adapt to potential risks to their operations from climate change impacts.

To reduce GHG emissions, fleets can follow several principles, each having its own sweet spot, depending on fleet characteristics, mission, and needs:

- *Conservation* completely eliminates GHG emissions for each VMT reduced. However, a fleet manager can only chop so much meat off transport demand before reaching bone.
- *Improving fleet efficiency* is likely the most effective GHG-reduction approach in the fleet manager's toolbox. Like conservation, every drop of fuel saved reduces GHG emissions 100 percent. Often, opportunities for savings are substantial. Fleets can improve efficiency by acquiring more fuel-efficient vehicles, including right-sizing vehicles to mission, and adopting operating behaviors such as less vehicle idling,

more efficient routing, driving during non-congested periods, reducing weight carried by vehicles, and regular tune-ups.

- *Replacing petroleum use with lower-carbon alternative fuels* can be done almost fleetwide. However, for first-generation alternative fuels, such as corn-based E85 and B20, the GHG reductions are limited to roughly 20 percent. Better opportunities will emerge with second-generation fuels, such as cellulosic E85, as well as EVs, which can provide almost a complete reduction in GHG emissions.
- *Shifting to lower-carbon transportation modes* such as mass transportation offers a niche solution in metropolitan areas and for specific trips.
- Fleets can also help reduce GHG emissions from the electricity generation sector by *integrating EVs in smart-grid applications*. By using their excess storage capacity, EVs can help spread out zero-emission renewable electricity generation to meet grid demand.

The impacts of climate change on temperatures, precipitation, severe storms, and sea levels will also have consequences for fleet operations. Fleet infrastructure, such as depots and fueling centers, may be vulnerable to these impacts. We can also expect reduced reliability and availability of some transportation infrastructure.

Many of these climate change impacts seem far away, but recent trends suggest that some are growing faster than projected. Proactive fleet adaptation planning will be far more effective and cost-efficient than responding to climate change impacts as they happen.

One final point is the notion of the culture shift. More than any other area, vehicles are where the average American can take the most active role in contending with climate change.

For the momentum to swing the nation toward such a culture, where a new vehicle is admired for what it provides rather than what it costs, the fleets will need to lead the public consciousness by example.

---

[1] Department of Transportation Bureau of Transportation Statistics, 2008 figures.

[2] National Research Council, Transportation Research Board, Board On Energy and Environmental Systems, *Effectiveness and Impact of Corporate Average Fuel Economy (CAFE) Standards* (Washington, DC: National Academies Press, 2002).

[3] Clean Air Act, 42 USC. 7545(o)(1).
[4] US Environmental Protection Agency (EPA), *Greenhouse Gas Emissions from the US Transportation Sector 1990–2003*, EPA 420 R 06 003, March 2006.
[5] US Department of Transportation (DOT), Federal Highway Administration, Office of Highway Policy Information, *Travel Monitoring and Traffic Volume,* http://www.fhwa.dot.gov/policyinformation/travelmonitoring.cfm.
[6] US Department of Energy (DOE), Federal Energy Management Program (FEMP), *Guidance for Federal Agencies on EO 13514 Section 12, Federal Fleet Management*, DOE/GO-102010-2960, April 2010.
[7] US DOE, FEMP, *EO 13514 Comprehensive Federal Fleet Management Handbook*, DOE/GO-102011-3353, July 2011.
[8] National Highway Traffic Safety Administration (NHTSA), *DOT, EPA Set Aggressive National Standards for Fuel Economy and First Ever Greenhouse Gas Emission Levels for Passenger Cars and Light Trucks,* DOT 56-10, April 1, 2010.
[9] US DOE, FEMP, *Guidance for Federal Agencies on EO 13514 Section 12, Federal Fleet Management.*
[10] David Greene and Steven Plotkin, *Reducing Greenhouse Gas Emissions from US Transportation* (Arlington, VA: Pew Center on Global Climate Change, January 2011).
[11] Joel Lovell, "Left-Hand-Turn Elimination," *New York Times Magazine*, December 9, 2007, http://www.nytimes.com/2007/12/09/magazine/09left-handturn.html.
[12] US Postal Service (USPS), *Sustainability Report 2010.*
[13] US DOE, *Estimated New US EPA miles per gallon (MPG) for 2011 Ford Fusion (4 cyl, 2.5 L) and 2011 Ford Fusion Hybrid (4 cyl, 2.5 L)*, http://www.fueleconomy.gov.
[14] Azure Dynamics, *balance™ hybrid electric vehicle: Specifications and Ordering Guide*, SPC500985-B, March 2011.
[15] US DOE, *Estimated US EPA MPG for 2011 Ford Fusion FFV (4 cyl, 3.0 L), 2006 Ford Fusion (6 cyl, 3.0 L) and 2011 Ford Fusion FWD (4 cyl, 2.5 L)*, http://www.fueleconomy.gov.
[16] Linda Gaines, Anant Vyas, and John Anderson, Argonne National Laboratory, Center for Transportation Research, *Estimation of Fuel Use by Idling Commercial Trucks*, Paper No. 06-2567, January 2006.
[17] US DOE, Clean Cities Program, *Clean Cities Alternative Fuel Price Report,* July 2011.
[18] US DOT, *Transportation's Role in Reducing US Greenhouse Gas Emissions, Volume 1: Synthesis Report*, April 2010.
[19] National Renewable Energy Laboratory (NREL), *FleetAtlas*, http://maps.nrel.gov/fleetatlas.
[20] NREL, *TransAtlas*, http://maps.nrel.gov/transatlas.
[21] US DOE, *Alternative Fuels & Advanced Vehicles Data Center (AFDC)*, http://www.afdc.energy.gov/afdc/locator/stations/.

[22] US DOE, FEMP, *Executive Order 13514, Federal Leadership in Environmental, Energy, and Economic Performance: Comprehensive Federal Fleet Management Handbook*, July 2011.
[23] DoD, Under Secretary of Defense (Acquisition, Technology and Logistics), *DoD Report to Congress of Use of Biofuels and Measures for Increasing Such Use in FY07–12*, November 2007.
[24] Julian Bentley and others, *Opportunities to Increase DoD's Biofuel Use in the Non-Tactical Vehicle Fleet: Further Evaluation and Implementation of the DoD Biofuels Congressional Study* (McLean, VA: LMI, October 2010).
[25] Virginia, *Silver Line (Washington Metro)*, http://www.virginia.dog.jaworzno.pl/p-Silver_Line_(Washington_Metro).
[26] Transportation for America, "Smarter transportation case study #8: Bus Rapid Transit Priority in Salt Lake City, Utah," October 18, 2010.
[27] ScienceDaily, ed., "Car Prototype Generates Electricity, and Cash," *ScienceDaily.com*, December 9, 2007.
[28] Jasna Tomic and Willett Kempton, "Using fleets of electric-drive vehicles for grid support," *Journal of Power Sources*, Vol. 168, March 2007, pp. 459–468.
[29] Kami Buchholz, "Universities using converted electric Scion xB for V2G research," *Automotive Engineering International Online*, August 10, 2011.
[30] The Climate Impacts Group and others, *Preparing for Climate Change: A Guidebook for Local, Regional, and State Governments* (King County, WA: Climate Impacts Group, King County, Washington, and ICLEI—Local Governments for Sustainability, September 2007).
[31] National Research Council, Transportation Research Board, *Potential Impacts of Climate Change on US Transportation* (Washington, DC: National Academies Press, 2008).
[32] The White House, *Executive Order (EO) 13514, Federal Leadership in Environmental, Energy, and Economic Performance*, Federal Registrar Vol. 74, No. 194 (8 October 2009), pp. 52115–52127.
[33] The White House Council on Environmental Quality, *Progress Report of the Interagency Climate Change Adaptation Task Force: Recommended Actions in Support of a National Climate Change Adaptation Strategy*, October 5, 2010.
[34] Pew Center on Global Climate Change, *State Adaptation Plans*, accessed September 13, 2011, http://www.pewclimate.org/what_s_being_done/in_the_states/adaptation_map.cfm.
[35] Climate, Energy and Commerce Advisory Committee (CECAC), *South Carolina Climate, Energy and Commerce Advisory Committee Final Report*, July 2008.
[36] Ibid.
[37] Ibid.
[38] National Research Council, Transportation Research Board, *Potential Impacts of Climate Change on US Transportation*, Special Report 290, 2008.

---

[39] National Research Council, Transportation Research Board, *Adapting Transportation to the Impacts of Climate Change: State of the Practice 2011*, Transportation Research Circular Number E-C152: (Washington, DC: National Academies Press, June 2011), p. 32.
[40] Ibid.
[41] Heather Lowe, "Preparing for Climate Change," (presentation, Climate Change Adaptation Strategies Workshop, Denver, CO, November 17, 2010).
[42] Pew Center on Global Climate Change, *Climate Change 101: Understanding and Responding to Global Climate Change*, January 2011.
[43] Martin Parry and others, *Climate Change 2007: Impacts, Adaptation and Vulnerability: Contribution of Working Group II to the Fourth Assessment Report of the Intergovernmental Panel on Climate Change* (Cambridge: Cambridge University Press, 2007).
[44] Peter Larsen and others, "Estimating future costs for Alaska public infrastructure at risk from climate change," *Global Environmental Change*, Vol. 18, Issue 3 (Elsevier, March 31, 2008), p. 442.
[45] Oregon Department of Land Conservation and Development (ODLCD), "The State of Oregon's Climate Change Adaptation Framework" *Oregon.gov* November 30, 2010, http://www.oregon.gov/LCD/news_2010.shtml.
[46] Ibid., p. 74.
[47] Ibid., pp. 75–76.
[48] Ibid., p. 79.
[49] Ibid., pp. 16–17.
[50] Ibid., Appendix 1b, p. 1/10.
[51] Ibid., Appendix 1c, p. 4/12.
[52] Ibid., Appendix 1c, p. 6/12.
[53] Mark Koetse and Piet Rietveld, "The Impact of Climate Change and Weather on Transport: An Overview of Empirical Findings," *Transportation Research Part D: Transport and Environment*, Vol. 14, Issue 3 (Elsevier, 2009), p. 17.
[54] Ibid.

Supply

The sunny morning is lovely, but the file on her desk casts a shadow. For the third time this month, the organization's suppliers are having trouble getting critical raw materials to the factories. It seems incomprehensible, but since storm surge increases along the East Coast forced the rerouting of Interstate 95, the transit of raw materials to processing facilities has become complicated and time-consuming. Something as simple as shipping from point A to point B now includes points C and D, and higher costs to boot.

Getting the Goods

We don't see it, and rarely do we realize it's there, but when it comes to goods and services, the supply chain is like the air we breathe. Look at any product; think about where it comes from. Think about where the pieces that make up that product come from. Think about the raw materials that make those pieces. Think of the machines that produce the raw material. That's the supply chain, and that's only a few steps backward through its links.

The US government recently announced a push to manage greenhouse gas (GHG) emissions in the federal supply chain. This isn't uncharted territory—it builds on the success of companies that have already done so—Walmart, Procter and Gamble, Timberland, and others. These organizations recognize the important relationship between their supply chain operations and a changing climate. The GHG emissions generated in the supply chain contribute to climate change and, thereby, profoundly affect operations across the

supply chain. The federal government's efforts will be led by the General Services Administration (GSA), which has issued a report outlining how the federal government can start collecting and using emissions information.

In recent years, organizations have paid more and more attention to the management of GHG emissions generated in the supply chain. The reason is simple: an item's production, storage, distribution, and service have become a multi-company process, in which third-party firms are often involved in managing discrete activities. These activities all have associated GHG emissions. Fuel use in transportation, energy use in storage, and manufacturing process emissions are all subject to the effects of a changing climate.

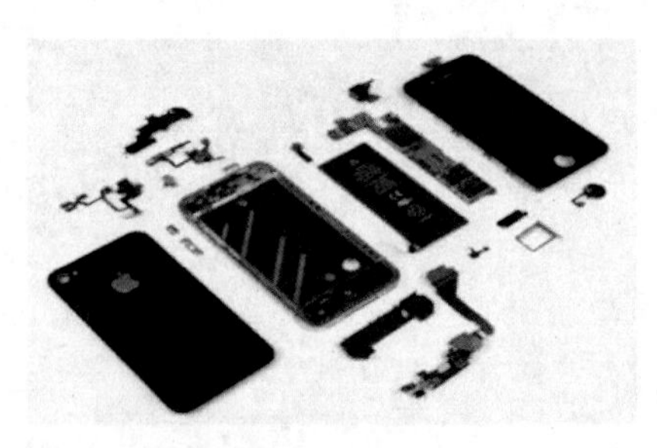

The supply chain for an Apple iPhone includes manufacturers that make the screen, case, battery, electronic memory, processor chips, and other components.

Manufacturing firms, for example, tend to specialize in certain areas, meaning multiple organizations make the individual components that are assembled into a final product. Apple's iPhone is the result of a battalion of suppliers marching in step to provide the pieces to a technological puzzle. (Apple's legendary secrecy is such that tracking the supply chain is the only way to get a peek at the company's upcoming products.)

These components all go to another company, which assembles them into an iPhone and sends them to distribution centers. This company has its own supply chain to consider. In fact, Apple employees never touch most iPhones.[1] So, as you can see, the GHG footprint of an iPhone has little to do with Apple (beyond product design) and much to do with activities in the supply chain. To reduce the footprint of this product, one would need to analyze the emissions of the supply chain.

For manufacturers (whether makers of consumer goods or high-tech providers), between 40 and 60 percent of a company's carbon footprint resides upstream in its supply chain—from raw materials, transport, and packaging to the energy consumed in manufacturing processes. Upstream represents every step before a product lands in the hands of a consumer. For retailers, that can be 80 percent.[2]

Other industries, including electronics manufacturers, see significant emissions from the use of their products downstream in the supply chain.

The point of all this is that significant efforts to reduce carbon in the supply chain require collaboration with supply chain partners. The same principles hold true for government vendors.

Supply chains are essential to the operation of almost any government agency. A supply chain provides the materials needed for operations; it facilitates the production and delivery of finished products to the end customer. History has shown that companies that are innovative in their supply chain design can take a leadership role in their industry. Examples range from Dell and HP to Walmart and Procter and Gamble. A former Nortel CIO summed it up nicely: "Companies don't compete; supply chains compete."[3] Government has the potential to operate in the same way.

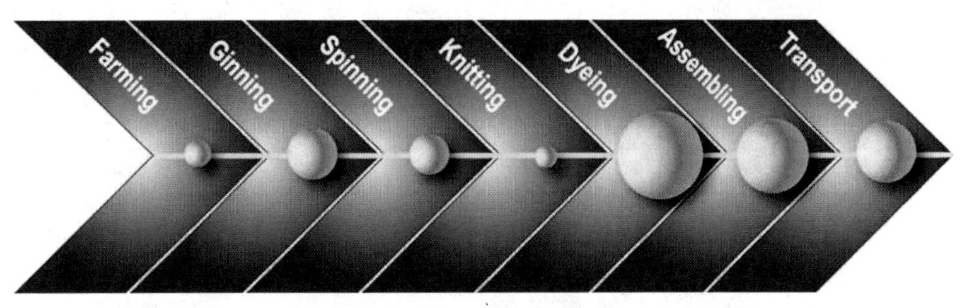

Proportional carbon emissions for a textile supply chain in Asia.

For public-sector managers, the supply chain is a critical lifeline to the ability to carry out the agency mission. Without a reliable source of materials and services, most operations would come to a standstill. Likewise, many agencies rely on the capability to deliver materials to a point of use to achieve their mission. The Department of Defense, for example, operates one of the largest supply chain infrastructures in the world, complete with maintenance centers, transportation assets, and numerous distribution centers.

But what is the best way to address climate change via the supply chain?

Clearly, mitigation and adaptation are the keys to any strategy. In a supply chain, they can help minimize impacts on the climate while securing business operations against disruptive changes. In other words, mitigation and adaptation create resilience—and that's good business.

Supply chain managers can mitigate GHG emissions by working with partners to reduce their own emissions. The alternative is to select partners with low GHG emission profiles in their operations or product sectors. Such mitigating actions are essential for reducing the overall climate change impact of the products an organization purchases and delivers.

Nevertheless, some climate change impacts are already occurring or appear unavoidable; risk is born from these.[4] They include higher temperatures, rising sea level, and changing weather patterns. Integrating adaptation strategies will help secure the supply chain against risks.

Mitigation and adaptation actions rely on a solid understanding of the supply chain network and a reliable capability for measuring GHG emissions.

The extended and global nature of supply chains means that roles are blurred. Who is the supplier when it can also serve as outsource manufacturer, logistics provider, distributor, or even customer? With that in mind, the term "partner" is used to describe external organizations that participate in the supply chain.

Mitigation

Studies show that companies that take GHG mitigation actions also tend to be business innovators.[5] Supply chain mitigation can lead to lower operating costs, often from energy-related savings, and higher quality across the supply chain.[6] Therefore, working with suppliers and supply chain partners to reduce GHG emissions is not just an environmentally preferable way to operate—it can genuinely improve the performance and quality of supply chain operations.

A GHG mitigation strategy for the supply chain can help manage three things:

1. Cost
2. Risk and operational disruptions
3. Reputation and brand.[7]

This strategy can also lead to a competitive advantage, for example, by creating a more viable cost structure or building a stronger brand. Often, leading organizations look for ways to use GHG mitigation actions to build business opportunities, not simply lower GHG emissions.

That's why mitigation in the supply chain is an increasingly common practice. The Carbon Disclosure Project (CDP), a United Kingdom–based organization that partners with shareholders and corporations to disclose the GHG emissions of major corporations, worked with 57 companies to survey their suppliers' carbon management capabilities. Of these, 45 percent, representing a cross section of industry, are currently reporting some aspect of their supply chain emissions; 86

percent are working with their suppliers to improve carbon management.

Interestingly, 56 percent said that within 5 years they will actually start deselecting suppliers that do not meet their carbon management requirements.

The survey claims an impressive sample size. Together, the companies surveyed have more than 1,850 suppliers around the world. Of these, 32 percent have detailed GHG emissions-reduction targets toward which they are working, and 69 percent have a board committee or executive responsible for climate change strategy.[8]

The basis for successful supply chain mitigation is a structured, continuous process. Mitigation generally involves measuring GHG emissions in the supply chain, including domestic and international operations, and taking actions to reduce them. Figure 7-1 illustrates a general process flow for this activity.

Figure 7-1. The Supply Chain Mitigation Process

A supply chain cannot be effectively mitigated through a single short-term project. Mitigation requires a continuous process that has the capability to drive positive change in the supply chain and the flexibility to adapt to changing supply chain operations.

Organizations need to know their starting point. The GHG emissions in the supply chain cannot be effectively reduced without first knowing where to start, meaning a baseline of the current emissions in the supply chain is critical. One can be established using a variety of methods, including screening (described later in this chapter).

When collecting emissions data from supply chain partners, stakeholders will need a method for allocating emissions from a corporate level to the specific emissions attributable to their supply chain. Allocation approaches are detailed in emissions accounting standards.

With the baseline as the foundation for the supply chain mitigation program, you can begin to understand where to take reduction actions and how much action needs to be taken.

The baseline tells you where you are, but an effective mitigation program also needs a target that will define where you want to go. This is a way to link the supply chain program to organizational program goals so if the organization goal is to reduce emissions, the supply chain target can reflect a contribution toward that goal.

A target can also double as an effective communications tool. It tells everyone, internal and external, the goals of the reduction program. It documents how aggressive your organization is in reducing supply chain emissions. Targets are also flexible. An organization can tailor them to portions of the supply chain with higher emissions, or address where it is most likely to reduce emissions. These may not be the same portions of your supply chain.

With a baseline set, and targets established, you can start taking action to mitigate, or reduce, GHG emissions. The best mitigation actions fit in the overall GHG management strategy and are appropriate for the capabilities of your supply chain partners. Mitigation actions can even include engaging these partners to build a better GHG management capability.

Action must be followed by monitoring. Supply chain emissions must be regularly measured to track progress toward the established goals. On the other hand, the complexity of building a GHG inventory at the corporate or product level means that measurement can only be done annually in most cases; more frequent measurements are simply too difficult. The *Greenhouse Gas Protocol: A Corporate Accounting and Reporting Standard* affirms annual measurement as realistic and preferred.[9]

Effective mitigation relies on a structured approach. We can all learn from companies that have succeeded in mitigating emissions. From their examples, we offer five recommendations for successful supply chain mitigation:

- Set a mitigation strategy with specific, measurable objectives.
- Screen and focus.
- Leverage GHG reporting standards.
- Tailor reduction actions to strategy and partner capability.
- Continuously mitigate.

Set a Mitigation Strategy with Specific, Measurable Objectives

Recommendation: *Set a mitigation strategy that establishes objectives and aligns supply chain actions with organizational GHG mitigation objectives.*

CHALLENGE

Every ship needs a rudder. Without a strategy, an organization risks inconsistencies and conflicts, which lead to lower confidence in partners, higher program costs, and overall poor results.

Managing GHG emissions in the supply chain involves multiple decisions, from the standards for measurement to how to engage partners in achieving reductions. Failing to engage partners in a way that makes them comfortable can make them less likely to participate in a mitigation program. Decisions made outside an overarching plan result in inconsistent or conflicting actions in your mitigation program.

SOLUTION

Mitigation Potential	★★
Operational Impacts	★★★
Financial Impacts	★★
Feasibility and Timing	★★★

An effective supply chain GHG emissions mitigation program needs a guiding strategy, which should be a dynamic document you regularly modify with new information and business priorities.

Before embarking on a mitigation program, you need to understand the context of GHG management in your supply chain and, from that, the right approach for mitigation. This starts with an understanding of the overarching organizational GHG-management strategy. A well-defined strategy must cover two elements: mitigation objectives and supply chain scope.

Once mitigation objectives are set, the next step is to define the supply chain or supply chains where you will mitigate GHG emissions. Defining the supply chain involves identifying its scope and products, along with the geographies to be included. This decision should flow from the mitigation strategy.

A supply chain mitigation analysis generally can have two basic scopes: cradle to grave and cradle to gate. A cradle-to-grave approach looks at both portions of the supply chain—the upstream, up to raw material extraction, and downstream, all the way through a product's use and disposal (Figure 7-2).

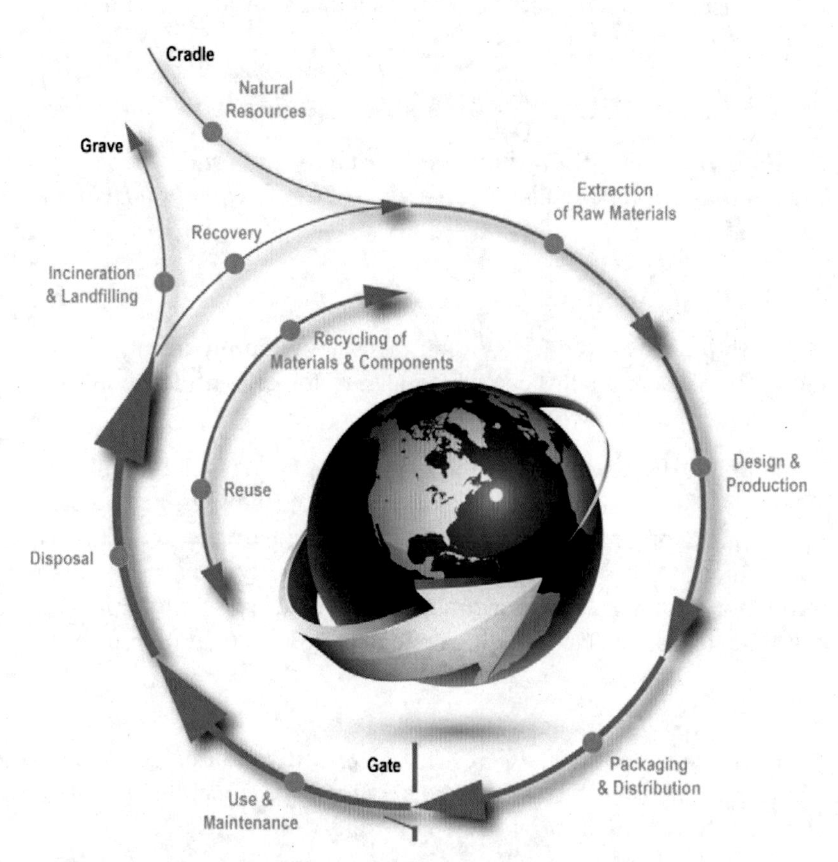

Figure 7-2. Cradle-to-Grave Elements of a Product Life Cycle

Cradle to gate captures the supply chain from raw material to a specific point in the supply chain; typically, this point is when the product is leaving your organization's property.

In reality, the scope of a mitigation program should fit with your ability to influence emissions and the scope needed to achieve your objectives. For some organizations, this can mean including only Tier 1 or even a select subset of Tier 1 suppliers in the mitigation program (Figure 7-3).

Figure 7-3. Two Tiers of a Simple Supply Chain

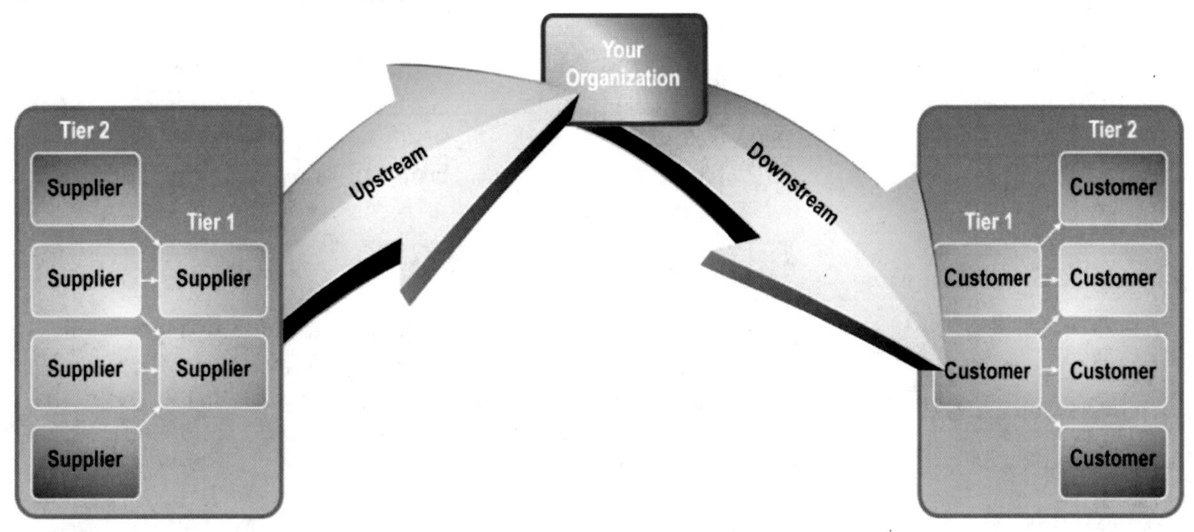

Organizations often use a cradle-to-gate approach if they are only interested in the emissions impacts of the materials they buy. In most cases, this means that they are taking a risk management or cost-efficiency approach to GHG mitigation. For brand management purposes, an organization may want to capture the full life cycle down to the customer. This allows it to verify it is measuring and managing all of its associated GHG emissions.

To decide how far upstream or downstream to manage GHG emissions, consider the practicality of capturing emissions data from partners or customers with which you do not have a direct relationship. In most cases, organizations directly conduct business with their Tier 1 suppliers and customers (Figure 7-3). However, the most GHG-intense activities in a supply chain can take place several tiers away from your organization, even at the point of your product's use or disposal. Despite this, taking mitigation actions where you don't have direct control is still possible. This can be accomplished through coordination with Tier 1 suppliers, education and outreach programs, and other methods.

EXAMPLE

The US federal government is embarking on a program to manage GHG emissions in its supply chain. GSA leads this effort, which is based on the sustainability and climate change strategy outlined for the entire government in President Obama's Executive Order 13514.[10] The president's plan clearly articulates an overall strategy of positioning the federal government as a leader in GHG management.

Furthermore, it is intended to serve as an example to the rest of the US economy.

To support this leadership strategy, GSA developed a suggested approach to managing government supply chain emissions. GSA has focused on working with industry to measure and reduce Tier 1 supplier GHG emissions. In its feasibility study, GSA includes defined outreach and training efforts for industry to improve capabilities for accurately measuring GHG emissions. GSA also identifies tools for making reductions.

Eventually, GSA hopes the program will be able to support full cradle-to-gate emissions management.[11] It expects suppliers and the government to see cost savings from the program that would go far in demonstrating leadership through mitigation.

Screen and Focus

Recommendation: *Supply chain GHG mitigation should start by screening an organization's supply chain to uncover GHG-intense processes, allowing mitigation efforts to focus on them.*

CHALLENGE

Including Tier 2 suppliers and beyond can significantly increase the number of organizations involved. Engaging all of these organizations in meaningful GHG mitigation actions is simply not practical in most cases.

Supply chains often include hundreds, sometimes thousands, of direct (Tier 1) suppliers and partners. Hewlett-Packard (HP), for example, has identified about 400 contracted manufacturing suppliers, 80 percent of them outside of the Americas.[12] Similarly, Ford Motor Company has more than 1,400 production-related suppliers in more than 4,400 locations in 60 countries. This is in addition to more than 9,000 non-production-related suppliers.[13] When companies start mitigating supply chain GHG emissions, they often struggle in scoping the program.[14]

SOLUTION

Mitigation Potential	★★
Operational Impacts	★★
Financial Impacts	★★
Feasibility and Timing	★★★

Reduction is often a question of quality or quantity. Say that an organization has 100 suppliers. It can go the route of quantity in reducing emissions from the source—for example, seeking and achieving reductions in 75 of those partners, resulting in a robust, impressive number to tout.

The reality, however, is that the 25 suppliers that you didn't get reductions from actually make up 98 percent of your supply chain's entire GHG emissions output. In this case, the search for quality actually would lead you to targeting fewer partners—but achieving greater gains in the process.

Effectively mitigating GHG emissions in the supply chain requires the wise use of resources—people, tools, and funds. Organizations or processes in the supply chain that emit few GHGs do not offer significant mitigation opportunities. The same can be said of those already subject to aggressive GHG reduction programs. Your resources must be applied to areas that will reduce GHG emissions.

The phased approach to GHG mitigation can help. The most common strategy is to start with a general measurement approach that allows you to gain an understanding of the GHG emissions in the supply chain. You can then target future activities based on the initial profile.

Focusing your actions means understanding where the GHG-intense activities are in your supply chain. Starting your mitigation activities with a GHG emissions screening gives you the opportunity to estimate the emissions throughout your supply chain. Granted, screening methods are not completely accurate, but they are still good for identifying where your GHG-intense activities are.

One GHG screening method is the use of an economic input-output life-cycle analysis, such as the Carnegie Mellon Green Design Institute's Economic Input-Output Life Cycle Assessment (EIO-LCA).[15] This approach combines US economic data with industry-specific GHG emissions; doing so allows you to estimate the GHG emissions associated with the financial output of an industry.

Because the EIO-LCA model links all of the industry connections across the US economy, the result is a very comprehensive estimate of the cradle-to-gate life-cycle emissions for a commodity. Where this falls short is in speaking to your particular supply chain. Specifically, these macro-level data lack details on your supply chain and your operations, including real-time data that will allow you to understand your supply chain's impacts in current terms.

This lack of specificity means that EIO-LCA is a good tool for screening supply chain emissions, but not for tracking reductions. It also doesn't provide direction as to where to focus reduction efforts. In addition, EIO-LCA data are collected on a national level, and national economic tables tend to vary in quality, detail, and timeliness.

No EIO-LCA data set is globally integrated, which does not mesh well with the global nature of today's supply chains.

All things considered, however, for most screening uses, the US data are a reasonable estimate for similar processes in other countries—so long as the technology or power sources do not significantly differ. If you are using the EIO-LCA data for non-US operations, you should take care to ensure that the data are a reasonable approximation for your target operation.

A second approach is to identify potential GHG emissions-intense activities in your supply chain through reference documents. A number of resources are available; they use a range of methods for identifying GHG emissions or energy-intense industries and processes.[16] In most cases, energy intensity is a reasonable stand-in for GHG emissions since most energy derives from fossil fuel, especially in the United States.

These tools can help identify processes in your supply chain that could be more GHG intensive, areas that may warrant mitigation action. However, there's a reason for the variety of tools: this isn't a one-size-fits-all process. This is especially true if you are already using partners that are environmentally and climate change conscious.

Once you have screened the supply chain and understand the point at which emissions arise, you can focus your resources on the most egregious offenders. The more GHG intensive an activity is, the more likely you are to find opportunities for emissions reduction. The exact portion of your supply chain on which you choose to focus is in large part governed by the mitigation resources you have available. A screen-and-focus approach helps ensure that these resources are effectively used.

EXAMPLE

Patagonia, a maker of outdoor equipment, wanted to understand GHG emissions and other sustainability drivers in its product supply chains. Rather than analyze the full life cycle of its entire product catalog, Patagonia screened its supply chains through a cradle-to-gate analysis of representative products. Today, by choosing one backpack, dress, jacket, etc., Patagonia can get a good idea of what drives GHG emissions in similar products. It now has a better perspective on the best actions for reducing the footprint of its products.[17]

Patagonia has taken the process a step further and is now using this information to influence future product design. It shares the details of its assessments with customers through a program called the "Footprint Chronicles."[18] As part of this program, it offers candid opinions on the sustainability of its products (Figure 7-4).

Figure 7-4. Patagonia's "Footprint Chronicles" for Its Nano Puff Pullover

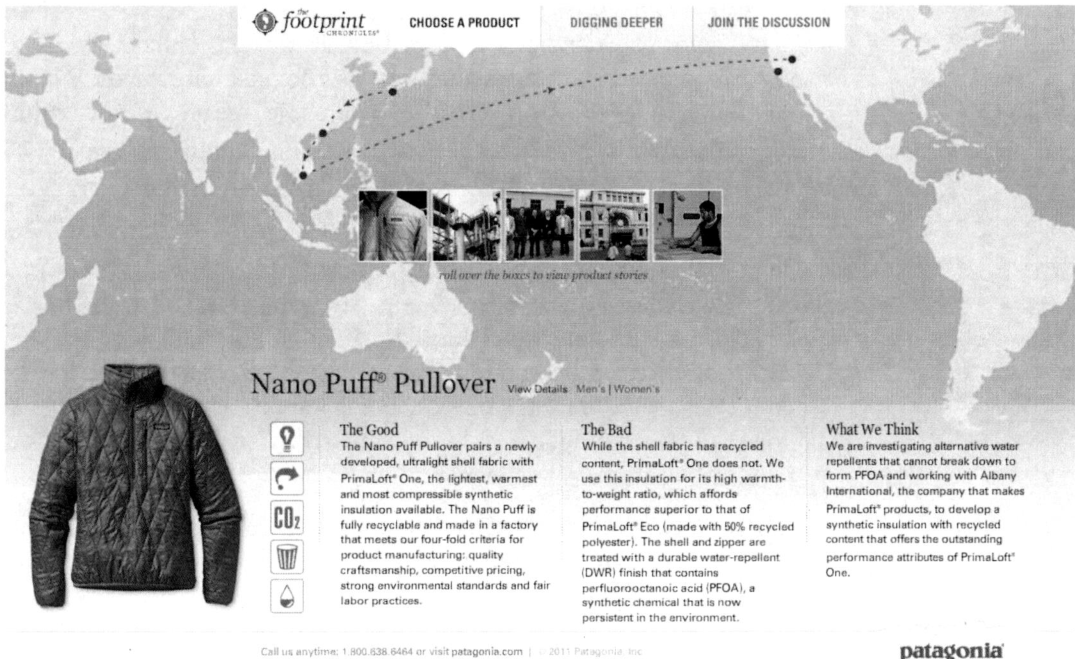

Leverage GHG Reporting Standards

Recommendation: *Eliminate significant guesswork by using reporting standards and rules for measuring GHG emissions that allow data users to interpret any reported GHG emissions information with consistency.*

CHALLENGE

Measuring GHG emissions can be a complex effort with many variables, and the reporting of these emissions is equally involved, with many decisions to be made. Among the issues to consider are the processes to include factors or sources to use in calculating emissions, and assumptions to make. The numerous decisions create a problem when it comes time for an organization to report its emissions.

The process is equally taxing on the report readers. Without a clear understanding of the assumptions and decisions being made, knowing how to interpret the report—or even gaining a perspective when comparing emissions inventories—is difficult. You can't take full advantage of a GHG emissions report unless you understand how the emissions were calculated.

SOLUTION

Mitigation Potential	★★
Operational Impacts	★★★
Financial Impacts	★★
Feasibility and Timing	★★★

GHG reporting standards resolve this issue. They allow for more consistent and reliable emissions data as a foundation for your supply chain mitigation program. They are a valuable resource for determining how to collect GHG emissions data from supply chain partners and provide a defined method for calculating emissions. They give you and your partners an unambiguous path to reporting.

Having a standardized method also means that the data you collect from multiple partner organizations are somewhat comparable because the same foundation is used to collect all of the emissions data. Comparability can be improved by adding some prescriptive parameters to the use of the standard. The comparability is not just limited to emissions between partners but also to emissions reports over time, which are useful for tracking emissions-reduction progress.

Four widely accepted GHG accounting standards are considered suitable for corporate (corporate value chain/scope 3 supply chain) and product reporting (Table 7-1).

Table 7-1. Four Widely Used Standards Relevant to Supply Chain GHG Emissions

Standard	Sponsoring Organization	Description
Scope 3 standard	Greenhouse Gas Protocol	Corporate-level supply chain measurement with other (Scope 1 and 2) corporate reporting protocols
Product life-cycle standard	Greenhouse Gas Protocol	Total life-cycle GHG emissions reporting
ISO 14044	International Organization for Standardization	Total life-cycle environmental impact reporting, covering impacts beyond just GHG emissions
PAS 2050 standard	British Standards Institution and Carbon Trust	Product emissions reporting

The organizations that develop these standards work closely with one another to ensure they are as compatible as possible.

Standards can drastically ease the burden of GHG data reporting and management on an organization and its supply chain partners. Standards promote consistent reports and GHG emissions data that can support comparability with additional prescriptive guidance such as product category rules. At the same time, using widely accepted methods for reporting allows your partners to reuse data and eliminates the need for creating unique reports for your program.

The other step is verification. In designing a mitigation and GHG reporting program for your supply chain, the verification you require from your partners is a means of ensuring the data you receive are valid. Organizations that plan to use GHG emissions as a procurement evaluation factor, comparing two bidders on the basis of their reported emissions, can use third-party verification as a way to ensure the emissions reports are valid. However, because it adds a step to the reporting process and can add significant expense, it can be a burden for partners, especially smaller businesses.

One thing, however, is critical: the decision to use verification and its associated methods should fit with your overall mitigation approach.

EXAMPLE

Several industry associations that understand the value of standards have gone as far as developing industry-specific standards based on widely used protocols.

The Beverage Industry Environmental Roundtable, for example, developed an industry-specific guidance protocol for GHG emissions reporting based on the GHG Protocol standards and other standards.[19] Several representatives from the soft drink, brewing, and distilling industries took part in developing this sector guidance, which modifies the basic GHG Protocol Corporate Standard requirements. The result is a tailored system specific to beverage industry processes. Organizations use this approach internally and with partners to develop accurate GHG emissions reports consistent across the industry.

Another example is in the electronics industry, where stakeholders recognized the value of using standards to collect consistent information from their suppliers. A working group within the Electronics Industry Citizenship Council (EICC)—comprising HP, IBM, and others—led development of GHG emissions reporting standards that focus on a common measurement approach. These standards also promote the sharing of best practices. The EICC also developed a reporting protocol using a third-party or secure database to store emissions data. Using this process, HP started collecting GHG emissions data from suppliers in 2008 and, a year later, started requiring reduction goals.[20]

Acknowledge partner capability

Tailor Reduction Actions to Strategy and Partner Capability

Recommendation: *Focus on selecting mitigation actions that align with the organization's strategic objectives as well as partner capabilities for the greatest chance at success.*

CHALLENGE

An organization has numerous mitigation actions from which to choose when reducing GHG emissions in the supply chain.[21] Deciding among them is sometimes the most difficult decision it faces.

Actions for mitigating supply chain emissions fall into two primary classifications. Combined, they can provide the necessary tools for implementing mitigation actions in your supply chain.

The first is business actions, including procurement and preferential sales. These actions provide a direct business incentive, positive or negative, to a partner to reduce its GHG emissions.

The second is non-business actions, such as training and outreach. Non-business actions include outreach to supply chain partners to help them achieve GHG reductions. These include a wide range of things, such as introducing partners to free sources for GHG management practices or investing directly in partner GHG reduction initiatives. Doing so reaffirms a commitment to the effort, while strengthening existing partner relationships.

With this wide variety of options, organizations need to select the mitigation approach most effective in each particular supply chain. Consider, for example, an approach that is too aggressive or requires too much of your partners. This, from the very start, will likely cause resistance to the program, or worse, its complete failure. On the other hand, an approach that is not aggressive enough will result in difficulty achieving the programmatic goals.

Mitigation Potential	★★
Operational Impacts	★★★
Financial Impacts	★★
Feasibility and Timing	★★

SOLUTION

The right approach will fit into the strategic objectives of an organization's mitigation program and take into account the capabilities of partners to report and reduce GHG emissions.

The best course of action for reduction can only be determined after you have identified the organizations or processes in your supply chain to target for GHG emissions-reduction actions. The most effective reduction actions take into account the overall goals of your mitigation program as well as the capabilities of your partners.

In dealing with partners, procurement incentives are the most common business action used in GHG mitigation. You can use procurement in several different ways. For example, you can include GHG emissions as criteria for evaluation when selecting partners, pushing your business toward partners with lower GHG emissions.[22]

You can also use product labels (third-party product certifications) as a criterion for procurement, specifically those that list product GHG emissions. An example would be restricting purchases in a product category to only certified products. Because product certifications are numerous, and they vary in quality, an organization must be careful to choose only those that are reliable. One resource for understanding the various labeling programs is the Ecolabel Index,[23] which is the largest global directory of ecolabels; the Ecolabel Index currently tracks 428 ecolabels in 217 countries and 25 industry sectors.

The approach depends on the relationship with the partner and the mitigation strategy. Most commonly, training is offered on GHG management to partners in the form of guidebooks, online training, or in-person classes; it is then supplemented by a list of available resources to assist in GHG management.

Table 7-2 shows examples of some common GHG reduction actions; almost all reduction actions are a variation on these.

Table 7-2. Common Supply Chain GHG Reduction Actions

Mitigation Action	Description
Education	Provide educational resources to partners to help them understand GHG mitigation approaches.
Reporting	Request partners to report on GHG mitigation actions; reports can include mitigation plans as well as GHG emissions inventories.
Emissions targets	Require partners to achieve specific GHG emissions targets, often with incentives for meeting or exceeding the targets.
Investment	Invest resources with partners to help identify and implement GHG reduction technologies or processes.
GHG management teams	Send a team of GHG management experts to the partner facility to identify GHG reduction opportunities.
Supply chain optimization	Assess and redesign your supply chain to minimize GHG emissions.
Product redesign	Assess and redesign your products to leverage lower GHG materials and design factors.

The methods you use to reduce supply chain GHG emissions should flow from your mitigation strategy. The development of this strategy includes identifying how aggressively GHG emissions reductions are to be attempted along with the overall objective and the time frame for meeting that objective.

Again, sometimes it comes down to an organization's size and kind of clout it wields. Larger organizations with significant purchasing power in the supply chain have the ability to use stronger methods to ensure the capture of reductions. They can, for instance, require participation in a GHG mitigation program to keep business partners.[5]

On the other hand, smaller organizations that do not have significant purchasing power must use other means to capture reductions, as partners may not be amenable to strong demands from a small customer. Many companies end up working with a mix of smaller and larger partners, with an approach tailored to the relationships with specific partners.

Because mitigation actions are the means for reducing GHG emissions, the action you select must fit with the objective you are trying to achieve. Figure 7-5 illustrates the partner engagement and reduction potential of the various activities.

Figure 7-5. Greater Partner Engagement Can Lead to Greater GHG Emission Reduction

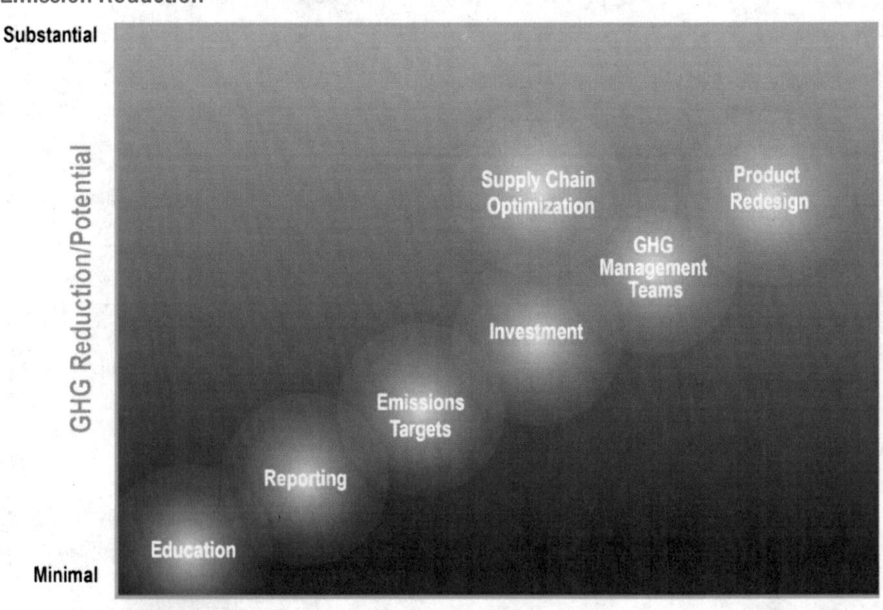

Whichever methods you select, you need to be able to link initiating that action and achieving the goal. Essentially, you should be able to logically connect the mitigation action to an impact in your supply chain (GHG emissions reductions, partners that are more knowledgeable, etc.) that will advance your organization toward its objective. Demonstrating this link reinforces the role of mitigation actions in your overall strategy and helps set intermediate performance targets based on these actions. It's also good management because stakeholders who understand why they're being asked to do something are more likely to buy in and remain engaged.

Of course, supply chain mitigation isn't done on an island; GHG emissions in the supply chain require partners to take emissions management and reduction actions as well. But, because companies do not all have the same capabilities when it comes to GHG management and mitigation expertise, you should consider the capabilities of your partners to support the actions.

A crucial first step is to work with your partners to understand their capabilities. Do they understand what GHG emissions are? What are the sources of GHG in their organizations? Have they completed a GHG inventory? Do they have GHG management plans? The more advanced a partner is, the more advanced the mitigation actions you can take with it.

You must be prepared to educate partners that do not report or manage GHG emissions on GHG management and the goals of your program. This can happen in a variety of ways, ranging from pointing them to GHG management resources from the Environmental Protection Agency, Department of Energy, and non-governmental organizations, to conducting on-site workshops. The type of training you give your partners depends on your program goals. You may want to roll out more training opportunities to help ensure the success of an aggressive reduction plan.

As partners build their capability, you can take mitigation actions that fit their current capabilities. This way, you can be mitigating the GHG emissions in your supply chain while GHG management capability is increasing.

EXAMPLE

HP and Ford are two companies known for effective supply chain GHG management and for demonstrating a clear link between strategy and actions.

HP focuses reduction actions in two areas. For most HP products, most of the GHG is emitted when consumers use the product; therefore, HP looks for product design alternatives to reduce energy consumption and make the product more recyclable. Redesigning products for better GHG emissions performance allows HP to realize substantial reductions in the overall GHG footprint in the supply chain. HP also directly engages suppliers in measuring and mitigating GHG emissions. The company finds that direct involvement with its suppliers results in better responses and understanding of barriers these suppliers need help in overcoming.[24] This then becomes HP's opportunity to provide this help.

Ford takes a different approach, using a targeted data-collection effort to understand its suppliers' capabilities for managing and reporting GHG emissions. In 2010, Ford engaged suppliers using two methods, the CDP supply chain questionnaire and the GHG protocol Scope 3 standard road test. Ford targeted 35 suppliers, which together represented close to a third of its $65 billion in annual procurement spending in 2009.[25] Ford found that half of these suppliers previously reported GHG emissions, illustrating a clear capability; however, their measurement and management strategies varied widely. Like HP, Ford found that direct supplier engagement is the best method for managing supply chain GHG emissions.[11]

Continuously Mitigate

Mitigate and learn

Recommendation: *Design your mitigation program to be continuous and perpetual.*

CHALLENGE

Mitigation should be a continuous journey that searches for new ways to lower supply chain GHG emissions. A common pitfall is to think of a mitigation program as a single round of reporting GHG emissions and identifying emissions-intensive activities in the supply chain. This can be especially true in an environment that favors annual budgets and quarterly financial reporting; in this setting, discrete, one-time programs are a natural fit. However, the goal of mitigation is to reduce the impact your supply chain has on climate change and build climate change resilience into your supply chain. This requires a perspective of 10 years or more.

SOLUTION

Mitigation Potential	★★★
Operational Impacts	★★
Financial Impacts	★★
Feasibility and Timing	★★★

Effective mitigation means continuously reducing GHG emissions in the supply chain by redesigning products and processes and leveraging emerging technologies.

An effective mitigation process is not a one-time activity, so it must allow for adjustment to new business conditions. GHG mitigation programs often set targets for reductions over the course of several years, similar to other environmental management system improvement actions, but much can change over that time frame, including products, partners, customers, technology, policy, regulation, and executive leadership. In addition, actual mitigation results may be better or worse than projected. All of these factors should result in adjustments to the mitigation program.

Adjustments to your targets should not be taken lightly, however. When people see frequent, significant changes to a program and its targets, they can lose faith. Frequent adjustments make the program appear unfocused. Target changes should go through a deliberate approval process with senior managers to ensure they are appropriate and remain in line with corporate mitigation goals.

The timeline for targets should also remain steady. If your target is for supply chain GHG emissions to reach a certain level in 3 years, then every year a new 3-year target should be set. This will keep people focused on continually improving performance.

With a continuous mitigation process, you can effectively manage your supply chain partner emissions and minimize the GHG emissions associated with your organization's supply chain. With a repeatable process, you and your partners will know what to expect from the program, and tracking progress will be simple and effective.

EXAMPLE

GSA has designed its supplier GHG management program as a continuous improvement process. The program is built on working with suppliers to understand GHG emissions in the federal supply chain and uses this information to guide future actions.

The implementation plan proposed by GSA covers two phases. The first is primarily a learning phase focused on understanding emissions in the supply chain and developing government capabilities for managing emissions. The second phase focuses on working with agencies and supply chain partners to accurately report emissions, set emissions targets, and encourage GHG management in the supply chain.

Both phases endorse annual GHG emissions reporting from supply chain partners and government agencies and encourage using each emissions report to identify reduction opportunities.[7]

Adaptation

As discussed throughout this book, current scientific consensus is that climate change impacts have begun, and the reasons why they're happening aren't as important as how they will affect us. Despite all our mitigation efforts, your supply chain operating environment will change. Higher sea levels and changing weather patterns, including more severe weather, will impact the ability of you and your supply chain partners to deliver products and services to your customers.[26]

The Association of British Insurers estimates that a 2°C increase in global temperature could increase financial property losses due to floods in Great Britain by 8 percent and property losses due to typhoons in China by 20 percent.

Adapting your supply chain to a changing climate is a matter of understanding the impacts climate change will have on your specific circumstances. With this understanding, you can take the appropriate actions to become more resilient, minimizing the impact on your supply chain and your customers. This becomes a question of managing risk in the supply chains as a means for approaching adaptation.

Supply chain risk management is an established business process for building a resilient supply chain secured against disruptions. Disruptions can include natural disasters, quality issues, financial failures, or other events. Figure 7-6 shows a recommended supply chain risk management approach. Modifying this risk management approach to focus on climate change–related risks can be an effective adaptation strategy.

The approach in Figure 7-6 is a systematic method for identifying climate change–related risks in your supply chain. By understanding the risks, you can prioritize them and take actions to address them. This can secure your supply chain against climate change impacts. The most important aspect of Figure 7-6, however, is that it emphasizes that this is a continuous process. This means that your adaptation approach will likely change with new information learned about climate change impacts, forcing the process to begin anew. Last, like all risk management programs, your climate change adaptation program should include a response element to define specific actions you will take in the event of unforeseen or unpredictable climate change, such as severe weather events.

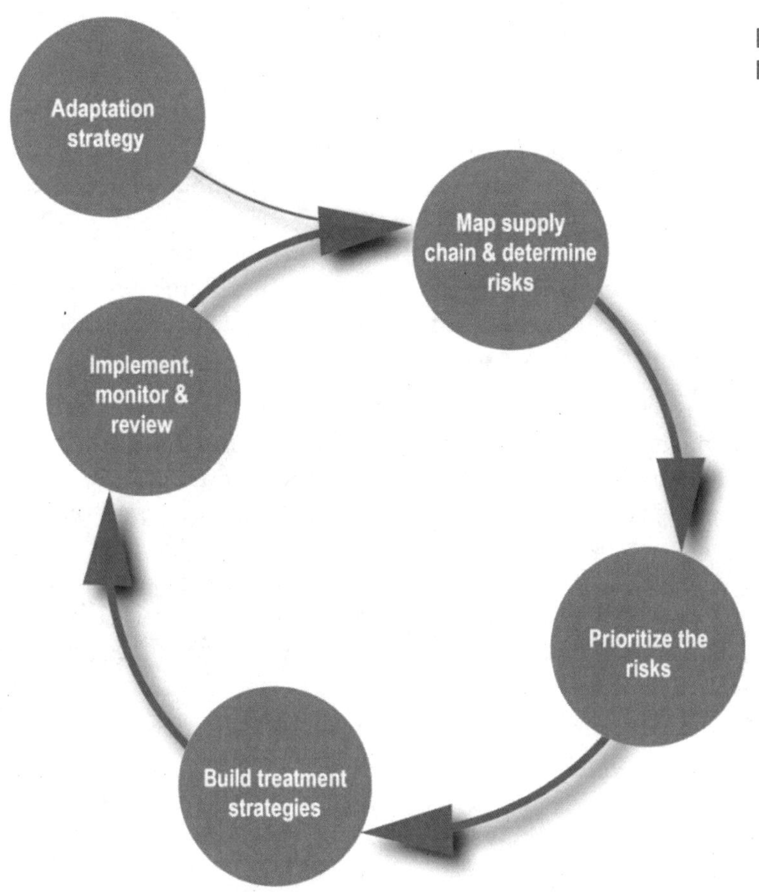

Figure 7-6. Supply Chain Risk Management Approach

Adaptation within the supply chain involves protecting your operations from events that are unexpected in either timing or impact. You can do the following to successfully adapt:

- Use a risk-based approach.
- Set an adaptation strategy.
- Identify and prioritize risks.
- Treat in accordance with your strategy.
- Continuously adapt.

Use a Risk-Based Approach

Recommendation: *Use a risk management approach that incorporates risk from regulation and climate.*

CHALLENGE

Climate change impacts are widely varied and uncertain. Adapting your supply chain to a changing climate ultimately means protecting your supply chain operations from the potential disruptions climate change can create.

As discussed in previous chapters, the wide range of risk sources identified by the Intergovernmental Panel on Climate Change includes changing weather patterns, increased drought, increased flooding, and many others. The range of uncertainty surrounding these predictions with regard to their timing and severity is just as wide. In fact, the highest risk to industrial operations is probably from extreme weather events rather than gradually rising temperatures, but we say probably because we just don't know.[27]

Managers would prefer not to deal with uncertainty when asked to make investment and strategic decisions, hence, the challenge. On the one hand, you don't want to waste your resources protecting your supply chain against something that never happens; on the other hand, you don't want your supply chain to be disrupted by an event for which you could have prepared. Supply chain adaptation requires an approach that helps you address the most significant risks your supply chain faces so that you can apply your resources effectively.

SOLUTION

Adaptation Potential	★★★
Operational Impacts	★★
Financial Impacts	★
Feasibility and Timing	★★★

Adapting your supply chain to climate change means managing your climate change–related risks (Figure 7-7). Supply chain risk management practices have been in place for many years and offer a good structure for this adaptation approach.

Using a risk management approach for supply chain adaptation begins with understanding the risks that climate change presents to your supply chain. A changing climate can disrupt normal supply chain operations in many ways. Each of these presents a risk to your operations. The good news is that many of these risks already exist, so you have an opportunity to plan for them while considering how a changing climate might (as predicted) increase their frequency or severity.[25]

The specific risks your supply chain faces are driven by the configuration of your supply chain. Climate change risk falls into two categories: regulatory and climate.

Regulatory risk is the risk of new regulations changing the cost structure of your supply chain, how you may manage your supply chain, or even materials you may use in your products.[28]

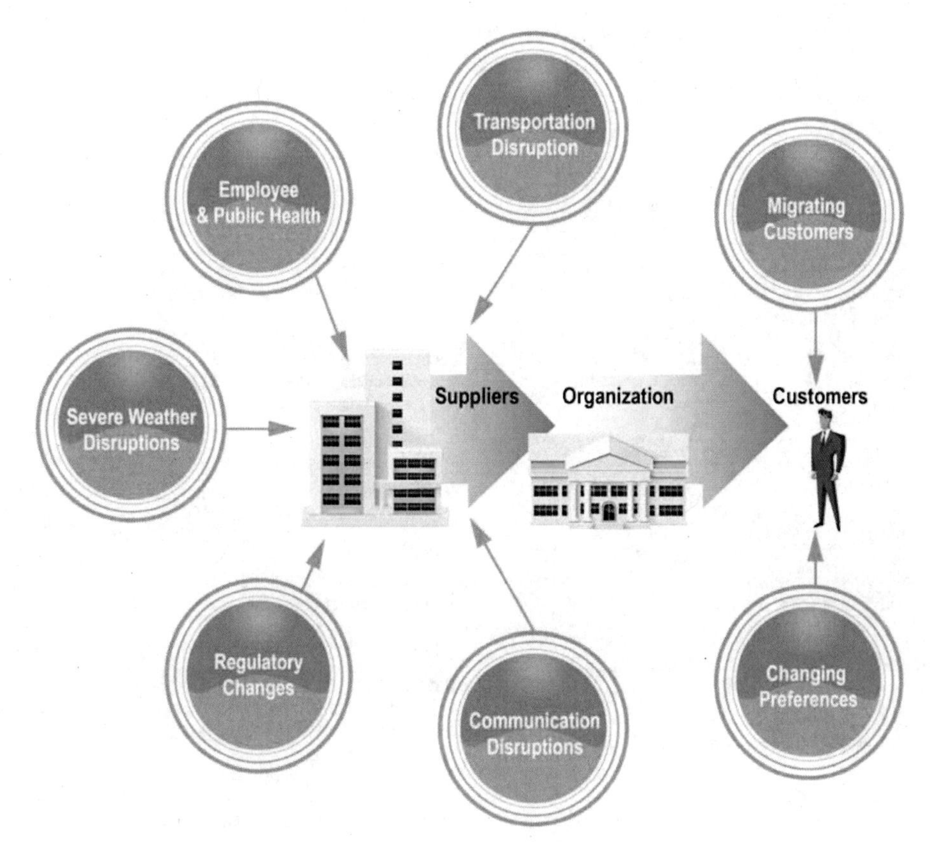

Figure 7-7. Overview of Climate Change Supply Chain Risks

Regulatory risks can be country specific. This means that new regulations may affect only one country where your supply chain operates. However, with the precedent of climate change treaties, a globally harmonized set of regulations could emerge that impacts your entire supply chain.[29] For example, since energy production is a significant source of global GHG emissions, regulations that alter the cost structure of energy production could arise.

The other category of risk is climate. These are risks resulting from the physical changes to the environment due to climate change. These risk events can disrupt the supply chain by interrupting production, disrupting transportation, or changing customer preferences.[3,26] Table 7-3 highlights examples of supply chain climate change risks and lists other chapters where they are detailed.

Table 7-3. Climate Change Risks

Climate Change Risk	Impact	Chapter
Public health degradation	Employee absence, reduced productivity, and restricted interactions	2
Communications disruptions	Lean and just-in-time systems particularly vulnerable	3
Water shortages	Reduced availability of water-intense products,increased costs, and impaired support from communities	4
Land use change	Possible changes in seaports, transportation routes,agricultural or industrial production locations	4
Severe weather	Interrupted business operations and reduced customer service	5
Infrastructure failure	Transportation, storage, production, and communication disruptions	5

Understanding the risks that a changing climate presents will help you adapt your supply chain. Customers and shareholders are asking organizations to address potential climate change impacts, and investor-associated organizations are applying similar pressure, so addressing these risks has internal and external benefits. Organizations such as the CDP and Ceres, in conjunction with recent Securities and Exchange Commission (SEC) requirements, are expecting companies to disclose these risks, compelling companies to better understand climate change.[30]

EXAMPLE

Companies like Levi Strauss & Co., Coca-Cola, and Sierra Nevada Brewing pay close attention to water supplies for both their facilities and their partners. For example, Levi Strauss assessed its supply chain for climate change risks and realized that water shortages could severely impact its cotton partners.[31]

Levi Strauss recognizes that cotton production is at risk due to various climate change impacts, depending on the growing location. In fact, the company's senior vice president for the supply chain blogged, "In India, China, and Nicaragua, this means water shortages.

In Vietnam and Bangladesh, on the other hand, it means too much water—in the form of flooding."[32]

Set an Adaptation Strategy

Recommendation: *Establish a clear, effective adaptation strategy that clarifies the climate change risks you are concerned about as well as the amount of risk you are willing to accept.*

Prepare adaptation plans

CHALLENGE

Much like supply chain GHG mitigation, adaptation can take many forms, ranging from low-effort preparedness to significant supply chain restructuring. Selecting how to adapt in your supply chain needs to be guided and directed.

Without clear objectives and guidance on adaptation, you can end up applying resources in suboptimal ways and overlook actions that will secure your supply chain against a changing climate.

SOLUTION

Adaptation Potential	★★
Operational Impacts	★★★
Financial Impacts	★★
Feasibility and Timing	★★★

Effective adaptation requires a clear adaptation strategy. A good adaptation strategy should also clarify the investment you are willing to make in treating the effects of climate change.

Your overall adaptation strategy should set the foundation and the goals for your adaptation program. Some programs may focus on responding to risks once they appear likely, while others may be more proactive and look for innovative operational concepts to adapt to climate change.

The right adaptation strategy for your supply chain should be based on how aggressive your corporate adaptation strategy is and how tolerant your business is of disruptions. Guidance comes from your organizational strategy and adaptation objectives. Typically, this strategy, like the mitigation strategy, involves managing three things:

1. Cost
2. Risk and operational disruptions
3. Reputation and brand.[8]

In any case, your adaptation strategy should be tied closely to a pair of factors: your organization's overall supply chain strategy and its accompanying risk management strategy. Aligning these strategic elements will ensure coordination across your enterprise. It will allow the sharing of resources to achieve organizational objectives. Without coordination at the strategic level, different organizations may take conflicting actions that create more chaos than resilience.

Any adaptation strategy needs to recognize that many climate change impacts on the supply chain are uncertain or unpredictable. Your strategy needs to include provisions for proactive risk management—identifying potential risks and taking action before the risk occurs—and reactive risk management—building a capability to react quickly when a risk occurs. Your strategy should define the balance between these two approaches. A more proactive approach will require more resources to treat potential risks before they occur. Conversely, a more reactive approach requires fewer resources but would result in a less resilient supply chain because adaptation focuses on reacting after events.

In developing a supply chain adaptation program, you need to research your organization's continuity of operations, business resiliency, and other risk management plans. More and more organizations are including supply chains in their corporate risk management plans. In addition, many stakeholder groups are encouraging or, in the case of the SEC, mandating that corporate risk programs address climate change risks. Climate change adaptation can tie into these efforts to take advantage of existing processes and resources, and avoid duplication of effort.

The adaptation strategy and associated program establish the priorities and resources for adaptation. With these elements set, you can build a supporting approach to guide your actions.

EXAMPLE

Levi Strauss based its strategy on its cotton consumption because 95 percent of its products are cotton based. As a result, it consumes about 0.5 percent of the world's cotton production; in 2007, the company used 289 million pounds of cotton in its products. Yet Levi Strauss does not have much information on where its cotton is sourced, since cotton sourcing is primarily left up to individual fabric mills.[33]

Levi Strauss's adaptation strategy focuses on managing the cost and availability of its cotton supply as well as projecting a sustainable brand image. In a life-cycle assessment of its 501 Jeans product, Levi Strauss found that cotton production represented 49 percent of water consumption across the life cycle. With cotton sourced from areas such as the Middle East and Africa that are subject to water scarcity, the company understood how the risk of a water shortage could impact the availability of cotton.[34]

Identify and Prioritize Risks

Assess adaptation risk

Recommendation: *Identify the climate change–involved risks to your supply chain and prioritize them to determine those most likely to significantly impact your supply chain. Focus actions on the risks that present the greatest expected danger.*

CHALLENGE

Every organization has limited resources to spend on treating risks. As discussed earlier (Table 7-3), the IPCC identified numerous ways climate change can impact our environment and, thus, your supply chain, but offers them only with regard to the level of certainty expressed by the scientific community.

Managers and stakeholders are responsible for protecting the supply chain against climate change disruptions while using limited resources. Deciding where to apply those resources requires an identification and prioritization method designed to determine the potential disruptions you should address to optimize your adaptation actions.

SOLUTION

Adaptation Potential	★★★
Operational Impacts	★★
Financial Impacts	★★
Feasibility and Timing	★★★

Identifying risks in your supply chain requires an understanding of that supply chain, including the products handled, partners involved, and geographic locations of operations and transportation routes. For these reasons, mapping the supply chain and understanding the current operations is an important part of an adaptation program.

The Supply Chain Operations Reference (SCOR) model is a structured approach to mapping a supply chain. It produces a geographic and process map of operations.[35] This method uses the standard structure of the SCOR model to quickly map the supply chain and understand the operations at each point. The SCOR model also includes a defined approach for managing supply chain risk.[36] Mapping the supply chain provides a geographic context to your operations, and many climate change impacts will vary considerably on the basis of geographic location and type of operation.[37]

When mapping the supply chain, remember the purpose of adaptation. To avoid wasting resources, you want to bypass detailed mapping of items or components in your supply chain that are widely commercially available—unless they are particularly sensitive to climate change (such as agricultural products). These products are widely available from multiple sources, so their supply is unlikely to be disrupted by climate change. The focus should be on the more strategic products in your supply chain and the components that go into them.[19]

Risks can often be identified by asking a few simple questions about your supply chain:

- How will a changing climate impact the physical locations in your supply chain?
- How will a changing climate impact your supply chain's ability to operate?
- What specific locations or assets are in locations particularly vulnerable to climate change, and what impact will that have on your operations?
- Is the demand for the products and services you offer sensitive to a changing climate?
- Could future regulations or industry standards related to climate change impact your supply chain?
- How will a changing climate affect your workforce and its ability to operate as normal?
- How will a changing climate affect your access to utilities (water, electricity, communications, etc.) and their cost?[28]

For the probability and impact assessments, you should use a simple scale to define these factors in relation to risk. Precision is what you might be inclined to draw from this process, but the more critical result is understanding.

It may be tempting to be more precise in your estimates, but the intent of this activity is to understand and pinpoint the location of your risk exposure. A simple scale will support rapid assessments that highlight the exposure to risk without requiring detailed analysis.

Once you have identified the climate change risks, you need to prioritize them. This starts with identifying the risks with the greatest potential to affect your supply chain. Figure 7-8 shows a common and effective scheme for prioritizing risks.

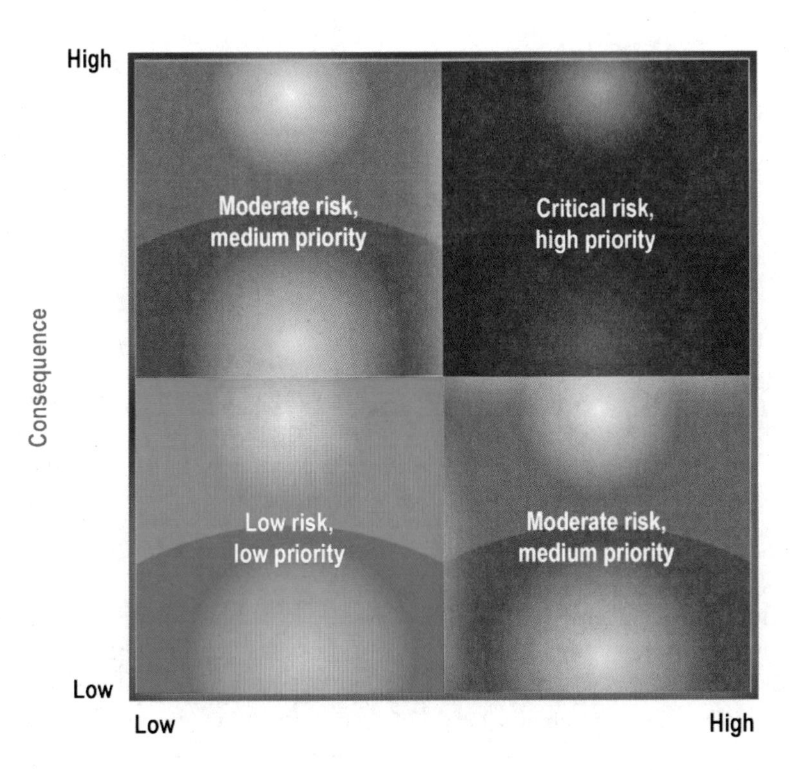

Figure 7-8. Risk Prioritization Scheme

The first priority for treatment actions should be to target those risks that have a high likelihood and that may significantly affect your supply chain operations. Likewise, the risks that are less likely and that have a minimal impact potential should be a lower priority for treatment; in fact, they should be addressed only if their occurrence would be troublesome.

This prioritization approach allows you to compare various risks that will have different impacts. This way, you're addressing the most significant threats to your business rather than those that are simply the most visible.

With the risks prioritized, the adaptation team should select which risks to treat and take appropriate action.

EXAMPLE

Levi Strauss started its climate change risk assessment with a cradle-to-grave, life-cycle analysis of two of its products: 501 Jeans and Dockers Original Khaki pants. The analysis looked at GHG emissions, energy consumption, and water consumption across the product line from raw material to final disposal. The company found that the largest climate change impacts occur in the customer use of the products due to the multiple wash and dry cycles the products go through; however, the biggest risk was in the water consumption. That's because, as ads for Levi's will tell you, its products are based on cotton, and cotton production is a very water-intensive process. With 95 percent of its products based on cotton, this was a clear priority for the company to address.[38]

Treat in Accordance with Strategy

Adapt as planned

Recommendation: *Appropriately adjust to the risks that climate change poses for an organization's supply chain, including the strategic alignment of your supply chain–appropriate treatment actions to ensure they lead directly to adaptation objectives.*

CHALLENGE

Without an overarching objective, adaptation actions can be haphazard and apply resources against the wrong threats, leaving the supply chain exposed to disruptions and wasting scarce resources. Effective adaptation programs need a guiding strategy to help select appropriate actions. They must aim toward organizational adaptation objectives and use resources as effectively as possible.

SOLUTION

Adaptation Potential	★★★
Operational Impacts	★★
Financial Impacts	★★
Feasibility and Timing	★★

An adaptation program is most effective when it has a unified purpose and is aligned with overall adaptation objectives. Organizations in general have the capability to adapt to a changing climate, and these adaptations are more powerful when coordinated.[37] The specific treatment actions you choose should be tailored to your operational requirements, capabilities, resources, and objectives.

Obviously, actions such as redesigning your supply chain network are more involved than developing a monitoring plan. One consideration is the effort required to implement a treatment action, which will allow for selecting the best action for a given risk. This should be followed by a business case to demonstrate the value of each treatment action.

At the same time, the more proactive the risk treatment action, the more valuable it is in terms of averting disruptions. While response plans are relatively low cost to create, their execution implies that a disruption has already occurred or is imminent. In these cases, you're banking on quick response, which, at best, will only reduce the impact of an event; even the most successful response still occurs as a result of a disruption to the supply chain.

It is a question of paying up front or paying later with interest. Restructuring your supply chain may be more difficult but can fully eliminate a potential risk; for example, moving a facility out of a flood zone eliminates the risk of flooding. Table 7-4 shows a sample of treatment actions.

Table 7-4. Example of Treatment Actions for Climate Change Adaptation

Treatment Action	Description
Network redesign	Relocation of one or more activities in the supply chain to move operations away from potentially adversely affected geographic locations, possibly including elimination or addition of facilities
Partner change	Selection of a new partner that does not have the same risk exposure
Product or material change	Selection of products or materials that meet the customer requirements or design criteria needed and represent a lower climate change risk
Targeted investment	Investment in assets or improvements to protect facilities or distribution channels from disruption
Response plan	Development of a response plan for risk events to minimize the impact on the supply chain
Monitoring plan	Development of a plan to monitor the business or physical environment for risk precursors so proactive actions can be taken

Selecting a treatment approach is not simply a matter of reducing risk. The approach selected needs to support your adaptation strategy and your strategic objectives.[39] Your business strategy might favor working closely with key suppliers, so, while a partner change approach might best reduce risk, it severs a key long-term business relationship.

EXAMPLE

Two companies are working proactively to minimize the risk that water shortages would represent to their supply chain.

Levi Strauss is working to help cotton farmers in Brazil, the Middle East, and Africa through the Better Cotton Initiative.[40] The goal is to enable farmers to grow cotton using less water and less pesticide.

A Sierra Nevada Brewing Company barley field is harvested.

Using less water will make cotton farming, and thus Levi's cotton supply, more resilient to changes in precipitation and water tables—both are expected impacts of climate change. By working within a consortium, the company can broadly impact the cotton production industry, even when it lacks direct relationships with the growers. This program also supports Levi Strauss's marketing efforts as it seeks to capitalize on its reputation as a sustainable organization.[41]

Likewise, Sierra Nevada Brewing Company has identified water shortages as a potential risk to its agricultural inputs. As a result, the company has started monitoring the water issues surrounding its Canadian suppliers to identify potential drought or water shortages before these events impact the supply of grains. At the same time, Sierra Nevada is reviewing other local sources for its raw materials in the event there is a shortage from its primary source. By taking a proactive, albeit low-key, approach to monitoring its supply chain, Sierra Nevada is minimizing the potential climate change impact on its ability to produce beer.[42]

Continuously Adapt

RECOMMENDATION

Adapt and learn

Recommendation: *Base your organization's supply chain adaptation program on a continuous risk management approach, recognizing the dynamic nature of climate change risks and adaptation methods and technologies.*

CHALLENGE

As stressed earlier, with regard to risk, the ultimate effects of climate change are uncertain. There is consensus on some effects, such as rising sea level and changing weather patterns, but the when, where, and how of these effects is uncertain. We don't know the timing of climate change impacts, the exact locations that will be affected, and how severe these impacts will be.

The effects of climate change transpire slowly, but many treatment options also take a long time to implement. Treatments such as relocating operations and using new materials can take well over a year to go from planning to completion. You need to make sure your adaptation plan allows enough time for you to proactively alter your supply chain ahead of a changing climate.

Adaptation Potential	★★★
Operational Impacts	★★
Financial Impacts	★
Feasibility and Timing	★★

SOLUTION

Uncertainty means supply chain adaptation cannot be a one-time event.[25]

Risk management does not stop with treatment actions. Supply chains are constantly changing, and the ongoing impacts of climate change are not fully understood. For these reasons, a continuous risk management program should be in place, including regular reviews of risk profiles. The result of these reviews should be regular reprioritization of risks and adjustments in mitigation plans. The reward for this vigilance is a more secure supply chain protected from significant climate change impacts.

You also need to continuously review your adaptation process and update your treatment approaches. Continuous adaptation planning and action will ensure that your supply chain can continue to operate effectively.

When you take a risk management approach, continuous adaptation has two elements: identifying and treating risk and monitoring the environment. With regard to risk, identification must be repeated regularly because as our climate changes, the effects of the change will become more certain. We'll also gain access to better data through regular reviews as the latest science and observed events will yield more findings that warrant consideration.[25]

The analysis of risks in the supply chain should also be updated regularly. This should happen, ideally, whenever you make changes to your supply chain. Organizations and the business they conduct will change over time, leading to changes in suppliers used, customers served, and technologies employed. When an organization makes a change in its supply chain, it should revisit its adaptation plan. That is the time to assess how change affects the way you are adapting your supply chain.

Growing Egyptian cotton is very water intensive.

Finally, consider what ongoing monitoring can provide. Many climate change impacts can be predicted with ongoing monitoring, especially the ones that aren't instantaneous. Events such as water shortages, migrations, and increased water levels do not happen overnight. You should monitor conditions that surround the predictable risks. Doing so allows you to identify whether changes are occurring so that you can take action—ideally, before they disrupt your supply chain.

EXAMPLE

Through its work with the Better Cotton Initiative, Levi Strauss is working with cotton producers, cotton consumers, and other partners to continually identify new methods to improve cotton production. It's specifically targeting lower water consumption and lower pesticide use. The company notes that factories are making its jeans adhere to a strict set of water quality requirements. Meanwhile, manufacturers and suppliers have been given a list of harmful chemicals that they cannot use.

Levi Strauss's business hinges on access to a reliable, affordable supply of cotton for its products, and the company understands that a one-time fix will not provide the resilient supply of cotton it needs.[43] Its approach has been cited by leaders in the business community, including Aron Cramer, CEO of Business for Social Responsibility, who said, "The Levi Strauss & Co. code of conduct was 'the code that launched a thousand codes.' It was the first of its kind for a multinational company in any sector and our first significant step toward developing a comprehensive responsible global sourcing program."

Final Thoughts

A climate change–resilient supply chain will grow. This growth begins with taking actions to mitigate your supply chain's contribution to climate change. It will evolve into something that allows your organization to adapt your supply chain operations to the impacts of climate change. Between this first step and the destination is an approach that allows you to take a leadership role in reducing the emissions generated in your supply chain. It enables your organization to assist your partners in protecting your long-term operational interests. However, to ensure you are progressing in a common direction, these actions need to be tied closely to your enterprise climate change strategy.

The GHG emissions impact of an organization is often greatest in the supply chain of the products used. This means that a comprehensive mitigation program must include the supply chain. Organizations should start pinpointing the GHG emissions in their supply chain and build mitigation programs to reduce those emissions. As more and more organizations conduct supply chain mitigation programs, GHG management within a supply chain will become more of a business norm, making supply chain GHG management easier for everyone.

At the same time, climate change impacts are here. You should start planning for adaptation today and hope your actions can minimize any potential disruptions. While the most significant impacts are projected to appear gradually, some actions take years to implement, and you can start with those right now. Redesigning your supply chain to build resilience against risk is one of those actions.

Finally, remember that climate change is not a static issue. Both mitigation and adaptation in the supply chain require a continuous process to account for changes in operation and climate science. Mitigation and adaptation programs can become normal parts of your organization and its business activities. The process includes a culture shift that requires leadership and communication. However, correctly done, with engaged partners and stakeholders along for the ride, it will help ensure mitigation targets are met and that your supply chain is better secured from climate change impacts.

---

[1] Pete Abilla, "The Apple iPhone Supply Chain," *www.shmula.com*, January 18, 2007, http://www.shmula.com/the-apple-iphone-supply-chain/304/; Texyt, "iPhone: Who's the Real Manufacturer? (It isn't Apple)," June 29, 2007, http://texyt.com/iphone+manufacturer+supplier+assembler+not+apple+00113; David Barboza, "Supply Chain for iPhone Highlights Costs in China," *New York Times*, July 5, 2010, http://www.nytimes.com/2010/07/06/technology/06iphone.html?pagewanted=all.

[2] Chris Brickman and Drew Ungerman, "Climate Change and Supply Chain Management," *McKinsey Quarterly*, July 2008.

[3] Tim Searcy, "Companies Don't Compete; Supply Chains Compete," *The AnswerNet Call Center Blog*, November 24, 2008, http://answernet.wordpress.com/2008/11/24/companies-dont-compete-supply-chains-compete/.

[4] Gary Guzy, "Climate Change: Is It Changing Your Supply Chain Risks?" *Supply Chain Brain*, http://www.supplychainbrain.com/content/index.php?id=1218&no_cache=1&tx_ttnews[tt_news]=3156.

[5] The Aberdeen Group, *The Sustainable Supply Chain* (Boston, MA: December 31, 2010).

[6] Ernst & Young, *Climate Change and Sustainability: Five Areas of Highly Charged Risk for Supply Chain Operations*, EYG No. FQ0019 (2010).

[7] Ibid.

[8] A.T. Kearney, *Supply Chain Report 2010* (London: Carbon Disclosure Project, 2010).

[9] Pankaj Bhatia, Janet Ranganathan, World Business Council for Sustainable Development (WBCSD), *The Greenhouse Gas Protocol: A Corporate Accounting and Reporting Standard* (Washington, DC: March 2004).

[10] The White House, *Executive Order 13514, "*Federal Leadership in Environmental, Energy, and Economic Performance*"*, *Federal Registrer* Vol. 74, No. 194 (October 8, 2009), pp. 52115–52127.

[11] US General Services Administration, *Executive Order 13514 Section 13: Recommendations for Vendor and Contractor Emissions* (April 2010).

[12] Hewlett-Packard, "Reporting and Managing Emissions from the Supply Chain" (presentation, Pew Center on Global Climate Change Energy Efficiency Workshop II—Supply Chain Strategies, October 23, 2008).

[13] Jonathan Newton, "Greenhouse Gas Management in Ford's Global Supply Chain," (presentation, The Climate Registry Western Climate Policy Forum, Seattle, WA, March 10, 2011).
[14] Jennifer Shafer and Taylor Wilkerson, *Barriers to Greenhouse Gas Accounting in the Supply Chain* (McLean, VA: LMI, September 2008).
[15] Carnegie Mellon University, The Green Design Institute, "EIO-LCA: Free, Fast, Easy Life Cycle Assessment," *http://www.eiolca.net.*
[16] For ways to reduce energy consumption in energy intensive processes see US Department of Energy (DOE), Energ-IntensiveProcesses:Addressing Key Energy Challenges Across U.S. Industry, March 2011, http://www1.eere.energy.gov/industry/intensiveprocesses/pdfs/eip_report.pdf.
For industry GHG emissions, see
US DOE, Manufacturing Energy and Carbon Footprints, http://www1.eere.energy.gov/industry/rd/footprints.html, and US Environmental Protection Agency, "2012 Draft U.S. Greenhouse Gas Inventory Report, http://www.epa.gov/climatechange/emissions/us inventoryreport.html.
For estimation resources and emission factors, see
Greenhouse Gas Protocol, www.ghgprotocol.org.
For cradle-to-gate GHG emissions estimates for many industrial processes see Product Ecology Consultants, http://www.pre.nl/content/simapro-lca-software, and Econovent Centre Portal, http://www.ecoinvent.ch/.
[17] Maya Ramaswamy and Taylor Wilkerson, *Case Study: Patagonia* (McLean, VA: LMI, 2008).
[18] Patagonia, "The Footprint Chronicles," http://www.patagonia.com/us/footprint/index.jsp.
[19] Beverage Industry Environmental Roundtable, *Beverage Industry Sector Guidance for Greenhouse Gas Emissions Reporting*, January 2010, http://bieroundtable.com/files/Beverage_Industry_Sector_Guidance_for_GHG_Emissions_Reporting_v2.0.pdf.
[20] "Reporting and Managing Emissions from the Supply Chain", Presentation, HP, October 23, 2008.
[21] World Economic Forum, *Supply Chain Decarbonization: The Role of Logistics and Transport in Reducing Supply-Chain Carbon Emissions* (Geneva: 2009).
[22] cKinetics, "Exporting Textiles: March to Sustainability 2010," *Free-Press-Release*, May 14, 2010, http://www.free-press-release.com/news-report-outlining-initiatives-to-quantify-supply-chain-climate-risk-predicts-thrust-by-2011-1273884525.html.
[23] Ecolabel Index, http://www.ecolabelindex.com/.
[24] Maya Ramaswamy and Taylor Wilkerson, *Case Study: Hewlett-Packard* (McLean, VA: LMI, 2008).
[25] Ford Motor Company, "Greenhouse Gas Emissions," *Sustainability Report 2010/11*, http://corporate.ford.com/microsites/sustainability-report-2010-11/issues-supply-environmental-emissions.
[26] Association of British Insurers, *Assessing the Risks of Climate Change: Financial Implications* (London: ABI, November 2009).

[27] Martin Parry and others, *Climate Change 2007: Impacts, Adaptation and Vulnerability: Contribution of Working Group II to the Fourth Assessment Report of the Intergovernmental Panel on Climate Change* (Cambridge: Cambridge University Press, 2007) pp. 357–390.
[28] Andrew J. Hoffman, *Climate Change Strategy: The Business Logic Behind Voluntary Greenhouse Gas Reduction*, Paper No. 905 (Ann Arbor, MI: Ross School of Business, November 2004).
[29] Jonathan Lash and Fred Wellington "Competitive Advantage on a Warming Planet," *Harvard Business Review*, March 1, 2007.
[30] John Sterlicchi, "How to undertake a climate risk assessment," *BusinessGreen.com*, February 17, 2010.
[31] Anne Field, "Perils of Climate Change," *Treasury & Risk*, July-August 2009 Issue.
[32] Amy Leonard, "Cotton and Climate Change," *Levi Strauss & Co. Unzipped*, May 25, 2010, http://www.levistrauss.com/blogs/cotton-and-climate-change.
[33] Colleen Kohlsaat, *Journey Toward Sustainability* (presentation, AAFA Environmental Committee Meeting, February 17, 2009).
[34] Tod Gimble, *Levi's 501 Cradle-to-Grave Study: Minimizing the Environmental Impact of your Jeans* (presentation, Levi Strauss & Co., 2009).
[35] Supply Chain Council, "What is SCOR?" http://www.supply-chain.org /scor.
[36] Lisa Harrington and others, *X-SCM: The New Science of X-Treme Supply Chain Management* (New York: Routledge, 2010).
[37], Wilbanks, T.J., P. Romero Lankao, M. Bao, F. Berkhout, S. Cairncross, J.-P. Ceron, M. Kapshe, R. Muir-Wood and R. Zapata-Marti, 2007: Industry, settlement and society. Climate Change 2007: Impacts, Adaptation and Vulnerability. Contribution of Working Group II to the Fourth Assessment Report of the Intergovernmental Panel on Climate Change, M.L. Parry, O.F. Canziani, J.P. Palutikof, P.J. van der Linden and C.E. Hanson, Eds., Cambridge University Press, Cambridge, UK, 357-390.
[38] "Journey Toward Sustainability", presentation, Colleen Kohlsaat, Levi Strauss and Company, February 17, 2009.
[39] "Climate Change Strategy: The Business Logic Behind Voluntary Greenhouse Gas Reduction", Andrew J. Hoffman, Ross School of Business Working Paper Series, November 2004
[40] Better Cotton Initiative, http://www.bettercotton.org.
[41] Karen Rives, "When Jeans Turn Green," *America.gov*, January 12, 2011, http://www.america.gov/st/energy-english/2011/January/20110112134236 nirak0. 2347681.html.
[42] Cheri Chastain, Sustainability Coordinator, Sierra Nevada Brewing Company, personal communication with author, June 11, 2009.
[43] "Journey Toward Sustainability", presentation, Colleen Kohlsaat, Levi Strauss and Company, February 17, 2009.

Security

It is June 3, 2020. Intelligence analysts brief the Director of National Intelligence on the projected decline in Russian energy production due to damage to key pipelines from unanticipated permafrost thaw in Western Siberia.[1]

Within a year, this decreased capacity is influencing global energy market speculation—contributing to skyrocketing energy costs. These costs are already up 300 percent over the past decade due to growing demand, failed biofuel crops, and widespread carbon tax measures. By 2024, slow but steady sea level rise is increasing the natural hazards threats for the growing populations in low-elevation coastal zones. The resultant flooding risks and higher sea surface temperature are fueling an unprecedented growth in humanitarian assistance and disaster response requests to the US, NATO, and international community during the typhoon and hurricane seasons, particularly in fragile regions or states with unrest and conflict.[2] With US military capabilities already eroded after a decade of budget cuts, these new mission demands are further taxing US foreign assistance and military forces already deployed to assist partner nations with public health crises, chronic food shortages, and disruptions in commerce caused by shifting precipitation and temperature patterns.

Where All the Pieces Come Together

This scenario is not improbable. It is, in fact, a reasonable possibility, a depiction of the influences climate change could have on our national security. The US national security community, which includes foreign affairs, defense, and homeland security, is increasingly recognizing and considering how a changing climate could impact our national security capabilities and interests. Some still question whether climate change is a national security issue, but the number of leaders and professionals trying to determine how to pragmatically move forward on this issue is reaching critical mass. Stakeholders are trying to determine how to meet the numerous challenges presented by climate change, with finite budgetary resources. The questions they seek to answer are twofold. First, what greenhouse gas (GHG) mitigation opportunities have the most potential to enhance energy security and military effectiveness? Second, how can the US national security community better understand climate change, plan for its impacts, and integrate a resulting strategy into processes already in place? The ultimate goal is figuring out how to stack the deck in our favor—and this can only be done by understanding the risks to national security, adapt to them, and thrive—with finite time, staffing, and fiscal resources.

What makes this effort particularly enigmatic is where we stand today. Recent decades have been times of great change, during which expansion of US national security has taken us from a narrow focus on just traditional international relations, counter proliferation, and state-on-state conflicts to more frequent missions of counterterrorism, counterinsurgency, reconstruction, peacekeeping, humanitarian assistance, and disaster recovery.

Some argue that the two most crippling "attacks" against the United States—9/11 and Hurricane Katrina—illustrate the increasingly commonplace conditions-based threats, also referred to as "creeping vulnerabilities."[3] These vulnerabilities have made a paradigm shift necessary. Naturally, with evolving national security interests comes a debate over the relevance of any policy for environmental change and climate security. This debate sheds light on the value of energy security and GHG emission reduction and why our national security interests require that we pay more attention to them.

National security is a tricky subject to discuss in wider circles because much (if not all) of the actions taken stem from high-level national policy. However, if you consider the wider context of support for these policies, the list of potential stakeholders widens. Your interested parties now include suppliers and contractors working in support of the Pentagon. These groups can identify and support policies that are smart bets for energy security and climate change response. These private-sector partners can drive innovation, produce independent research, and generally expand their level of support to include climate change as a strategic interest of the national security community.

We present five recommendations for how the US national security community can achieve its energy security goals and enhance effectiveness; they also allow for cutting GHG emission liabilities. These recommendations primarily focus on the Department of Defense (DoD) as the largest federal energy consumer and GHG emitter, but also apply more broadly across similar activities at the Department of Homeland Security (DHS) and other federal agencies. They apply to stakeholders—those in the private sector who work with the federal security sector as well as local and state governments that are host to numerous facilities and strategic interests. They focus on establishing an energy and climate risk framework along with the associated metrics and energy-efficiency measures. They address renewable fuel options and propose a GHG mitigation impact assessment.

We also explore adaptation and the direct security implications posed by the impact of climate change. The five recommendations for adaptation elaborate on the climate scenario and policy needs, while proposing war-game exercises, capability assessments, and an early warning system for state fragility.

National Security Challenges and Interests: Constantly Evolving

US national security challenges have become increasingly complex. We once focused on rivalries with a single superpower or near-peers who sought to undermine our interests. Today, our national security focus is on a world with numerous centers of power, where a host of disparate and nuanced set of challenges are not easily resolved through traditional diplomacy and military "hard power" alone.[a]

Smart power means developing an integrated strategy, resource base, and tool kit to achieve American objectives, drawing on both hard and soft power.

—Richard L. Armitage and Joseph S. Nye, Jr.

A big part of our changing security strategy has been shaped by a decade spent in Afghanistan and Iraq. Our recent military and diplomatic efforts in those nations have reinforced the idea that fighting an unconventional enemy with conventional military forces and tactics is difficult at best. In 2005, after several years of combat operations, DoD officially acknowledged this shift by issuing DoD Directive

[a] The military often uses the term "kinetic action" to describe its broad range of combat or operational activities. These also fall under the label of "hard power" when used to address international conflicts. The complement is "soft power," where the focus is on values, cultures, policies, and institutions.

3000.05—an order that changed policy to give stability, security, transition, and reconstruction (SSTR) missions the same attention as traditional combat operations. Four years later, DoD Instruction 3000.05 expanded upon this by requiring a capacity to support US government missions across the full spectrum of conflict while bolstering civilian security, providing interim services, restoring local infrastructure, and delivering humanitarian relief.[4]

This policy evolution is best summed up in the 2010 *National Security Strategy*, where the concept of a "whole of government" approach ("govspeak" for effective interagency coordination and action) was first embraced. This strategy integrates "all of the tools of American power... [to] enhance international capacity to prevent conflict, spur economic growth, improve security, combat climate change, and address the challenges posed by weak and failing states."[5]

The *2010 Quadrennial Defense Review* (QDR) supported an integrated approach and urged the incorporation of intelligence, law enforcement, and economic tools for promoting stability.[6] The 2011 *National Military Strategy* takes a similar view, confirming the "enduring national interests," cited as

- *The security of the United States, its citizens, and US allies and partners;*
- *A strong, innovative and growing US economy in an open international economic system that promotes opportunity and prosperity;*
- *Respect for universal values at home and around the world; and*
- *An international order advanced by US leadership that promotes peace, security, and opportunity through stronger cooperation to meet global challenges.*[7]

These priorities reflect the evolving security policy and reemphasize the increased effort to view homeland security, military, and foreign affairs through a different lens. Climate change is just another in a series of 21st century X factors that could threaten American global security interests.

Environmental Change and Climate Security: Not a New Concept

Discussing the security implications of natural resources was a product of the Cold War, but it wasn't until the 1990s that these talks intensified to the point where policies for environmental change and climate security were considered. Any movement toward formalizing these discussions (understandably) took a back seat after the 9/11 attacks; at that moment, the US national security community shifted its focus to the global war on terrorism with an emphasis on counterinsurgency and SSTR missions.[8] In fact, this shift was so pronounced that all references to environmental security and climate change were absent from national security policy documents by 2005. Despite this policy change, many DoD combatant commands continued to recognize the relevance of environmental security initiatives, viewing them as a better way to address some aspects of their missions (especially in the early 2000s).[8,9,10]

Another shift began in 2007. That's when the non-profit research and analysis organization, CNA, issued a widely read report, *National Security and the Threat of Climate Change*, which reopened the climate change discussion within national security circles. Shortly after, policy guidance cemented climate change as a part of the discussion. The Fiscal Year 2008 National Defense Authorization Act included language requiring of the defense community "that the next national security strategy and national defense strategy include guidance for military planners on the risks of climate change, and that the next quadrennial defense review and examine capabilities the armed forces will need to respond to climate change."[11]

In 2007, a CNA report changed the dialogue in the national security community when it demonstrated that climate change multiplies threats in already unstable regions. Therefore, climate change must become an integral part of national security and defense strategic thinking.

More changes came in 2008. That year, the United States Africa Command (USAFRICOM) was established with the intent of applying a broadened approach to US security interests on that continent. New military attention on soft power issues in Africa included a number of climate-relevant considerations: water availability, natural resource scarcity (and abundance), climate change, and their interrelated security implications—which spurred research and application on the subject.[8]

Perhaps no opinion was as blunt, or as sobering, as one issued by the National Intelligence Council (NIC), which argued

> *[although] climate change is unlikely to trigger interstate war ... it could lead to increasingly heated interstate recriminations and possibly to low-level armed conflicts. With water becoming scarcer in several regions, cooperation over changing water resources is likely to be increasingly difficult within and between states, straining regional relations.*[12]

By 2010, national security leaders at the highest levels had joined the discussion. Among those speaking out on the issue was Director of National Intelligence, Admiral Dennis Blair, who asserted that climate change's intensification of challenges associated with water resources, natural disaster, and arctic change would "aggravate existing world problems" and strain US military resources over the next 20 years.[13]

Today, national security policy documents devote ample discussion to the issue. The 2010 QDR discusses rising demand for resources and quickly urbanized waterfronts, asserting that a combination of the effects of climate change, public health concerns, and tensions related to culture and demographics could "spark or exacerbate future conflicts" in several regions. The 2010 *Joint Operating Environment* ("the JOE") includes climate change in its list of the 10 key trends that will impact US military forces, citing potential natural hazards, coastal inundation, and new tensions related to the exploitation of natural resources as catalysts to watch.[14]

GHG Reduction via Energy Security: Addressing Mandates, Supply, and Price Volatility

Buoyed by what is now a half decade of policy support and robust discussion on all issues related to climate change and national security, the US government and some military services have action under way. In recent years, an examination of GHG emission liabilities has taken place as part of federal and military compliance planning and concerns about becoming more sustainable.

Until recently, federal action on GHG management and reduction had been undertaken on a strictly voluntary basis. That changed in October 2009 through the release of Executive Order (EO) 13514—Federal Leadership in Environmental, Energy, and Economic Performance, the first government-wide mandate for GHG emission inventory reporting, GHG reduction goals, and departmental strategic sustainability performance plans. This EO, issued by President Barack Obama, requires federal agencies (including DoD) to understand, account for, and reduce their GHG emissions. The federal government's largest energy user and GHG emitter is DoD, so in January 2010, DoD unveiled an aggressive target: reduce GHG emissions from installations and non-tactical vehicles by 34 percent by 2020.[b] Although this excluded tactical systems and weapons platforms (they are not subject to the president's EO),[15] DoD avowed in its 2010 sus-

[b] DoD's stated target would significantly exceed the government's goal, which is to achieve 28 percent reductions.

tainability report plan that even these areas can achieve the proposed reductions.[c]

Hanging over any federal policy is the push to reduce the country's strategic reliance on petroleum-based products. Foreign oil imports are at the crux, and a big part of reducing this foreign oil dependence is broadening strategic sources of supply (Figure 8-1). Not only would this improve the balance of trade, it would reduce exposure to the shocks of volatility in the world petroleum market. To this end, several pieces of legislation—including the Energy Policy Act of 2005, Energy Independence and Security Act of 2007 (EISA 2007), and National Defense Authorization Act of 2009—have set federal goals and expanded efforts to address the dependence on foreign-sourced oil. Accordingly, DoD policies and programs are increasingly incorporating energy security imperatives and activities (emphasizing energy efficiency and renewables, for example).[d]

Figure 8-1. US Petroleum Consumption, Production, and Import Trends

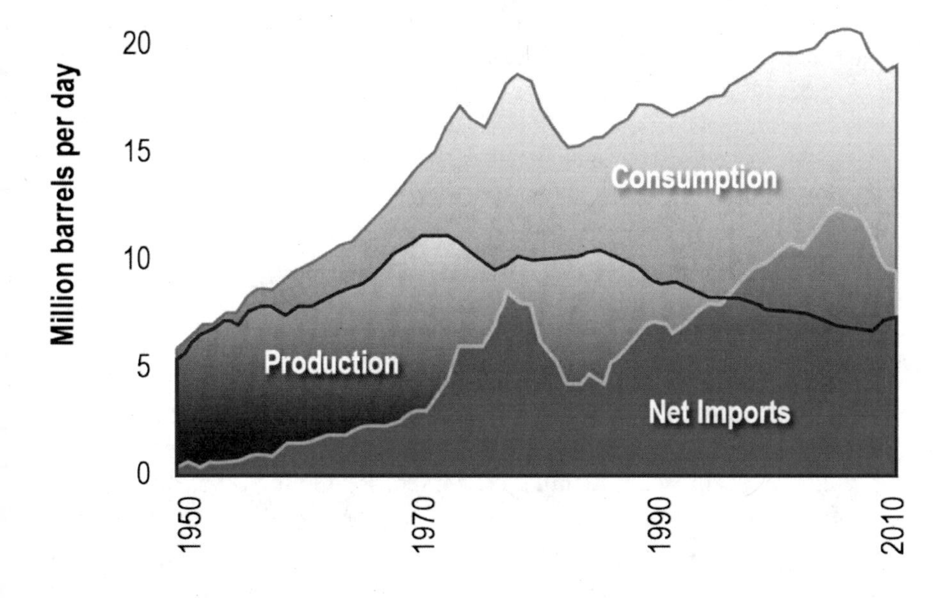

[c] The most compelling case for GHG inventories and reductions comes from those with "boots on the ground": combatant command logisticians and PFCs tasked with defending fuel convoys in-theater. In short, energy is a deadly serious business, especially for sustaining US military capabilities.

[d] Strong guidance on the matter was offered in the 2010 QDR, which urged DoD to address energy and climate changes, not only because of national security, but also because of the potential impact on strategic mission readiness.

A $1 increase in the price of a barrel of oil sustained over the course of a year can increase DoD fuel costs by $130 million ... these fuel price increases wreak havoc on our budgets.

—The Honorable Sharon Burke, Assistant Secretary of Defense Operational Energy Plans and Programs

For DoD, it's all about *energy security*. In defense circles, this means ensuring access to "reliable supplies of energy," while protecting and delivering "sufficient energy to meet operational needs."[16] One piece of the puzzle is the military's immediate focus on reducing operational energy logistics risks and in-theater fuel costs (in Afghanistan, for example); another aspect is the defense community's reliance on volatile petroleum sources, a strategic vulnerability for the military.

For example, the Navy alone experienced a jump in its annual fuel costs from $1.2 billion to $5 billion in just 1 year (FY 2008).[17] This is a budgeting nightmare. The challenge of near-term budgeting is the substantial price volatility in fuel contracts, spurred by two key factors: tightening global petroleum supplies and political instability in oil-producing nations. Meanwhile, in the long term, rising energy costs have a corrosive effect on military programs and capabilities, particularly with diminishing budget appropriations. Challenges abound if growing global demand for energy outstrips projected energy production capacity.[15]

DoD largely views GHG mitigation as a compliance issue, a cultural function of how it operates, but it sees energy security as a mission imperative. DoD, DHS, and other national security agencies need to consider GHG emissions while pursuing energy security projects that result in greater military effectiveness. This will make GHG mitigation an outcome along with greater capability and energy security.

GHG Mitigation Opportunities: Assembling the Ingredients

Energy security and climate change are increasingly relevant to national security stakeholders in ways that are not always obvious. The drive to reduce energy consumption can create new opportunities (energy security and combat effectiveness). On the other hand, advocating renewable energy projects abroad that lower GHG emissions can create new challenges for US national security interests (food security, for example). The recipe for success has a few key ingredients: support of leadership, clear goals and decision-making metrics, a well-articulated analysis framework, innovative technologies and approaches, and practical outputs designed to complement and enhance federal agency and military service capabilities.

How, then, can stakeholders cope with a future operational environment in which expensive energy coexists with GHG emission reduction mandates? They need to find opportunities for the national security community to *execute GHG mitigation* through actions that increase energy security and *enhance military effectiveness*. Our five recommendations can help the nation's security community make progress toward its energy security *and* GHG mitigation goals:

- Employ a risk-based energy and climate decision framework.
- Mandate programmatic energy security and GHG metrics.
- Expand the use of energy-efficient technologies and demand reduction approaches.[e]
- Expand renewable fuel use.
- Develop a smart-power/interagency assessment of climate mitigation impacts.

Employ a Risk-Based Energy and Climate Decision Framework

Recommendation: *Mandate the development and use of a framework that focuses on energy security and climate risk management. This will ensure the acquisition and sustainment of the military services' tactical systems, weapon platforms, and installations.*

Manage energy and GHG risks

CHALLENGE

DoD energy risk strategy has been lacking. In fact, it has operated without the standard measures of success any program needs, including policies, metrics, and governance.[18] The current increase in awareness of energy security needs and operational energy challenges—sparked by legal mandates, Defense Science Board reports, and lessons learned on the battlefield—has provided a boost. Still, more limited DoD-level policies on energy have left stakeholders with a jumble of military service–level energy security strategies, implementation plans, and technology investment. These do have a common thread, but they unfortunately lack strategic cohesiveness and unified purpose.

Several cases have illustrated this point. The coal-to-liquid controversy resulted in part from limited DoD guidance for addressing energy security coupled with a lack of consideration of GHG emissions. As a result, congressional actions have complicated recent efforts to adapt to changing energy needs and the pursuit of advanced renewable fuel opportunities.

In this particular case, the military began exploring the use of coal-to-liquid (fuel) and natural gas-to-liquid synthetic fuels as part of its

[e] See the "Structure" chapter for renewable energy and power generation efforts included in DoD and service energy security plans.

strategic energy security efforts (because of US domestic production and use). These technological processes, though, have been the subject of much debate and, in some circles, maligned.[f] The challenge with coal-to-liquid technologies is that the source materials and processes result in higher life-cycle GHG emissions than the conventional fuels they replace made from petroleum crude oil. One military service aggressively explored this alternative fuel option, and some in Congress disapproved.

The result was Section 526 of EISA 2007, which forbids US government agencies from procuring any fuel with life-cycle GHG emissions greater than that of conventional petroleum (except for test and evaluation purposes). This forced a complete reevaluation of the military's coal-to-liquid aspirations and complicated its current and future alternative and renewable fuel plans by limiting its flexibility in synthetic fuel purchases.[g]

In spite of this setback, efforts to prioritize and coordinate energy security initiatives are moving forward, including the recent establishment of a new position at the Pentagon—the Office of the Assistant Secretary of Defense for Operational Energy Plans and Programs. DoD also released its first Operational Energy Strategy. These are steps toward greater coordination of energy security strategies, programs, and efforts being implemented and planned by the military services.

Today, military service initiatives are exploring the use of advanced renewable fuels in tactical systems and weapons platforms but must now be aware of GHG hurdles (one danger area is in the suite of options where energy security and climate security offer divergent solutions). Supporters of renewable fuels must be prepared to address uncertainty over how to assess and balance several factors, including fuel life-cycle GHG emissions, water requirements, and land impacts.

Although this new position in DoD helps by filling the energy policy leadership and integration gap, helping to coordinate response to challenges, a clear DoD policy and a framework are needed. Such a framework must proactively incorporate GHG considerations when

---

[f] A *Washington Post* editorial published on June 18, 2007, included a headline labeling the coal-to-liquid debate as a "boondoggle."

[g] However, the military services' experience helped in establishing and institutionalizing the services' alternative fuel qualification and weapons platform certification process, which has enabled the rapid test, evaluation, qualification, and certification of the new hydrotreated renewable fuels (green F/A-18C/D Hornet, green A-10 Thunderbolt II "Warthog," etc.).

pursuing energy security efforts. But until a policy and mechanisms exist, the military's efforts to plan and manage for GHG liabilities, such as including them in its capabilities planning and programs, will likely remain inconsistent.

Another area where energy demands cannot be ignored is emerging and next-generation tactical systems. The planned operational lifetime for weapons platforms often ranges from 20 to more than 30 years (the B-52 fleet has been in service for more than 50 years). Energy-reducing technologies that influence sustainment costs and GHG emissions for coming decades will only slowly penetrate the entire fleet. Acquisition programs (as well as installation efforts) need frameworks capable of evaluating these types of life-cycle risks.

In these cases, concepts, components, and designs will lock in the energy demand of critical mechanisms. DoD requirements and acquisition policies are the factors that pull the most weight when engineers try to meet key performance parameters, and these factors ultimately drive the programmatic decision making. Because energy consumption is locked in early in the design process, so are the resultant GHG emissions. But because of how research and development progresses, and the time devoted, these early decisions remain a factor far into the future—in terms of longevity, today's concept could become the next B-52. When you add together time for design and development, and then include acquisition, the lifespan for such systems are measured in decades, similar to infrastructure.

SOLUTION

Mitigation Potential	★
Operational Impacts	★★
Financial Impacts	★★
Feasibility and Timing	★★★

DoD needs to accelerate the progress of its strategic energy security and operational energy strategies, such as the Operational Energy Strategy, and mandate the development, integration, and use of an *energy and climate risk management framework*. Doing so should be targeted toward efficiency.

Combining energy and climate in one framework offers more bang for the buck in addressing mission and compliance drivers. This would be accomplished through the issuance of a directive and instruction for acquisition and sustainment. Although proposed as a separate but complementary approach, this framework and any supporting policy documents would leverage existing structures, such as the DoD 5000 series of instructions and directives. These, in turn, address acquisition processes and military standards, such as MIL-STD-882D, "Standard Practice for system Safety," which provides direction for environment, safety, and occupational health (ESOH) risk management.

This new framework should integrate energy and GHG considerations into decision making by laying out policy guidance. It should identify institutional mechanisms, whether for programs or facilities. The main focus would be on across-the-board acquisition and sustainment, where early identification of energy and GHG risks (and the associated tradeoffs) would allow management throughout the program's life cycle. Energy security and GHG risks would then be incorporated into design-phase decision making (before so-called "Milestone A"). Existing technologies and life-cycle cost-effectiveness would see an expanded evaluation process.

This policy and framework should also apply to projects and installation-level decisions that have energy security and GHG implications. For example, energy efficiency and fixed-site renewable technologies[h] are often viewed as solutions to the interlinked challenges of energy and climate security, but the respective benefits and environmental risks of these technologies vary. Figure 8-2 shows a "tradespace"—the options and tradeoffs of choosing between energy security and GHG mitigation. It illustrates how many of these options can contribute to one, the other, or both ends.

The World Resources Institute (WRI) provides the following description

> *This chart compares the energy security and climate characteristics of different energy options. Bubble size corresponds to incremental energy provided or avoided in 2015. The reference point is the "business as usual" mix in 2025. The horizontal axis includes sustainability as well as traditional aspects of sufficiency, reliability, and affordability. The vertical axis illustrates lifecycle GHG intensity. Bubble placements are based on quantitative analysis and WRI expert judgment.*[i]

---

[h] For example, large, immobile solar photovoltaic facilities such as the one found at Fort Carson, CO.

[i] For specific details on the assumptions underlying the options on this chart, go to www.wri.org/usenergyoptions.

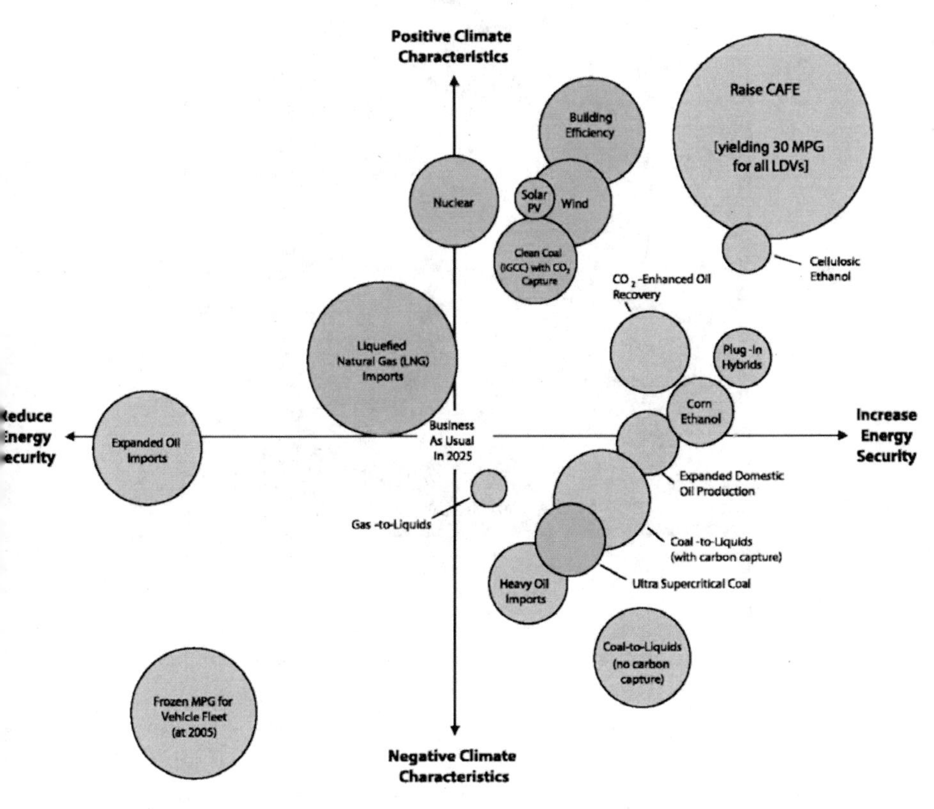

Figure 8-2. Snapshot of Selected US Energy Options Today: Climate and Energy Security Impacts and Tradeoffs in 2025

New policy guidance from the Office of the Secretary of Defense would provide the services with consistent policy and risk-based mechanisms for understanding these tradeoff implications and provide for analysis of alternatives. This could apply to weapon system programs, installation construction projects, and anything in between.

EXAMPLE

For years, MIL-STD-882D, "Standard Practice for System Safety," has guided program-wide risk management for ESOH. It has even been used by informed civilian agencies, such as NASA, in their efforts to prioritize (Figure 8-3) and manage both programmatic and institutional risks. This risk-based approach is used by acquisition program managers, who must prepare a programmatic ESOH evaluation to document and manage the risks they identify, along with regulatory compliance.[19]

Figure 8-3. Risk Matrix

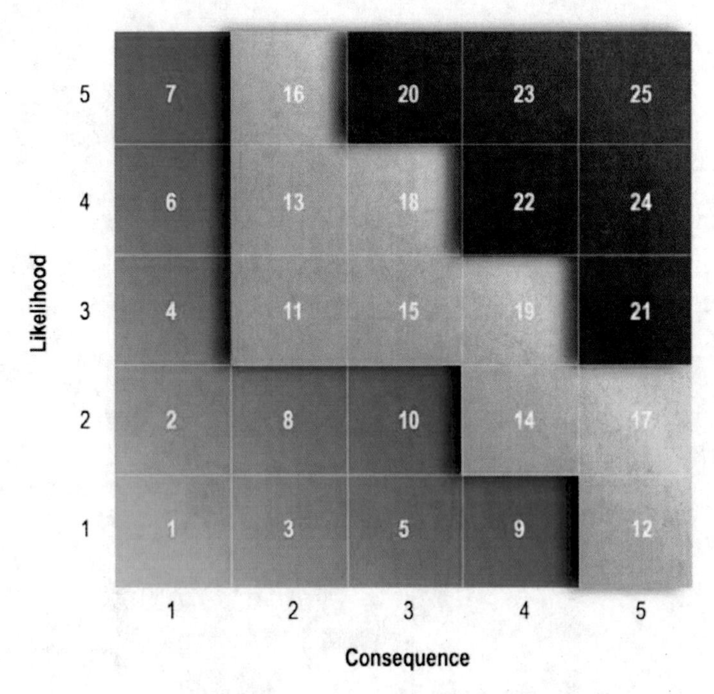

As shown in Figure 8-3, the likelihood and the consequence of a risk determine a rank (number in colored cells), which is then used to prioritize risk reduction and contingency planning. All risks, even those with a low rank (green), are put on a watch list. Those with intermediate (yellow) or high (red) ranks are also targeted for risk reduction. The highest risks also require a contingency plan.

The F-22 Raptor fighter program was one of the first weapons-platform programs to fully incorporate hazardous materials risk and pollution-prevention requirements using this approach.[20] The program team was able to reduce hazardous material use in the production aircraft after identifying the associated risks, and then used this information to manage and mitigate each area of concern. By prioritizing risks and reducing them through design choices, the program's ESOH liabilities, including material obsolescence, maintenance costs, and environmental compliance, were reduced over the weapons platform's life cycle.[j]

A similar framework, NASA Policy Requirement 8000.4A, "Agency Risk Management Procedural Requirements," mandates the use of both risk-informed decision making and continuous risk management

[j] See note g, this chapter.

practices. While traditionally used in space vehicle programs, NASA adapted and used this approach for institutional risk management purposes in recent years and has included risks such as encroachment and climate variability or change. This institutional risk approach is intended to increase the agency's sustainability. It can also help prioritize, focus, and guide NASA's energy and GHG mitigation, along with its climate adaptation efforts.[21,22]

Mandate Programmatic Energy Security and GHG Metrics

Mandate energy and GHG metrics

Recommendation: *Mandate that federal agencies with security interests (DoD and DHS) incorporate metrics for energy security (and the resultant GHG intensity) into (1) tactical system and weapon platform retrofits and (2) evaluations and decisions related to acquisition programs.*

CHALLENGE

Over the past 5 years, all branches of the military have developed energy security strategies. They've explored opportunities to reduce their fuel consumption, diversify their energy portfolios, deploy alternative energy technologies, and establish unique goals and objectives. Several challenges remain, one is the establishment of consistent policy; another is the use of metrics associated with energy security strategies to identify options for and areas of investment that achieve their stated objectives. This should be done, regardless of whether it involves a tactical system, weapons platform, forward operating base, or installation.

DoD provided a joint requirements perspective (both climate and energy security) by adopting guidelines for energy efficiency for acquisition programs, but it has not yet mandated its use across the board. Furthermore, there are no defense-specific requirements or metrics for GHG intensity. These are necessary for determining the alternatives and their implication for achieving multiple DoD and service goals. Developing baselines for program alternatives and managing progress are challenging when you're operating in an environment of limited applications and metrics that aren't standardized.

SOLUTION

Mitigation Potential	★
Operational Impacts	★★
Financial Impacts	★★
Feasibility and Timing	★★★

Specific metrics are necessary to assess the energy and GHG implications of options and decisions. If something is not measured, or has no defined requirements, then it usually is not effectively managed. For weapon platform acquisition and maintenance, we recommend expanding existing mandates that involve integrating energy efficiency (and the resultant GHG intensity). The same should be true for installation energy metrics that analyze available options. At a minimum, *energy efficiency must be a required parameter*. By embracing standardized, integrated metrics in these contexts, efforts to shift from first-cost to life-cycle effectiveness and sustainment (which provides value to taxpayers) will be strengthened.

The key focus would be to target early points in both the design phase for new systems and any retrofit plans for existing ones and leverage these checkpoints for what they offer: opportunities to gauge how effective a new system will be for both energy efficiency and security.

In the medium and long terms, requirements mandated by policy can combine with metrics to help accelerate innovation. This can be accomplished by inserting technological advances into weapons platforms as well as military construction and infrastructure programs. The risk of not having clear metrics and requirements is that key opportunities will be missed. The result would be failed efforts to enhance mission capabilities, minimize fuel consumption, and reduce GHG emissions.

EXAMPLE

Since 2004, DoD has employed mine-resistant, ambush-protected (MRAP) vehicles in Iraq. These vehicles were designed to better protect troops against improvised explosive devices (IEDs). They succeeded in enhancing force protection—exactly what they were designed to do. MRAPs, more than your standard in-theater transport, have stronger, heavier armor capable of deflecting these deadly attacks.

Of course, there was a tradeoff. In the case of MRAPs, increased armor protection meant lower fuel efficiency, decreased mobility, and diminished maneuverability.[23]

In April 2007, the Office of the Under Secretary of Defense for Acquisition, Technology and Logistics released a memo announcing a new DoD policy regarding how the department measures the cost of fuel. This analysis, which considers not only the direct purchase cost of fuel, but also any associated factors, was to be included in all assessments of available alternatives for all tactical systems and weapons platforms. The intent was to decrease fuel intensity, increase energy efficiency, and enhance US military capabilities—the very areas in which MRAPs were lacking.

On December 8, 2008, this policy was incorporated in DoD Instruction 5000.02 and has since informed all major defense acquisition programs. Today, they are aware of the tradeoffs because of metrics measuring mission effectiveness. For the MRAPs, the lesson that decreased exposure to IEDs would increase energy supply line vulnerability doesn't have to be learned in the field.[24]

Since then, metrics that consider the entire cost of fuel are increasingly influencing the design and acquisition of next-generation MRAPs, where the goal of lowering the vehicle's weight is balanced with the need to protect.

This mandate has even spurred innovation, including the development of new technologies like the structural blast chimney (SBC). Developed by Hardwire LLC under a Defense Advanced Research Projects Agency program, SBCs vent blast energy from underneath the vehicle upward, thereby amplifying protection but also reducing vehicle weight and increasing fuel efficiency.[25]

Expand the Use of Energy-Efficient Technologies and Demand Reduction Approaches

Recommendation: *Where feasible, expand the use of energy-efficient technologies and approaches in tactical systems and weapon platforms to decrease GHG emission liabilities, while increasing energy security and mission effectiveness.*

CHALLENGE

US tactical systems and weapons platforms have an energy shackle that becomes stronger and more burdensome if coupled with inefficient technology. Inefficient technologies are the cause of waste and high sustainment costs. In the previous MRAP example, emerging wartime needs to protect troops from IEDs in Iraq resulted in the quick development of MRAPs along with the "up-armored" High Mobility Multipurpose Wheeled Vehicle (commonly known as the Humvee). But the accelerated development and deployment of these vehicles resulted in transports that were less fuel efficient, which, in turn, indirectly increased the vulnerability of troops and in-theater lines of communication because more fuel convoys were needed. Energy and GHG metrics can only take you so far; they do not necessarily equate to technology decisions or how such systems are used.

The reason is that energy inefficiency is not just a function of technology, but also of *technology use*. Combat and contingency operations are largely driven by external factors, but there are recurring impacts on *how our military operates and trains*. This is one area firmly within our control, where modifications for promoting greater energy efficiency can be made and result in reduced GHG emissions. Remember, GHG reductions are not and should not be the primary consideration in a combat zone. However, failing to act on operational energy lessons learned in Iraq and Afghanistan are lost opportunities. Lessons learned have pointed us toward tactics that can enhance combat effectiveness, maximize solider safety, and, as an added benefit, reduce GHG liabilities.[k] In some ways, the recent DoD Operational Energy Strategy could be looked upon as a de facto operational GHG mitigation strategy, something to certainly consider.

SOLUTION

Mitigation Potential	★★★
Operational Impacts	★★
Financial Impacts	★★
Feasibility and Timing	★★

Enhancing energy efficiency and reducing demand is one of the largest energy security and GHG mitigation opportunities for DoD, DHS, and the federal government. For the last decade, DoD and its branches have better understood the threat of unstable petroleum supplies to in-theater efforts to connect deployed troops with their supply base. They understand how price volatility can wreak havoc on budgets. Every unused gallon of fuel reduces the requirement for new energy and, in turn, its associated logistics burden and risks, costs, and, again, avoided GHG emissions.

A key step in achieving this is the identification and pursuit of opportunities that maximize combat readiness (more training) and warfighter safety (fewer convoys in a combat). We recommend expanding the military's existing efforts to insert energy-efficient technologies into new tactical systems and weapons platforms programs and retrofits. Initiatives to reduce demand in operational planning and training should be expanded. Great opportunities are available in the form of technology enhancements, operational efficiency, and training.

[k] Just ask the Army and Marine Corps Energy "Tiger Teams"—groups of experts assigned to investigate or solve technical or systemic problems in the field—they provided ample lessons learned from recent in-theater engagements.

In the context of weapons platforms, opportunities are here today and on the horizon. Clearly, the tendency is to look to the future for solutions, and yes, in the coming decades, technological advances inserted into, for example, military aircraft, will provide key opportunities for reductions in fuel consumption and GHG emissions.[l] Energy efficiency-enhancing technologies for airborne and naval weapon platforms can eventually be incorporated into new designs or during refitting programs. However, we don't have to wait for future breakthroughs. For instance, the energy consumption of aviation weapon platforms can be reduced today using technologies that already exist. Improved aerodynamics, lightweight materials, and power plants that run with higher efficiency all provide these opportunities now.[26] One example is the use of advanced laminar coatings that could reduce drag and increase fuel efficiency by 16.5 percent.[27,m] Innovative sign concepts can have dramatic impacts: Boeing's blended wing body, for example, could reduce fuel consumption by as much as 32 percent compared with today's conventional airframes.[28]

X-48B fuel-efficient blended wing body aircraft.

Existing energy-efficient technology insertion and retrofits can likewise benefit Navy and Coast Guard surface ships. These include more efficient power plants and concepts, advanced propeller designs, improved hull designs, and novel low-resistance hull coatings. Engine and hull coating upgrades may be options for vessels requiring major refits.[26,29] The Navy and Coast Guard can also expand their examination of opportunities to improve prime mover efficiencies,[n] including combined diesel and gas turbine plants as well as hybrid electric drives.[29]

In-theater, the Army and Marine Corps have been studying energy-efficiency technologies to use in their tactical systems and at their forward operating bases (FOBs) as well as how to supplement these technologies with renewable power generation to further reduce de-

---

[l] In this case, there will also be ways to enhance mission capabilities (greater range and loiter time).

[m] New technologies, materials, and approaches should also carefully screen and consider novel risks to performance, economic, health, or other environmental factors.

[n] A "prime mover" is a machine that transforms energy from thermal, electrical, or pressure form to mechanical form, typically an engine or turbine.

Forward Operating Base Salerno in the mountains of Afghanistan.

mand. They recognize that these are real opportunities for running fewer convoys. Not only does this reduce dependence on in-theater supply lines, but it also increases combat effectiveness and enhances the safety of the warfighters who would otherwise be protecting these convoys.

One area being pursued is the optimization of generator efficiency at FOBs, not only in-theater but also back home for training. US ground forces have learned that they can do a better job conserving fuel and reducing maintenance labor hours if they accelerate the adoption of larger, more-centralized generators at *enduring* FOBs.[o] In addition to this tactic, they should aggressively re-equip combat and training units with lighter, more fuel-efficient tactical generators, rather than keep their older generator sets. In the midterm, improved small-scale, localized power, referred to as "microgrid power management," along with storage technologies, should be pursued because it reduces risks to supply routes, increases energy security, and reduces GHG emissions.

Operational changes (and the technologies that enable them) also present immediate opportunities for efficiency gains and reduced fuel consumption, which can also result in better energy security and lower GHG emissions. For example, ocean-deployed military surface vessels are already paying heed to the routes they take to destinations and how fast they move along these routes. By optimizing operating speeds and using Global Positioning System–enabled route planning, vessels can benefit from immediate reductions in fuel use. Navy and Coast Guard vessels should also focus on enhancing and managing energy and power conversion, similar to shoreside heating, ventilation, and air conditioning (HVAC) and power efficiency and management approaches.[29,p] We recommend expanding these energy management and optimization efforts as a means for increasing capabilities, meeting energy security objectives, mitigating fuel costs, and reducing GHG emissions.

These solutions focus on in-theater approaches, but other approaches can be applied stateside. We recommend accelerating efforts to ana-

[o] In 2008, the Bush Administration began characterizing any Iraq bases as 'temporary' or 'enduring' rather than 'permanent.' The label has stuck.

[p] This is similar to the discussion in the "Structure" chapter, where energy managers in federal and commercial buildings have increasingly been looking at the operations and technologies used in their facilities. For instance, they are upgrading HVAC systems with more intuitive information management systems and monitoring for performance and energy use.

lyze training programs and optimize the use of training aids such as simulators. Optimizing training will still provide the desired results: maximizing performance and combat effectiveness. Optimization simply allows other goals to be reached, including energy security and GHG reduction.

One opportunity already deployed focuses on the increased use of weapons platform simulators.[30] Doing so provides a means for potentially enhancing training effectiveness while reducing weapon system wear and tear, fuel consumption, and GHG emissions (though the relative reduction tradeoff needs to consider the sources of electricity used in such training).[q]

EXAMPLES

The Navy has been a model for envisioning and preparing for operational changes and technology to achieve energy efficiency.

The USS *Makin Island* (LHD-8), a *Wasp*-class amphibious assault ship, has been fitted with an electric auxiliary propulsion system that reduces fuel consumption during low-speed operations, which can be up to 75 percent of the vessel's operations.[29] The results were immediate, and impressive. On its first voyage alone, this technology saved the Navy $2 million just in fuel, and this technology is expected to save $250 million over the vessel's life cycle.[17]

Meanwhile, 2012 will see the beginning of a pilot program to bring hybrid electric drive system technology to the Navy's USS *Arleigh Burke* (DDG-51), the lead ship of the *Arleigh Burke*–class of guided missile destroyers.[31] Following this pilot program, a full launch of DDG-51 with this technology is anticipated in 2014.[17]

In addition to weapons platform technology upgrades, the Navy has several operational efficiency efforts under way, including studying and adopting operational practices for its naval aviation platforms and ocean-deployed military surface vessels. In 2012, the Navy will begin piloting a new smart navigation technology called "Smart Voyage Planning Decision Aid." This technology is designed to match the mission and ocean conditions with the most efficient route.[31]

Two additional efforts are already taking place in the Navy's aviation program. These cases involve cold truck refueling and SMART—

[q] Lively debate continues about simulators' contributions to enhancing or degrading training effectiveness. Study into these approaches continues, but many factors (such as simulator fidelity, quality, and use in training) complicate such arguments.

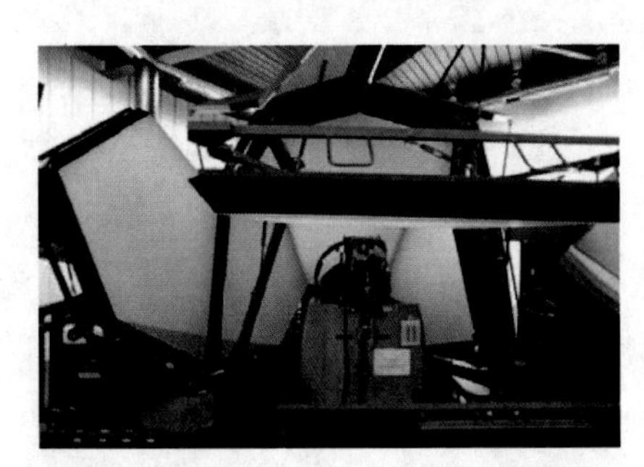

Flight simulator use is expected to increase and helps to reduce fuel consumption during training.

short-cycle mission and recovery tanking. Cold truck refueling is the practice of not leaving the jet turbine running while refueling. Currently in effect, this system is already freeing up substantial budgetary resources due to fuel savings.[30] SMART tanking produces results along the same lines: it has demonstrated a 65 percent fuel burn reduction for carrier-based, naval aviation air wings.[30]

Complementing these existing operational approach examples are efforts to reduce flight hours during training. Simulator use is currently limited to training and proficiency in flight preparation, emergency procedures, tactical repetition, and mission rehearsal. That said, expansion of simulator activities is expected to take place over the next decade.[30] Since 1998, the Navy has increasingly explored, analyzed, and expanded its use of simulation "trainers" for its aviators, allowing increased familiarization and training within its many rigorous and comprehensive training programs.[32]

Expand Renewable Fuel Use

RECOMMENDATION

Use renewable fuels

Recommendation: *Enable and expand renewable fuel adoption, purchase, and utilization in the various military branches.*

CHALLENGE

DoD tactical fuel consumption (three-quarters of which is jet fuel) is the US government's largest consumptive use of petroleum-based fuel. Although advanced "drop-in" renewable fuels are already being qualified for use in tactical systems and certified weapons platforms,[r] they have not been produced on a commercial scale domestically and therefore lack demonstrated business models. As a result, they are not currently available in sufficient quantities and, for the time being, are expensive to procure.

Adoption is tricky, however, because of evolving standards. The Environmental Protection Agency's (EPA's) Expanded Renewable Fuel Standard (RFS2), mandated by EISA 2007, defines renewable fuels on the basis of their life-cycle GHG emissions. This sets a high bar for next-generation cellulosic biofuels, since EPA is mandating that these fuels have life-cycle GHG emissions 60 percent lower than those of conventional fuels.

[r] "Drop-in" refers to those that are designed for easy insertion and immediate use in existing systems without modifications. In the United States, renewable fuels are defined as having life-cycle GHG emissions at least 20 percent less than the conventional fuels they replace.

Because of RFS2 mandates, production of ethanol, biodiesel, and some advanced renewable blended fuels is expected to skyrocket through 2022. However, these mandates do not include the production of jet fuel, and the bulk of these conventional biofuels, including biodiesel, are banned for use in DoD tactical system and weapons platforms because of risks involving incompatibility, maintenance, and operation.

How About Installation Energy?

See the *"Structure"* chapter for energy efficiency and renewable energy recommendations applicable to installations.

The military's efforts largely focus on the advanced drop-in biofuels from non-food feedstocks and their use in existing equipment, but there are production volumes and costs barriers for the next few years. Limited availability and high costs are expected to last for at least the next few years.

Interagency, public-private partnership efforts are under way to accelerate the development of this domestic industrial base.[s]

SOLUTION

Mitigation Potential	★★
Operational Impacts	★★★
Financial Impacts	★
Feasibility and Timing	★

DoD should continue to expand its use of renewable fuels in its tactical systems and weapons platforms. Doing so supports a number of strategic policy objectives and military organizational goals related to energy security and GHG emission reduction.

A 2011 DoD study suggests that use of advanced renewable fuels can be expanded through two actions: first, by enabling renewable fuel use by the military, and second, by cultivating partnerships in industry and government.[33]

For the first, DoD must continue its current technical certification efforts for weapons platforms and tactical systems, and must, in fact, accelerate them to ensure it is permissible to use drop-in renewable fuels. For the second, DoD must build upon partnerships with other US government agencies, including the Departments of Energy (DOE), Agriculture (USDA), and Transportation, along with EPA and private industry to adopt and implement supply chain road maps. The key function of these road maps is to support the cost-effective production and procurement of advanced biofuels (from non-food feedstocks in the long term), which are critical to meeting the existing alternative fuel use goals of the military services.

[s] In June 2011, the Departments of the Navy, Energy, and Agriculture signed a memorandum of understanding to support and accelerate industrial base capabilities for the commercial scale production of cost-effective, drop-in biofuels utilizing Defense Production Act authority.

Each military branch, along with the Coast Guard, should expand efforts to certify major tactical systems (tanks, Humvee, etc.) and weapons platforms (aircraft, vessels, etc.) in anticipation of commercially available, renewable fuels, including the rapidly emerging alcohol-to-jet pathways.

Once qualified for use, renewable fuels lead to a several percentage point reduction in the GHG emissions of tactical systems and weapons platforms. For instance, one of the Air Force energy security goals is to have alternative fuel blends account for 50 percent of its US fuel consumption by 2016.[t] The military services' objectives are to continue the qualification of these fuels and certify the next-generation of drop-in renewables for use in Air Force aircraft and ships no later than 2016.

DoD already has in place formalized partnering agreements with DOE; we recommend DoD expand this partnership to allow the development of commercial-scale "operational demonstrations" of these advanced renewable fuel production systems. At the same time, USDA, DOE, and DoD, including the Navy, should continue to nurture industry relationships related to non-food feedstock biofuels. This will help to establish vertically integrated supply chains for renewable operational fuels, particularly those that use non-food feedstocks (to avoid exacerbating food security concerns).

DoD leadership should emphasize that driving these efforts and supporting legislation that helps create private-public partnerships to make drop-in fuels available on a commercial scale—at acceptable cost—is a strategic national security imperative.[u]

EXAMPLES OF DROP-IN RENEWABLE FUEL TESTING

Hydrotreated renewable fuels have largely been vetted throughout the military's technical communities. These are renewable fuels that use hydrogen to pull off oxygen through hydroprocessing technology. The result is a jet fuel, considered "green," that can be blended with conventional fuels. Because there are no noticeable impacts to performance or changes to the equipment, it keeps fuel usage at the status quo from an operational standpoint.

---

[t] Advanced drop-in renewable fuels are 50/50 blends; the EPA requires a minimum of 20 percent GHG reduction over conventional petroleum. With the US Air Force's 50 percent use goal, these would equate to at least a 5 percent reduction in stateside GHG emissions from their jet fuel emissions.

[u] Ideally, this would establish long-term contracting authority and add jet fuel to EPA's volumetric mandates.

In 2010, the Air Force and Navy flew planes fueled by hydrotreated renewable fuels. The Air Force made the first military flight, flying an A-10 "Warthog" running on a 50/50 blend and conventional JP-8 jet fuel. The Navy performed a similar supersonic demonstration with an F/A-18 Super Hornet, appropriately nicknamed the "Green Hornet." Both demonstrations flew on a mixture of biomass-derived hydrotreated renewable jet fuel (HRJ) and conventional jet fuel where the HRJ feedstock oil was produced from camelina. Camelina is a member of the mustard family, not generally considered a source of food oil (and often referred to as a weed).

The "Green Hornet" can use a 50/50 blend of camelina hydrotreated renewable fuel and conventional JP-8 jet fuel.

The demonstration is considered a key step toward the Air Force goal of using 50 percent alternative fuels by 2016.[34] The Navy's efforts, meanwhile, represent the first wide-scale aviation test program for evaluating the performance of a 50/50 biofuel blend in supersonic operations, another critical gate clearing the F/A-18 E/F through its entire flight envelope. Such testing has since expanded to other aircraft and tactical systems throughout the various branches for advanced renewable fuel certification efforts.

Develop a Smart-Power/Interagency Assessment of Climate Mitigation Impacts

Recommendation: *Jointly develop a new assessment, done every 4 years, that will allow a better understanding of climate mitigation tradeoffs and impacts in the future security environment (FSE). It could be called the "Quadrennial Assessment on Defense, Development, and Diplomacy (QA3D)."*

Assess mitigation security tradeoffs

CHALLENGE

Although the Department of State and US Agency for International Development (USAID) are responsible for sustainable development programs, DoD has a stake in understanding the security implications of these foreign policy activities.

The 2010 *Quadrennial Diplomacy and Development Review* (QDDR) outlines the US foreign affairs community's new emphasis on the crosscutting challenge of energy, natural resources, and climate.[35] It repeatedly emphasizes the role of energy and natural resources as key enablers in meeting US foreign policy objectives. These, of course, are ultimately intended to increase global stability.

Current biofuel production requires far more water than other energy types.

In many ways, this is a policy response to increased international criticism of the US and developed world's perceived fault and inaction to curb GHG emissions.[3] Such criticism, and the resultant policy response, comes at a time when biofuels and renewable energy technology are being promoted for "green development" programs that enhance energy security and reduce GHG emissions.

However, the oft-cited remedy for addressing rapid growth in energy demand is the technological leapfrogging to renewable energy, something that also promotes sustainable development. This path offers a tremendous opportunity for development, energy security, and GHG reduction, but it also needs to be considered in the context of the overall impacts of these efforts on natural resources (Table 8-1) and local political dynamics. Such programs could have unintended consequences that challenge US national interests.

Table 8-1. Comparison of Water Consumption by Energy Types

Energy Type	Water Consumption (m³/MWh)
Solar	0.0001
Wind	0.0001
Gas	1
Coal	2
Nuclear	2.5
Oil	4
Hydropower	68
Biofuel (first generation)	178

Foreign policy practitioners recognize the now prominent debate on food security versus biofuels. America makes a prodigious contribution to global food production, but our own biofuel-related energy security efforts can also have a big impact on food prices.

The food-versus-fuel debate becomes more nuanced when looking at the impacts on the pieces that make up agriculture. Agriculture is a huge consumer of not just energy, but also of energy-intensive inputs (fertilizer, pesticides, and fuel) and water. The water sector is also expected to be impacted by climate mitigation activities.

The dynamic availability of water and its necessary role in energy and agriculture production represents a challenging interconnection between climate change and national security. Although energy security is a key priority, the NIC reminds us that it is not the only factor to consider in our domestic and foreign activities. The interrelationships between food, water, and climate security warrant thoughtful consideration of their potential interactions and how those interactions might generate unintended consequences.

SOLUTION

Mitigation Potential	★
Operational Impacts	★★
Financial Impacts	★★
Feasibility and Timing	★★★

A strategic baseline and recurring assessment—the QA3D—is needed to understand the opportunities for, and implications of, green development, as well as its indirect and unintended national security implications. Granted, the military already participates in several of these assessments (the QDR, for one), so a separate, jointly developed assessment focusing on national and international climate mitigation's impacts on national security may seem redundant. However, this baseline and recurring assessment is critical in the specific areas addressed.

We propose this review take a twofold approach: first, establishing a National Security Council–coordinated assessment of the impact the nation's international GHG mitigation efforts would have on the global security environment; second, requiring DoD and the State Department to prepare, along with USAID, the QA3D 1 year before their other required assessments (the regularly scheduled QDRs and QDDRs). In such an interagency environment, the departments can develop a joint assessment in an informed, coherent, and deliberate manner to supplement the *National Security Strategy*, along with informing their respective QDRs and QDDRs. Through this process, US foreign affairs efforts can be tailored to support institutional capacity building for sustainable decision making. This can result in the development of resilient communities, nations, and regions.

These would be tremendous opportunities to realize national security benefits. Furthermore, developing partnerships that promote informed, renewable energy strategies and projects can result in cleaner energy, meet development needs, and effectively support GHG emissions reduction.

Climate Security Implications

In recent years, national security and climate change have been viewed in the context of direct impacts—sea level change, floods, drought, and even drastic "abrupt climate change" scenarios. Climate change, and the fear and uncertainty surrounding it, is already impacting international dialogues, something that can be expected to intensify in the coming decades, particularly if impacts manifest.[12]

Concerns over resource scarcity and access to new resources are already influencing international decisions, prompting nations to take action in their own interests. Today, they're jockeying for position to ensure access to natural resources such as energy, minerals, water, and agricultural land. So important is it to Russia that it established itself as an "energy superpower," renewing its claim to potentially lucrative Arctic mineral rights by asserting that an underwater ridge near the North Pole is really part of Russia's continental shelf.[12,36]

Conditions that contribute to and cause environmental mass migration may spur new tensions; these could lead to intervention, either by individual countries or under international banners.[37] At a minimum, humanitarian responses stemming from or attributed to climate-driven hazards, environmental stress, and natural resource competition will rise. This intensified demand will require effective multilateral partnerships.[38]

Tensions over New Borders and Ocean Exclusive Economic Zones

Border disputes and tension over access to and management of natural resources are nothing new. What we've not seen is how climate change could amplify these conflicts as areas become newly accessible. Debates over Antarctic ice have long been framed in the context of exclusive economic zones (EEZs) but are now expanding northward with the potential of arctic transit corridors and claims of territory over new natural resource discoveries.[38]

In international law, EEZ relates to the law of the sea (of which the United States is not currently a party). An EEZ is a sea zone in which special rights for exploration and marine resource use are held by a single sovereign state, extending from the seaward edge of the state's territorial sea out to 200 miles from its coast. A number of nations are perpetually engaged in disputes over EEZ rights.[v]

As mentioned, Russia is promoting claims that the arctic seabed, and the resources contained within, is an extension of its continental shelf. Canada, Denmark, Norway, and the United States have national interests to protect within the Arctic Circle (Figure 8-4). Potential

[v] In fact, the legal position of ice in international law in the Northern Hemisphere differs from that in the Southern Hemisphere according to Emily Crisps in "The Legal Status of Ice in the Antarctic Region."

economic factors associated with changes to these areas are also uncertain, as shifts in arctic ecosystems could influence the viability of lucrative and important fishing industries. The result of such changes, where fisheries and other ecosystem services within national EEZs or international waters experience climate-driven shifts, could, in the future, contribute to greater tensions between nations due to declining or newly accessible resources—regardless of type.

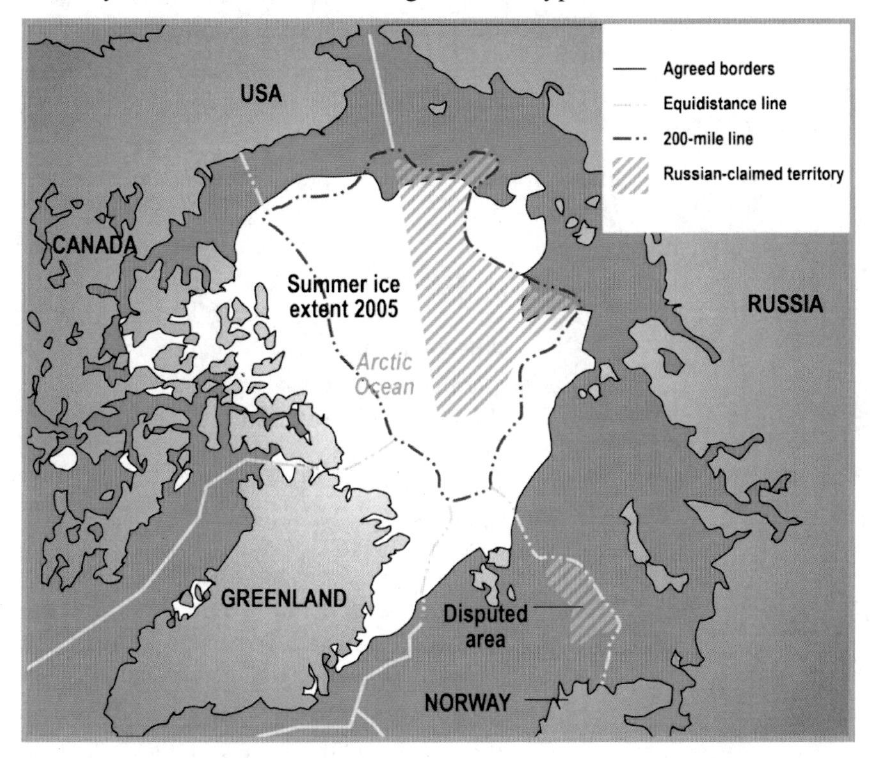

Figure 8-4. Melting Sea Ice within the Arctic Circle Breeds Disputes about National Borders

Natural Hazards

Humanitarian responses are frequently spurred by natural hazards and disasters, both prolonged and sudden. Droughts, floods, and hurricanes are the images that spring to mind when discussing the national security implications of climate change. Earthquakes are not linked to climate change but provide a context for how humanitarian responses unfold.

Over the last two decades, the US national security community has coped with the demands of humanitarian assistance and disaster response (HADR). The demands themselves have steadily grown, as has the percentage categorized as "complex emergencies" (emergencies with security concerns or violence) and those requiring US military involvement.[2]

We are still uncertain of the direct role climate change has in the intensity and frequency of natural disasters such as storms and floods, but it will likely spur some changes in how the US government assesses and responds to natural hazard risks.[37] Even "normal" storm and hurricane events could see intensified storm surge impacts through merely modest sea level rise. These would be felt particularly in coastal nations and megacities with growing urban populations—meaning that the demand for HADR from the US and international community would likely increase.[39]

Water Stress

The National Academies of Science, the Intergovernmental Panel on Climate Change (IPCC), and NIC suggest that climate change will degrade freshwater resources in quantity, quality, or both. Changes in temperature and hydrological regime will certainly challenge many water-limited areas.[w] Even in places where absolute water quantity and management remain sufficient, changes to the timing of precipitation could not only increase famines, floods, and fire, but also impact the success or viability of agriculture and ecosystems.

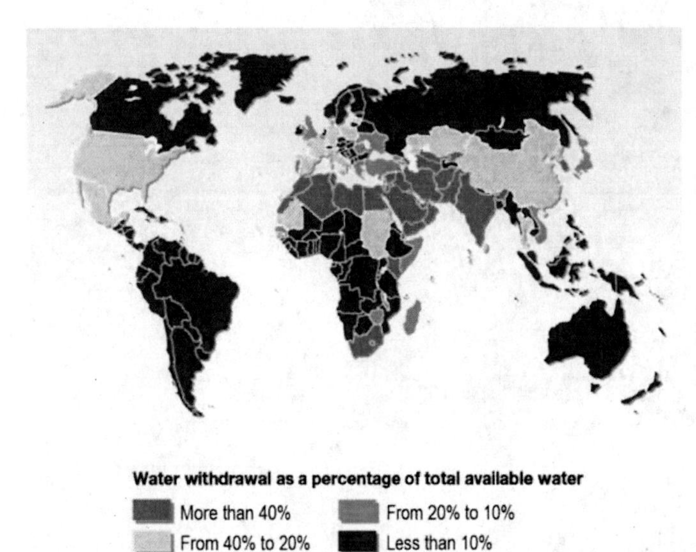

Over 2.8 billion people will face water stress by 2025.

These changes are not limited to countries currently under water stress, which occurs when the demand for water exceeds the available amount during a certain period or when poor quality restricts its use. The NIC's technical resources indicate that 21 countries (with 600 million people) are currently water or cropland stressed, and this group is anticipated to grow to 36 countries (including 1.4 billion people) by 2025.[12]

[w] "Hydrological regime" refers to changes over time in the rates of flow of rivers and in the levels and volume of water in rivers, lakes, reservoirs, and marshes. It is closely related to seasonal changes in climate. In regions with a warm climate, the hydrological regime is affected mainly by atmospheric precipitation and evaporation; in regions with a cold or temperate climate, the air temperature is a leading factor.

Food Production Declines

Food insecurity can have significant humanitarian and development implications and may contribute to greater fragility, instability, potential state failure, crises, and migrations. Food production and distribution are often at the nexus between environmental factors and human security domains, with impacts on national, regional, and local agriculture production, food transportation, and distribution.[x]

Accordingly, something as simple as temperature and precipitation shifts could intensify chronic food security challenges even though the location and severity of famines largely depend on local weather and the region's resilience. Efforts to expand alternative fuel sources could play a role; the rapid expansion of biofuel feedstock production from food cultivation has often been cited as a contributing factor to the economic food scarcity experienced globally in 2007–08.[12]

International Mass Migration

Increased natural hazards or chronic natural resource scarcities could lead to mass population migrations.[37] The aftermath of Hurricane Katrina exemplifies the scale and difficulties of such mass migration, even in the United States.

While increasingly Hobbesian conditions abroad may not result in classifying similar migrants as environmental or climate refugees,[y] increased numbers of refugees can certainly generate human suffering and exacerbate internal and international tensions. Chronic natural resource challenges can likewise have a staggering human cost. For example, the ongoing crisis in Darfur has been cited as a conflict driven by environmental change as well as somewhat by climate change.[z]

---

[x] It is both a function of agricultural output (environment and economic) and distribution (socioeconomic).

[y] Thomas Hobbes argued in his 1651 book *Leviathan* that the state of nature dictates that "during the time men live without a common power to keep them all in awe, they are in that condition which is called war; and such a war as is of every man against every man" (*Leviathan*, chapter. XIII)—in other words, anarchy.

[z] This controversial subject found a prominent voice when U.N. Secretary-General Ban Ki-moon published in the *Washington Post* an Opinion piece titled, "A Climate Culprit In Darfur" (June 16, 2007).

Increased Fragility and Instability, and Failed States

Desperate Somalis ignore military forces and rush toward food supplies.

Current environmental security writings downplay the potential for interstate "environmental wars" but are open to the idea that shocking changes to the availability of natural resources, whether chronic declines or abundance, could contribute to increased state fragility.[12,40] The NIC anticipates that climate change will "exacerbate resource scarcities."[12] Over the past 5 years, US policymakers have focused on state fragility, viewing it as a practical application of human security.

State fragility involves the relationship between the state and its citizenry and is generally gauged using a suite of broad indicators designed to track security, political, economic, and social sectors. Highly fragile states usually parallel those on the list of unstable and failed states.

DoD and development practitioners are increasingly concerned about the impact that degrading natural resources have on fragility, particularly its social and economic components, which are indicators of future political and security condition declines, and eventual instability. Given the climate change impacts on water, food, and energy, countries with limited resilience or adaptive capacity could become more fragile and weak, and eventually fail.

Climate Adaptation Opportunities

Approaches to fragility reduction are intersecting with the efforts of those building climate change resiliencies in nations, regions, and local communities. This convergence reveals new adaptation opportunities and synergies. Proactive US government planning, partnership, and peace-building activities take on new urgency through a clearer understanding of climate impacts. However, the growing assortment of needs and constrained budget resources necessitates thoughtful interagency approaches. It also calls for the engagement of non-traditional capacities to effectively protect our national security interests and achieve national military objectives, such as to "Deter and Defeat Aggression" and "Strengthen International and Regional Security."

These new challenges in adaptation also come with new opportunities. As adaptation planning and technical transfer expand within the US development community, the result is collaboration with the defense sector, particularly when a security situation is compromised in complex environments. Climate adaptation offers a venue for international engagement, knowledge exchange, and technology transfer.

Efforts to better realize this potential should include the following:

- Develop climate scenarios for planning.
- Issue conflict–prevention policy.
- Prepare climate scenario interagency war games.
- Mandate service-level climate change capability assessments.
- Develop a conflict-instability-fragility early warning system.

Develop Climate Scenarios for Planning

Recommendation: *Develop global, geospatially explicit, climate model–informed scenarios with explicit ranges, uncertainties, and natural resource impacts to meet national security planning needs over the next 5, 10, 25, and 50 years.*

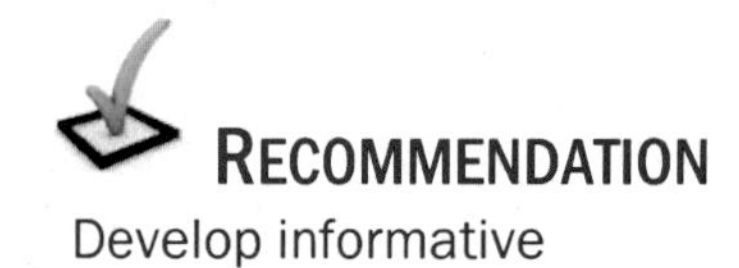

Develop informative scenarios

CHALLENGE

Science and scenarios are disconnected from what is useful to the military. On the one hand, climate scientists and modelers have generated plenty of literature and model data and continue to expand their knowledge while acknowledging the uncertainties. On the other hand, national security leaders, such as retired Major General Richard Engel of the NIC and Rear Admiral David Titley of the Navy's Task Force Climate Change, lament the scarcity of climate scenarios that provide usable inputs for planning and capability assessments.[41]

The military already prepares in the face of many uncertainties, including a changing future security environment and tactical situations on the battlefield. Warfighters can only effectively prepare if they know the possible worst cases, expressed in understandable terms of reference. Climate scientists, in partnership with the national security user community, can help provide these. Although products such as the National Research Council's *Warming World: Impacts by Degree*[42] are a step in the right direction, military leaders have noted the need for more useful models, impact assessment, and localized information.

SOLUTION

Adaptation Potential	★★★
Operational Impacts	★★
Financial Impacts	★
Feasibility and Timing	★★

Collaboration between DoD and the National Oceanic and Atmospheric Administration (NOAA), as suggested by Rear Admiral Titley, would help address these gaps.[41] We recommend taking this a step further: mandate that DoD, DHS, State, USAID, and other federal agencies develop global planning scenarios with explicit ranges, extremes, uncertainties, and natural resource impacts to meet national security planning needs. The US Interagency Climate Change Adaptation Task Force could serve as the responsible party for facilitating its own recommendation to "ensure scientific information about the impacts of climate change is easily accessible." This should function as a venue to leverage the capabilities and applications across the federal government.[43] These scenarios should be developed using the best science currently available and revisited after every IPCC assessment report update.

Before incorporating the science, coordinated efforts should be made to determine the baseline information required for the homeland security, defense, diplomacy, and development communities. The focus should then be on the development of climate scenario products for planning purposes. For example, lessons should be gleaned from the current Air Force Minerva Energy and Environmental Security Initiative.[44] Such scenarios could benefit from the following:

- 5-, 10-, 25-, and 50-year planning horizons
- A range of anticipated cases (from best estimates to extremes)
- Downscaling to resolutions appropriate for various national security community needs.

In marrying any current climate science "best estimates" with national security user needs, US government scenarios need to be jointly prepared with stated uncertainties. They should be used for capability assessments, updating global future security environments[45,aa] and contingency plans. They have further use in programmatic and installation risk management and adaptation. These scenarios could also be leveraged in US foreign assistance efforts as baseline stress tests to identify vulnerabilities and build greater resilience with regions or local communities.

EXAMPLE

Since 2009, DoD's Strategic Environmental Research and Development Program (SERDP) has been funding projects to adapt regional and downscaled local models to create more accurate military climate impact and adaptation approaches.

Many military training and power projection bases, particularly naval installations, are located in coastal areas, and their supporting infra-

[aa] Funded by the DoD Minerva program, the Climate Change and African Political Stability (CCAPS) Program is studying and pioneering approaches to better understand the implications of climate change on FSE on the African continent, including humanitarian disasters and conflict.

structure was often built using outdated design assumptions about climate and natural hazard regimes. Over time, rising sea level and exposure to intensified storm surges have made these facilities more vulnerable (Figure 8-5). This becomes a national issue when you consider that DoD's operations rely on the ability to use this supporting infrastructure to provide energy, water supply, waste treatment, transportation, and other essentials.

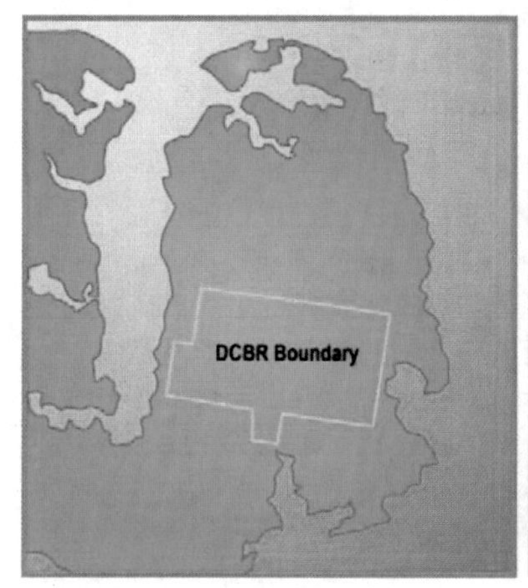

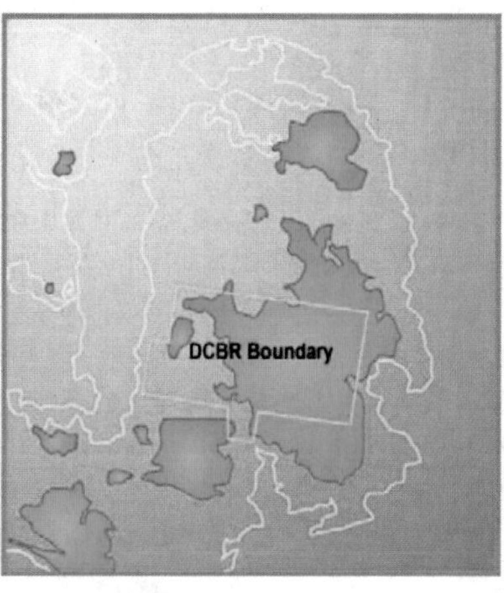

Figure 8-5. Change in Land Cover at Air Force Dare County Bombing Range (DCBR) in North Carolina Due to Sea Level Rise Projected by 2100

In response, SERDP has been supporting efforts to model the impacts of sea level rise, develop a vulnerability assessment framework, and integrate with asset and mission capability risk approaches. A number of bases are already included in this effort, including Eglin Air Force Base, FL; Naval Station Norfolk and Hampton Roads, VA; Naval Base Coronado, CA; Marine Corps Base Camp Lejeune, NC; and Marine Corps Base Camp Pendleton, CA. These ongoing projects seek to better understand how vulnerable installation capabilities might be and what to do about it.

Eventually, the outcomes of these assessments will help identify what military installation planners need to know, what the existing climate models can provide, and the information gaps that need to be filled.

Issue Conflict-Prevention Policy

Recommendation: *Issue a DoD directive and instruction on fragility-reducing, conflict-prevention policy (augmenting existing DoD Instruction 3000.05) that mandates the coordinated development of state fragility, natural security, and climate security strategies.*

Mandate conflict-prevention strategies

CHALLENGE

Today, the reality is that the current capabilities of the US national security community may not meet future demand for HADR or missions that result in stability, security, transition, and reconstruction.

There is little discussion of capabilities and strategies in the context of "natural security," a term defined within the Center for a New American Security (CNAS) concept paper as ultimately meaning "sufficient, reliable, affordable, and sustainable supplies of natural resources for the modern global economy … [it covers both] consumption and consequences."[46]

This is especially true given climate change threat amplifiers; it is especially worrisome when you include increases in environmental complexity. Up until 2005, the US military's foremost priority was preparation for and execution of combat operations. However, the real-world demands of stabilizing operations forced a rapid shift in this thinking; the result was DoD Directive 3000.05.[4] This policy stated that combat operations and stability missions were both primary mission responsibilities for the US military.

Combatant commands recognize and proactively pursue conflict-prevention missions as much as possible, but DoD policy and US governmental priorities still do not explicitly reflect this necessity. Given climate change–induced impacts, we assert that non-traditional security concerns will likely expand and require proactive conflict and fragility prevention activities.

The QDDR reflects such a shift in the foreign affairs community, but this also requires DoD and the military services to rethink the scope of their role—including the capabilities needed (engagement brigades)[3] and preparation for missions. Retired General Anthony Zinni, Former Commander of US Central Command, is purported to have said, "I have two missions: warfighting and engagement. If I do engagement well, I won't have to do the warfighting."[47] Those who share this view would presumably welcome a proper response.

SOLUTION

Adaptation Potential	★★
Operational Impacts	★★
Financial Impacts	★★
Feasibility and Timing	★★

DoD should issue a directive and instruction on conflict prevention and engagement to complement the DoD Directive 3000.05 focus on warfighting and stabilization operations as well as the declared State Department and USAID policies.[4] A primary directive in this new DoD policy should be to mandate development and integration of strategies for fragility, natural security, and climate security.

The instruction should also provide guidance on how the military should integrate these crosscutting strategies into its program planning, budgeting, and execution activities. Although the policy for stabilization activities requires a focus on early warning signs indicating instability, policies for conflict prevention should leverage the state fragility concept adopted by the defense and development communities.

This policy mandate should actively support the military-to-military engagement objectives of regional combatant commands to build and broaden capacity with foreign militaries. Similar interagency approaches to multilateral partnerships should be more effectively leveraged with competing near-peer powers and allies.

Another goal should be pursuing bilateral agreements that allow national security–critical objectives to be met. These should be framed in the context of fragility reduction and climate adaptation. These agreements should focus on economic development and technological collaboration in ways that offer strong incentives for partners to engage in bilateral or broader multilateral collaborations. For example, this approach could build on current engagement objectives and efforts in the USAFRICOM and US Southern Command areas of responsibility.

Prepare Climate Scenario Interagency War Games

Recommendation: *Use climate scenarios for plans, war games, and exercises for humanitarian response and stabilization missions, along with geopolitical conflict-prevention responses with US government interagency partners.*

CHALLENGE

Two primary challenges face those who would integrate and "operationalize climate change"[bb] and its effect on national security interests. First, the defense, diplomacy, and development communities would seem to have few, if any, scenarios that are consistent and scientifically grounded, yet relevant. Second, the lack of such scenarios for assessing the future security environment has made it difficult for leaders, analysts, and capability planners to understand broader US national interests and how they are affected.

[bb] "Operationalizing climate change" is not about standing up a new program but rather making it part of the current institutional planning process, exercises, and thought processes.

This first challenge will be partially addressed by our suggestion to develop scenarios, but national security policymakers and practitioners must still work to integrate and operationalize these recommendations.

SOLUTION

Adaptation Potential	★★
Operational Impacts	★
Financial Impacts	★★
Feasibility and Timing	★★★

Building on the previous recommendation to develop more robust US government climate scenarios, the next logical step is to organize climate scenario interagency war games. Not war games in the traditional sense, these are "climate change–induced conflict response games," in which the use of defense assets is only one among several options.

This program must comprise individual and joint DHS, DoD, State, and USAID war-game exercises, where the goal is to plan for, prioritize, and rehearse responses to likely and extreme climate change impacts. Initial scenario-driven operational areas of focus might include the following:

- Geopolitical tensions and conflicts
- Stabilizing missions and peacekeeping
- HADR (foreign and domestic).

A number of organizations already develop scenarios and support war-game exercises—including the US Army Training and Doctrine Command, Army War College, Air Force Air University, and Federal Emergency Management Agency—so these efforts can be leveraged to jointly develop phased war-game and exercise plans.

These exercises are a well-established mechanism but need to focus more on the unique convergence of fragility, natural resources, and climate impacts. They also need to be performed internally and in an interagency context; this will best engage partners in smart-power activities. These war games should focus on working through the issues, roles, coordination, and capabilities impacted by climate change.

Exercises must also be expanded to include combined operations with our allies. The resulting lessons learned should aid in consensus building and technology transfer, such as through venues like the NATO Science Programme,[cc] and they could facilitate opportunities

[cc] The NATO Science Programme initially promoted collaboration between scientists in NATO countries. When the Cold War ended, the Programme was opened up to scientists and experts from non-NATO countries.

to advance environmental intelligence and fragility-reduction approaches.

EXAMPLE

Several exercises have already explored the energy, resource, and climate change concerns facing national security practitioners. In May 2005, a high-profile war game called Unified Quest 2005 engaged hundreds of active-duty and retired military officers at the Army War College in Carlisle, PA. Participants closely modeled the asymmetric warfare faced in Iraq to craft an Army war strategy in preparation for a potential war.[48] One aspect the participants explored was the role of natural resource access. Their scenario addressed the control and cutoff of energy resources to Europe. One notable result of this scenario, as noted by a participant, was that the "average American will be pretty upset when gas hits $8 per gallon."[48]

The Unified Quest war-game examined the impact of natural resource access and discontinuities in energy supplies.

Other externally hosted scenario exercises have focused on the national security implications of climate change. In July 2008, CNAS hosted representatives from military organizations and businesses around the world who participated in an international climate change war game.[49]

The game was set in the year 2015 and offered both US and international participants a risk-free environment to work toward a "Framework Agreement on Managing Long-Term Climate Change." This was done in the context of climate projections for 2050 and 2100 and their implications.[49] In November 2009, the not-for-profit organization, CNA, developed and hosted a second, similar scenario war game. These activities illustrated the need for better information on climate change consequences, offered novel implications for planners and decision makers, and analyzed the capacity needed to respond to climate-related threat multipliers.[49]

Mandate Service-Level Climate Change Capability Assessments

Recommendation: *Mandate department- and service-level capability assessments using guidance on the future security environment that includes a natural security component and climate scenarios.*

CHALLENGE

The socialization and exchange (the "what" and "how") required to develop energy and climate strategies for DHS and DoD still must take place through discussing *anticipated* climate impacts on doctrine, training, power projection platforms, and future force capabilities. They have never answered the whats and hows of the matter, even with studies that have already addressed climate change concerns and capabilities. For example, the 2010 QDR, while presenting DoD's assessment of future capabilities, highlights a need to develop "enterprise-wide climate change and energy strategies."

SOLUTION

Adaptation Potential	★★★
Operational Impacts	★★
Financial Impacts	★★
Feasibility and Timing	★★★

Military programs and planners already plan for anticipated mission capability requirements. In their acquisition and planning process, they use a host of reports and guidance, including the QDR, force structure guidance, assessments of future security environments, and any joint requirements. Using the proposed scenarios and climate war-games outputs, we echo something recently endorsed by Navy Task Force Climate Change: specific capability assessments for climate change–driven geopolitical, stabilization, and conflict-prevention missions. For DoD, these capability assessments could potentially be coordinated through the Joint Requirements Oversight Council, which is a part of the DoD acquisition process. A Defense Science Board study focusing on climate change impacts on Africa has given us a starting point.

The USCG CUTTER *HEALY* is one of only four ice-hardened vessels in the American fleet.

The service branches would perform the assessments as well as develop documents on needed initial capabilities, along with capability development. These efforts should focus on what the new or enhanced capabilities should be, at what level and appropriate timing for development. These capability assessments should include factors such as force mix, tactical systems, and weapons platform programs, as well as force projection and training installations.

EXAMPLE

A 2011 National Academy of Sciences study sought to identify and understand climate change and the naval capabilities that may be needed. This report offers a strategic, technically focused analysis of the naval capabilities required to respond to a rapidly changing Arctic. This is not a full Navy capabilities assessment and plan; it nonetheless provides a starting point for integrating anticipated climate change impacts and linking the necessary military capabilities, such as the need for "ice-hardened" vessels rather than "ice breakers."

Develop a Conflict-Instability-Fragility Early Warning System

Recommendation: *Develop interagency, joint, and military service–level requirements for risk-based, conflict-instability-fragility early warning and decision support systems, including natural hazard, natural resource, and climate components.*

RECOMMENDATION

Develop early warning systems

CHALLENGE

Discussions regarding climate change and national security quickly reveal many of the gaps in understanding regarding the connection between instability and fragility and the environment.

Over the past few years, numerous academic studies and US governmental research and development efforts (NIC, national labs, Army, etc.) have examined how conflict, instability, and fragility are assessed. They've looked into how to develop early warning systems that incorporate environmental factors, including climate change.[40] Unfortunately, while several instability predictive models and fragility early warning approaches exist, most have not incorporated natural resource or environmental factors (not for lack of research, though). Furthermore, they are often built for specific uses and not devised in a joint context that would include strategic, operational, and tactical levels.

Early warning systems must be risk based to effectively inform decision makers, but no such system is available yet. USAID developed an instability-fragility approach and is examining opportunities to incorporate natural resources and environmental factors. However, the incorporation of such factors is still pending.

SOLUTION

Adaptation Potential	★★★
Operational Impacts	★★
Financial Impacts	★
Feasibility and Timing	★

In general, early warning is less about prediction than it is a monitoring of trends. It provides a timely warning of events that have implications for US national security interests (signposts, red flags, etc.).

We recommend the development of an interagency, joint, and military service–level conflict-instability-fragility early warning and decision support architecture that leverages existing systems and begins to address identified gaps.[50] Conflict and state instability are directly relevant to national security. Over the past 5 years, the Army and USAID have adopted the fragility concept in their doctrine and increasingly into decisions. Fragility measurement and early warning systems have been rapidly emerging amidst increased academic and policymaker use of the term.[51]

USAID has gone further to develop an instability and fragility analysis approach at the country level. Fragility and natural resource strategies should build on these advances to include human and natural system interface sectors such as agriculture, water, and energy. These offer a familiar, mission-relevant frame of reference.

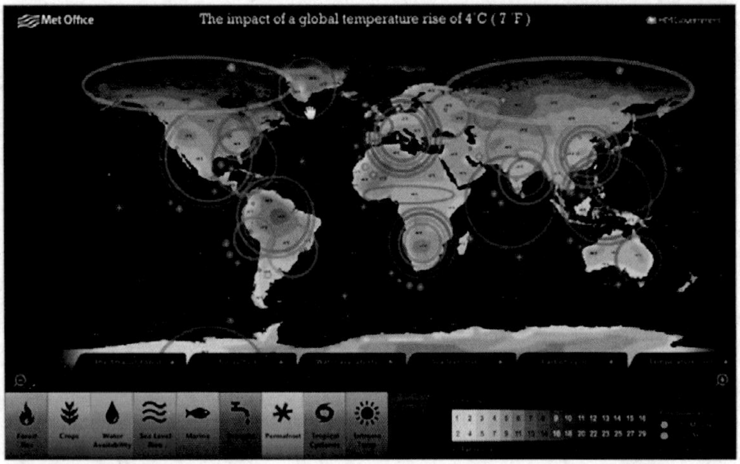

An interactive web site created by the UK Met Office illustrates some impacts that may occur if the global average temperature rises by 4°C.

The aim is to provide useful situational awareness and forewarning that can be used for conflict-prevention planning and proactive engagement. This architecture should use fragility as a bridging concept and incorporate natural hazards, natural resources, and climate components. We suggest that this early warning system leverage and integrate with DoD-funded climate vulnerability research, such as the projects of the Minerva Initiative at the University of Texas at Austin, University of North Texas, and Air War College, to the extent possible.[dd]

The US national security community is positioned to support development of decision support architecture for observed climate changes and implications. The US network of embassies, military attachés, and intelligence personnel is global and can provide timely reporting of intelligence. This intelligence should include more reporting on natural resource, environmental, and climate factors.

In a time of limited budgets, this interagency capability could be cost-effectively expanded with additional regional subject matter experts. These experts would supply physical, environmental, and climate analysis, while providing a common framework to avoid duplicative departmental efforts. The necessary data for such a system are becoming more available—they just need to be assembled correctly and in a meaningful way.

[dd] The Minerva Initiative is a broad effort focused on areas of strategic importance to US national security policy—certainly climate change and energy security would fit the bill.

EXAMPLE

Remotely sensed data from NOAA, NASA, and the National Geospatial-Intelligence Agency already provide geospatial monitoring and analysis of the changing natural resources and environment.

A well-known but more focused example of such an approach is the USAID Famine Early Warning Systems Network. This program was developed to identify international food supply failures. It assesses and warns of conditions that can lead to famine or other food insecurity in sub-Saharan Africa, Afghanistan, Central America, and Haiti.[52] The Famine Early Warning Systems Network utilizes a combination of remotely sensed satellite data from NASA and NOAA, coupled with a ground network, to monitor agricultural, market, and livelihood conditions, which all help to identify countries with emerging food security risks.[52]

Final Thoughts

Climate change and its implications for the future security environment represent an immense challenge for the US national security community. However, these same global trends and environmental changes also present a host of opportunities for our service members, civil servants, and citizen stakeholders.

Fortunately, many of the key elements are already in place for the homeland security, national defense, and foreign affairs communities to address the policy, management, and technical gaps associated with energy and climate change. The primary question now is how to move forward with finite resources and seize opportunities to leverage US leadership and vision in advancing security and resilience to climate change.

In conclusion, we offer the following:

- US military leaders increasingly realize that the current national security approaches are being undermined by rising energy costs and their indirect environmental impacts.
- The means achieving greater energy security is often tied to climate mitigation, so value-added benefits should be identified and aggressively pursued, even if the reasons for doing so are unrelated.
- The national security community needs to identify and pursue more solutions that result in "wins" on multiple fronts, including energy and GHG.

- Conflict prevention will expand as a mission in the future, so the foreign affairs and defense communities will need more proactive planning and coordinated action to effectively protect national security interests.
- Enhanced situational awareness and management of the national security risk portfolio is necessary to effectively plan and prepare for the future.

The national security challenges associated with climate change are vast, but the opportunities have never been greater to reshape the elements and use of US national power. US technological prowess, intellectual expertise, intelligence gathering and analysis, logistical supply and planning, and airlift/sealift capabilities require investment, yet leadership can reap favorable security returns. These outcomes can and should simultaneously address the emerging challenges and opportunities presented by climate change.

---

[1] National Intelligence Council (NIC), *Russia: The Impact of Climate Change to 2030: Geopolitical Implications*, CR 2009-16, September 2009.
[2] CNA and Oxfam America, *An Ounce of Prevention: Preparing for the Impact of a Changing Climate on US Humanitarian and Disaster Response*, (Alexandria, VA: CNA and Oxfam America, June 2011), p. 9.
[3] Shannon Beebe and Mary Kaldor, *The Ultimate Weapon is No Weapon* (New York: Public Affairs, 2010), p. 175.
[4] DoD, "Military Support for Stability, Security, Transition, and Reconstruction (SSTR) Operations," DoDD 3000.05, November 28, 2005. Superseded by "Stability Operations," DoDI 3000.05, September 16, 2009.
[5] The White House, 2010 *National Security Strategy*, p. 13.
[6] DoD, *Quadrennial Defense Review Report* (Washington, DC: DoD, February 2010), p. 13.
[7] DoD, *National Military Strategy of the United States* (Washington, DC: DoD, February 2011), p. 4.
[8] Jeremey Alcorn, *US Environmental Security: Understanding And Enabling It To Matter*, http://envsec.gmu.edu, 2008, pp. 95–96/.
[9] Carolyn Pumphrey, ed., *Global Climate Change National Security Implications* (Carlisle Barracks, PA: US Army War College, Strategic Studies Institute, 2008).
[10] Kent Butts and Curtis Turner, *Environmental Security and Cooperation Workshop: Building Governmental Legitimacy and Creating Conditions Inhospitable to Terrorism* (Carlisle Barracks, PA: US Army War College, Center for Strategic Leadership, September 2004), p. 1-2.
[11] Rymn Parsons, *Taking Up the Security Challenge of Climate Change* (Carlisle Barracks, PA: US Army War College, Strategic Studies Institute, August 2009).

[12] NIC, *Global Trends 2025: A Transformed World* (Washington, DC: US Government Printing Office, 2008), pp. 41, 54, 66.
[13] Dennis Blair, *Annual Threat Assessment of the US Intelligence Community for the Senate Select Committee on Intelligence*, February 2, 2010.
[14] US Joint Forces Command, *Joint Operating Environment (JOE)*, 2010, http://www.jfcom.mil/newslink/storyarchive/2010/JOE_2010_o.pdf.
[15] DoD, *Strategic Sustainability Performance Plan FY 2010*, August 26, 2010.
[16] DoD, *Quadrennial Defense Review Report* (Washington, DC: DoD, February 2010), p. 87.
[17] Alaina Chambers and Steve Yetiv, "The Great Green Fleet, The US Navy and Fossil Fuel Alternatives," *Naval War College Review*, Vol. 64, No. 3 (Summer 2011), p. 66.
[18] Defense Science Board, *Report of the Defense Science Board Task Force on DoD Energy Strategy "More Fight—Less Fuel* (Washington, DC: Defense Science Board, February 2008), p. 4.
[19] Defense Acquisition University, *Acquisition Community Connection*, https://acc.dau.mil/CommunityBrowser.aspx?id=234122&lang=en-US.
[20] Global Security, "F-22 Raptor Materials and Processes," *Globalsecurity.org*, http://www.globalsecurity.org/military/systems/aircraft/f-22-mp.htm.
[21] NASA, "Risk Management," *Environmental Management Division*, http://www.nasa.gov/offices/emd/home/risk_management.html.
[22] James Leatherwood, "NASA's Assessment Framework for Addressing Adaptation to Climate Change Impacts" (presentation, the National Defense Industrial Association (NDIA) E2S2 Conference, 2011).
[23] US Department of the Army, Environmental Planning Support Branch, *Programmatic Environmental Assessment for Fielding and Use of Mine Resistant Ambush Protected Vehicles at Army Installations in the United State,*. June 2009.
[24] NDIA, "Fully Burdened Cost of Fuel (FBCF)" (presentation, August 12, 2008).
[25] Kate Brannen, "Chimney' Deflects IEDs," *Defense News* (Gannett Government Media Corporation, 2010).
[26] David McCollum, Gregory Gould, and David Greene, *Greenhouse Gas Emissions from Aviation and Marine Transportation: Mitigation Potential and Policies* (Arlington, VA: Pew Center, December 2009).
[27] Greener by Design, *The Technology Challenge: Report of the Technology Sub-Group*, August 2001.
[28] Robert H. Liebeck, "Design of the Blended Wing Body Subsonic Transport," *Journal of Aircraft*, Vol. 41, No. 1, 2004, pp. 10-25.
[29] John Benedict, "Adapting Ship Operations to Energy Challenges—Overview" (presentation, JHU/APL, Adapting to Climate & Energy Challenges: Options for US Maritime Forces, March 29, 2011).

[30] CDR Daniel Orchard-Hays, "Aviation Energy Initiatives in the Fleet" (presentation, JHU/APL, Adapting to Climate & Energy Challenges: Options for US Maritime Forces, March 29, 2011).
[31] Glen Sturtevant, "Adapting Ship Operations to Energy Challenges," (presentation, JHU/APL, Adapting to Climate & Energy Challenges: Options for US Maritime Forces, March 29, 2011).
[32] CDR Scott Fuller, "Navy Flight Simulator Training," (presentation, JHU/APL, Adapting to Climate & Energy Challenges: Options for US Maritime Forces, March 29, 2011).
[33] DoD, *Opportunities for DoD Use of Alternative and Renewable Fuels: FY10 NDAA Section 334 Congressional Study*, July 2011.
[34] Samuel King Jr., "Air Force officials take step toward cleaner fuel, energy independence," *Inside Eglin AFB*, March 26, 2010.
[35] US Department of State, *Leading Through Civilian Power: The First Quadrennial Diplomacy and Development Review (QDDR)*, 2010, www.state.gov/s/dmr/qddr/.
[36] Yuri Zarakhovich, "Russia Claims the North Pole," *Time.com*, July 12, 2007.
[37] Jeffrey Stark, Christine Mataya and Kelley Lubovich, *Climate Change, Adaptation, and Conflict: A Preliminary Review of the Issues*, CMM Discussion Paper No. 1 (Washington DC: USAID, October 2009).
[38] National Academy of Sciences, *National Security Implications of Climate Change for US Naval Forces* (Washington, DC: National Academies Press, 2011).
[39] The June 2011 Oxfam America and CNA joint release, "An Ounce of Prevention: Preparing for the Impact of a Changing Climate on US Humanitarian and Disaster Response," assesses HADR requests in terms of need and type. Importantly, it presents these anticipated needs in the context of the current US and international HADR structures, capacities, and trends.
[40] Army Environmental Policy Institute, *Environmental Factors In Forecasting State Fragility* (Washington, DC: AEPI, June 2010).
[41] Lauren Morello, "Defense Experts Want More Explicit Climate Models," *New York Times*, June 24, 2010.
[42] National Research Council, *Warming World: Impacts by Degree*, 2011.
[43] The White House, "US Interagency Climate Change Adaptation Task Force," *Council on Environmental Quality,* http://www.whitehouse.gov/administration/eop/ceq/initiatives/adaptation.
[44] More information on the Air Force Minerva Energy and Environmental Security Initiative can be found at http://afri.au.af.mil/minerva/index.asp.
[45] Climate Change and African Political Stability, The Robert S. Strauss Center for International Security and Law, http://ccaps.strausscen ter.org/.
[46] Sharon Burke, *Natural Security* (Washington, DC: Center for a New American Security, June 11, 2009)"
[47] Kent Butts, "Water & Health: Security and Stability Partnerships" (presentation, Center for Strategic Leadership).

[48] Sandra I. Erwin, "Iraq Lessons Pervade Army War Games," *National Defense Magazine*, June 2005.
[49] Sharon Burke and Christine Parthemore, *Climate Change War Game: Major Findings and Background* (Washington, DC: Center for a New American Security, June 1, 2009).
[50] AEPI, *Environmental Change and Fragile States: Early Warning Needs, Opportunities, and Intervention*, 2011.
[51] Monty Marshall, *Working Paper: Fragility, Instability, and the Failure of States Assessing Sources of Systemic Risk* (Washington, DC: Council on Foreign Relations, Center for Preventative Action, 2008), p. 2.
[52] US Geologic Survey, *FEWS NET Data Portal*, http://earlywarning.usgs.gov/fews/.

Initiative

Even if you're on the right track you'll get run over if you just sit there.

—Will Rogers

None of what this book discusses is worthwhile if stakeholders or functional managers don't commit to pursuing these goals on behalf of their organization. Initiative involves the first step taken and the energy that drives it. It's a moment seized, an opportunity acted upon.

Today you're at a crossroads. You can continue your present course, and hope for the best. Or you can take the initiative to act on what's down the road. That's the argument that's been made for so long in regard to climate change. Unfortunately, it's not that simple for your organization.

Here are some things to keep in your back pocket as you decide on the right course of action for your organization:

- Many steps to mitigate greenhouse gases (GHG) will save money. Not all steps, to be sure, but finding ways to operate more efficiently should accomplish both ends.
- Some recommendations involve spending money now to save it in the future. These can be considered investments or, alternately, hedges against great losses if certain events unfold. They are insurance payments of a sort.
- Contribute toward national and international goals. Make the world better for America's children and grandchildren. Invest in the future on their behalf. Be part of the solution.

- Gain public relations benefits. Be recognized publicly as a forward-looking organization aware of important social trends and willing to take part in them.
- When the time comes to impose constraints through legislation or regulation, help attain credibility for your organization so that it can participate in the process.

After eight chapters covering seven functional areas, we offer up one final number for consideration: 70. That's how many challenges, solutions, and recommendations this book has presented. It's a dizzying array.

The Next Step

These 70 items have underlying themes worth reinforcing. Each represents a way of thinking broad enough in its approach to climate change to offer a framework for the specific actions it encourages. These apply to any functional manager, not just to someone in a given specialty.

Here's the challenge that 70 presents, however. No one can go at this alone. Having 70 recommendations for an organization to consider, spread across seven functional areas, is the quintessential opportunity for collaboration across silos, key in any organization's success, regardless of the challenge it faces. You might not be the lead for all of these initiatives, but you can support those who are and help shape strategies and solutions that benefit the organization as a whole.

Each chapter's recommendations were rated on the basis of four criteria: mitigation/adaptation potential, operational impacts, financial impacts, and feasibility and timing. But what do they say collectively?

Functional managers in the public and private sectors face an issue far too often pushed to the back burner. Individuals and organizations have approached the climate change challenge in the same way they've tackled every other concern facing them—with an eye on the efficient use of limited resources. This isn't a time for wasteful or wild spending, no matter how pressing the concern.

The themes that emerge in this book indicate that a response to climate change doesn't have to be wasteful or out of line with an organization's strategic goals. In fact, the opposite is true; the things people can do now to act on climate change risks are also what's good for any organization's health. They are efficient, proven, and just need to be supported and enacted by stakeholders in a way that makes sense for a particular situation.

Mitigation Themes

Five dominant themes emerged from this book's mitigation recommendations (Table 9-1):

1. Plan cooperatively.
2. Seek the financial benefits of GHG reductions.
3. Operate more efficiently.
4. Take direct action.
5. Conduct ongoing assessments.

Table 9-1. Summary of Mitigation Recommendations

Health	Information	Land	Structure	Vehicles	Supply	Security
Assess health impacts 1	Exploit remote work 2	Regulate local land use 1	Measure, report, reduce 1	Reduce transportation demand 2	Set mitigation strategy 1	Manage GHG risks 1
Use telehealth techniques 2	Increase operating efficiency 3	Sequester carbon in forests 4	Coordinate mitigation efforts 1	Improve fuel efficiency 3	Screen and focus 1	Mandate energy and GHG metrics 1
Reduce heat islands 4	Increase hardware efficiency 3	Reduce livestock emissions 4	Mitigate through procurement 4	Reduce fuel emissions 4	Leverage reporting standards 1	Expand energy efficiency 3
Advocate energy efficiency 3	Inform efficiency decisions 3	Adapt agricultural practices 4	Mitigate cradle to grave 4	Optimize transportation modes 4	Acknowledge partner capability 4	Use renewable fuels 4
Appraise mitigation effects 5	Connect to mitigate 1	Trade carbon credits 2	Mitigate through behavior 2	Use electric vehicles 4	Mitigate and learn 5	Assess mitigation tradeoffs 5

Note: The numbers in the boxes correspond to the primary theme of this recommendation.

Plan Cooperatively

There are more than a few partners out there to lean on when mitigating GHG—and several recommendations identified this. Efforts to mitigate *structure*-associated GHG, for example, are likely to provide opportunities for cooperation between users of that infrastructure and stakeholders in related types of infrastructure. Similarly, *supply chain* managers will want to seek the cooperation of their major suppliers if they wish to achieve meaningful reductions throughout the supply chain. This collaboration of functional managers involves planning and sharing goals and methods.

A functional manager should work with those holding their organization's purse strings; budget managers and chief financial officers need to support mitigation and adaptation planning.

Seek the Financial Benefits of GHG Reductions

Along those lines, a number of this book's recommendations highlight how GHG mitigation can be financially rewarding to an organization. For example, the "Health" chapter highlighted how the use of telehealth techniques can enhance access to medical information while saving resources otherwise devoted to this pursuit. Providing staff members remote connectivity to an office reduces the amount of office and parking space needed, increases employee productivity, and lowers labor costs in return for the flexibility and time savings that employees obtain. Organizing transportation demands better results in reductions in fuel consumption as well as wear and tear on fleet vehicles.

Carbon credit markets, as described in the "Land" chapter, also present opportunities. These markets can monetize mitigation achievements. Not all recommended actions will necessarily generate a financial return, but the ones that do offer the best of both worlds—syncing the organization's financial mission with its climate-related mission.

Operate More Efficiently

Operational efficiency is purely a good business model, and every organization should strive toward it. In the process, they can reduce GHG emissions. The "Information" chapter discusses curbing power use through centralized processing facilities. It also suggests finding greater operational efficiency through an enterprise architecture approach, examining critical interactions within an organization and looking for ways to organize processes more efficiently. The "Vehicles" chapter shows how organizations can adopt more fuel-efficient fleets. The "Security" chapter argues that operational energy efficiency pays off in both reduced costs and enhanced military effectiveness. And the "Health" chapter points out that opportunities exist to reduce the energy intensity of medical equipment.

The biggest issue with efficiency is positioning an organization to get there, a process that often involves an investment. Normally, this is a capital investment associated with more energy-efficient equipment, and frequently such equipment pays for itself over time—it also happens to help mitigate GHG emissions.

Enhancing an organization's ability to operate efficiently often results in reductions in the resources needed to produce output and subsequent savings in energy—and GHG emissions. Suffice it to say, the search for greater operational efficiency is stressed throughout this book as means for emission mitigation.

Take Direct Action

A number of recommendations advocate the use of equipment or processes that directly mitigate GHG. For example, the "Vehicles" chapter recommends substituting alternate fuels for conventional fuel and electric vehicles for a conventionally powered fleet. The "Land" chapter notes the sequestration opportunities available in forests or other lands. Similarly, altering agricultural practices to adopt less GHG-intensive methods ("Land"), diminishing the effect of urban heat islands ("Health"), and modifying infrastructure procurement practices ("Structure") are all examples of things that can be done that don't deviate from the reduction goal path.

Of course, a great many direct actions that could reduce GHG emissions aren't always sensible or financially feasible. This book avoids discussing these actions. Nevertheless, the available direct actions do work. They are effective because they are the path of least resistance. A manager or stakeholder who focuses on mitigation should be able to identify at least some that are well within the organization's financial limits.

Conduct Ongoing Assessments

The thought that goes into a stakeholder's mitigation decisions plays a big role in what's actually done; this theme reinforces the key pieces in these critical evaluations.

For example, the book recommends establishing baseline criteria that include GHG mitigation in investment or purchase decisions ("Health"). We also suggest delaying further effort until lessons are gleaned from initial mitigation actions ("Supply"). And we suggest assessing tradeoffs associated with mitigation actions ("Security").

Mitigation is more than a series of discrete actions. It involves an ongoing, fluid assessment process. It starts with initial planning, continues with measurement and judgment whether the actions worked, and infers how these efforts might be improved in the future. Thus, the evaluation of mitigation efforts is ongoing and prolonged. Stakeholders will find it effective to consider options for ongoing assessment, draw out any lessons learned, and change plans to ensure the most value for the resources deployed.

Remember, too, that mitigation actions are likely to become more strenuous over time as public policy gels and the actual threats and impacts of climate change become clearer. The trick is to benefit from early actions, learn as much as possible, and prepare for even more stressors.

Early actions like those this book recommends may involve energy fixes that pay for themselves. Later actions, on the other hand, are likely to involve real costs and become increasingly expensive.

Adaptation Themes

Recommendations for adaptation fall under the following six broad themes (Table 9-2):

1. Use climate models to understand projected change.
2. Identify specific threats.
3. Assess the associated risks.
4. Plan adaptation actions cooperatively.
5. Employ warning systems.
6. Take direct action.

Use Climate Models to Understand Projected Change

Several chapters ("Information," "Vehicles," and "Security") recommended that stakeholders and functional managers, particularly those in large organizations with installations in geographically diverse areas, familiarize themselves with the information from climate models. In doing so, they can learn how to use this information to assess implications for their organizations.

This means taking a step beyond their current understanding. By obtaining a basic grasp of the results and implications of these models, stakeholders can understand how they are expected to change over time. And they can become better versed in the geographic dispersion of such changes and the uncertainties surrounding them. Finally, they can better gauge the timing of climate change.

Table 9-2. Summary of Adaptation Recommendations

Health	Information	Land	Structure	Vehicles	Supply	Security
Anticipate acute events (2)	Address climate risks (3)	Improve coastal zoning (6)	Assess climate risks (3)	Change organizational thinking (4)	Use risk management (3)	Develop informative scenarios (1)
Prepare warning systems (5)	Use climate models (1)	Manage water availability (2)	Plan to adapt (4)	Identify upcoming challenges (1)	Prepare adaptation plans (4)	Mandate prevention strategies (6)
Anticipate chronic events (2)	Leverage social media (5)	Adapt forest management (6)	Implement warning systems (5)	Assess your vulnerability (6)	Assess adaptation risk (3)	War game climate change (1)
Address contamination threats (2)	Innovate through crowdsourcing (4)	Change agricultural practices (6)	Develop flood controls (6)	Create adaptation plans (4)	Adapt as planned (5)	Assess capability needs (3)
Expect mental stress (2)	Plan adaptation cooperatively (4)	Expect increased salinity (2)	Protect and harden (6)	Increase modal flexibility (6)	Adapt and learn (6)	Develop early warning systems (5)

Note: The numbers in the boxes correspond to the primary theme of this recommendation.

Such broad knowledge will help stakeholders as they think through the challenges climate change poses for the areas in which their organizations operate. Understanding the timing of such challenges provides a window into their relative likelihood for specific functions. Plus, if a manager chooses to directly use of one of these models, it will allow experimentation with various scenarios to see how much difference they make in the organization's future.

The knowledge gained from such understanding and use will assist managers in at least two respects. One is *the money*: the knowledge will show why investment in adaptation actions is necessary in protecting an organization's assets. Just as a manager might experiment with financial models to assess investment decisions, climate models help point to desirable adaptation actions. At a time of limited resources and greater scrutiny of expenditures, the ability to make a strong case for using these resources is a must.

At the same time, understanding removes some uncertainty with regard to *the timing*. Understanding and using climate change models enables a manager to better predict the point at which actions will become necessary—or at least are likely to become necessary. Climate change is a serious challenge, but realistically speaking, not everything needs to be done at once. Climate models can be used to understand when to make an investment and when to defer it. Models offer context as to whether to hedge bets due to too much uncertainty surrounding events. Understanding can also indicate whether action may simply be premature.

Of course, this flies in the face of how this book was crafted from the outset—to bring climate change to the non-scientific crowd. As a result, this is probably the least comfortable overarching theme you will encounter, because, honestly, who has the time or background to learn how to use information from climate models?

The problem is that adaptation challenges are so localized and so time-varying that it will be tempting for stakeholders to forego proper planning because they failed to properly characterize the risk at hand. But there's really no other way—an organization that operates worldwide needs detailed information for multiple places to properly adapt.

In the future, stakeholders are likely to look for solutions from experts drawn into an organization as needed. This cadre of available, capable analysts should increase in the future, but stakeholders still need to understand what they are buying into. And, as with any expert advice, *buyer beware*—you need to know enough to consider the advice in light of what is best for your organization.

Identify Specific Threats

Although stakeholders now have a need to understand climate models, and be able to use them, more is required. Assets must be protected, a process characterized by a second theme running through the adaptation recommendations (see "Health" and "Land," for example). The ability to translate general climate trends into discrete events that could threaten the assets of an organization is a must.

Whether the event is the effect of warmer weather, rising sea level, or extreme rainfall—it will require specific actions that involve not just understanding the course of such events, but also a strategy for adapting to them.

Inside, detailed knowledge of organizational processes and functions is required to identify specific threats to an organization. That knowledge may be spread around many corners of an organization, making coordination and cooperation a necessity. The organizational structure and hierarchy can help or hinder the process. High-level buy-in and leadership will bear fruit; efforts without this support may not succeed. Because the viability of the organization may be at stake, be prepared to be a part of the solution.

Assess the Associated Risks

Risk is everywhere, and in one manner or another, almost all the chapters (including "Information," "Structure," "Vehicles," "Supply," and "Security") recommend that stakeholders assess the *climate-related* risks that face their organizations. This assessment means understanding the probabilities associated with discrete events—which will likely change over time as further understanding of climate change is gained—and the vulnerability such probabilities present to the organization's assets should the risks materialize.

Risk assessments pop up in several chapters (the "Structure" chapter has the greatest detail) because they're a smart part of any plan. In this case, they're especially important when they focus on the risk associated with adaptation to climate change impacts. The important lesson is that a manager who engages in protecting his or her organization's assets against the consequences of climate change needs to understand the risks such change poses and the vulnerabilities exposed. An entire field is devoted to this, and it should be part of an organization's earliest considerations.

Plan Adaptation Actions Cooperatively

This theme has two segments. First, make planning part of your adaptation effort; it takes thought, time, and effort. Throughout the book, we recommend that managers lay out specific adaptation plans, though a few chapters ("Information," "Structure," and "Vehicles") caution that such plans will likely change as actions are taken and knowledge is gained.

Second, coordinate planning with others in your organization. The main point is that adaptation planning in one part of an organization is likely to be more effective if it is consistent with the adaptation plans of others who hold a stake in the overall success. Coordinating adaptation planning should make the effort more efficient and less costly.

Employ Warning Systems

More than half of the chapters explicitly recommend setting up warning systems to alert when one or more of the events predicted by climate models come to pass. These warning systems would seem to have the most benefit to health professionals, infrastructure managers, national security personnel, and land-use managers; however, any stakeholder who wants to know when and how much to invest in adaptation actions will gain from their implementation. Particularly relevant advice on this issue is given in the "Security" chapter. The broader point is that early warning has value for climate change adaptation planning, so investment in a warning system is likely a wise choice.

The role of social media is an interesting dynamic of warning systems, since social media is entirely reliant on people—who you're connected to and how you engage them. It can be applied in any setting.

The unique benefit of social media is that it can be used to establish an early warning system that leverages the separate knowledge of many different individuals (see the "Information" chapter). The hurdle, however, is developing the means and the capability to put such a system into play. Thus, when considering early warning systems, you need to examine how to cultivate and develop its widespread acceptance and use, in addition to making it an element of adaptation planning (see "Health" and "Structure").

Take Direct Action

This theme echoes the mitigation discussion: the path of least resistance often pays huge dividends. As with mitigation, a number of direct actions can be taken now to adapt to an organization's GHG emissions.

What sorts of actions? They are wide ranging, each requiring different knowledge and skills as well as varying levels of investment and commitment. They include improved coastal zone management, changed agricultural practices, modified forest management (see "Land"), increased flood control, hardening and other protections for facilities ("Structure"), and increased transport-mode flexibility ("Vehicles").

All seem sensible first steps toward adapting to the consequences of climate change. As events unfold and these consequences become more apparent, more such actions likely will emerge. Actions can be taken right now to adapt to climate change, but more actions will unfold over time, and these actions are likely to play a big part in an organization's ability to withstand whatever weather-related events occur. (The learning process for an organization is discussed in the "Supply" chapter.)

Author Ratings of the Recommendations

We've set a high bar by even presenting a recommendation. If it's in this book, LMI stands by it 100 percent. But each recommendation has its stronger areas. Throughout this book, recommendations have been rated using four criteria:

1. Potential
2. Operational impacts
3. Financial impacts
4. Feasibility and timing.

We can sum up these four criteria as "value." Today's decisions require a hard look at the resources available, the direction of an organization, and where the risks lie. By pinging on these four separate areas, an additional factor emerges for consideration: context—what fits best with an organization's values.

Obviously, the precise financial impact that implementing a recommendation might have on an organization at a particular location can't be provided within these pages. Instead, these ratings should be thought of relative to one another for a particular category within a particular chapter. Given those caveats, some things can be learned from applying these four criteria to the various recommendations.

However, we don't deny that some hold more promise than others. In Table 9-3, we array them by the total number of stars given by our functional experts. For example, if a recommendation receives 3 stars for each of the four criteria, it garners 12 stars in all, similarly for fewer stars (thus, an 11-star rating means it received 3 stars for each of three criteria and 2 for the other). Remember that the star ratings aren't comparable among chapters because the recommendations refer to different sorts of actions and that the relative importance of mitigation or adaptation potential, operational impacts, financial impacts, and feasibility and timing may differ among organizations. All things considered, some of these recommendations clearly have great potential.

Table 9-3. Total Numbers of Stars per Recommendation

Category	12★s	11★s	10★s	9★s	8★s	7★s	6★s	5★s	4★s
Mitigation	2	1	9	11	8	2	1	1	0
Adaptation	4	2	4	5	10	4	5	1	0
Total	6	3	13	16	18	6	6	2	0

For example, our authors give six recommendations 12 stars, meaning they fall into the *best* category for all four of our criteria. In our view, these six recommendations deserve careful consideration because they promise effective results at little or no financial or operational cost (in our authors' view, they will be highly cost-effective). Further, 38 of the 70 recommendations receive at least 9 stars, meaning they average 3 stars in one category and 2 or more in every other. This implies that by at least one criterion they are rated *best* while averaging *better* among the rest. Those 38 recommendations, too, hold great promise for cost-effective results.

We see from the table that of the 38, 15 are related to adaptation and 23 to mitigation. There are some reasons for the difference.

Mitigation

Table 9-4 shows that, in general, more of the recommendations within this book are rated "better" than "best" except with regard to feasibility and timing.

Table 9-4. Summary of Ratings for Mitigation Recommendations Using Four Criteria Important to Managers

Criteria	Good ★	Better ★★	Best ★★★
Mitigation potential	6	17	12
Operational impacts	6	17	12
Financial impacts	5	20	10
Feasibility and timing	3	13	19

Recall that, in this case, we're referring to technical feasibility and timing of implementation. Within this context, even a single-star rating means something could be completed within a 10-year planning horizon. For the most part, the technologies supporting these recommendations are mature and proven, meaning that, in these cases, the implementation timeline is 1 year or less. This reinforces a powerful point: some things can be done now, assuming the other criteria are favorable.

Will these recommendations work? Or more specifically, will they help stakeholders substantially reduce GHG emissions? Of the 35 recommendations for adaptation, 29 will provide at least moderate GHG reductions (meaning up to 20 percent of a baseline over 10 years). This is a substantial level of mitigation.

Are they affordable? That's the question on everyone's mind these days. Few ideas will get off the drawing board without an early pronouncement of fiscal soundness.

Of the 35 adaptation recommendations, 30 have little to no net financial impact on the organization; some even offer a positive net return. From this, the message is clearly yes; these are actions the organization can afford to take. The benefits of some are so clear that a manager may be penalized (internally or by the competition) for failing to act.

Will taking these actions impair operations? Our analysis suggests that 29 recommendations (of the 35) can usually be implemented with minor changes to business practices. Implementation, of course, may require some temporary resources and training, but a third of the recommendations can usually be implemented with existing staffing and resource levels and with minimal process changes. So these aren't calls for complete restructuring of an organization.

Which of these recommendations offer the most bang for the buck? We think all of the recommendations offer opportunities within an organization for real change. They are affordable, actionable, and appropriate for a meaningful time frame and will not unduly interfere with the organization's operation.

Overall, these recommendations offer many opportunities within an organization for real change. They are affordable, actionable, appropriate for a meaningful time frame, and will not unduly interfere with the organization's operation.

Adaptation

As with the mitigation recommendations, the adaptation recommendations were more likely to fall into the "better" category, not "best" (Table 9-5), in part because of the stringent requirements for earning a best-in-class distinction.[a]

Table 9-5. Summary of Ratings for Adaptation Recommendations Using Four Criteria Important to Managers

Criteria	Good ★	Better ★★	Best ★★★
Adaptation potential	1	19	15
Operational impacts	7	19	9
Financial impacts	14	14	7
Feasibility and timing	5	15	15

Consider, for example, what it takes to be rated "best" in adaptation potential. The recommendation must show promise to significantly reduce or eliminate vulnerability to likely climate change threats over the next 10 years—a tall order given the difficulty of knowing with any precision what those threats will be.

At the same time, even the recommendations rated "good" address climate change–related threats possible in the coming 10 years and offer important co-benefits.

That's why these are *no-regret* recommendations for adaptation potential. All but one of these recommendations fall into the "better" and "best" classes. They reflect the fact that the focus of this book is concentrated on adaptation challenges that the scientific community has identified as "highly likely."

Operational impacts of adaptation actions tend to be more intrusive than those of mitigation actions; they'll probably require organization-wide responses. The operational impacts of adaptation actions arise from the never-ending nature of the adaptation challenges. Rising sea levels can impact a facility *every single day*. Recurring high temperatures continue to debilitate workers, often for extended periods. Adaptation actions therefore need to take on these qualities of pervasiveness.

---

[a] Refer to the "Action" chapter for a reminder of these criteria.

Perhaps the biggest lesson that can be taken from an examination of Table 9-5 is that the financial impact of adaptation actions is not as positive as that of mitigation actions.

Financially, adaptation will test the organization. It will take careful planning and coordination to incorporate these actions into the organization's strategic plans to reduce the financial impact. An easy example of this is how climate change impacts or threatens major infrastructure—not only what's already in place, but also continued major investments and site selection decisions.

The alternative to this financial impact is not to adapt. Think about that for a minute. If an organization has good reason to expect these impacts on operations and fails to adjust accordingly, it would face severe negative impacts and real costs from the failure to adapt. Assessing the risks from climate change, and weighing the relative costs and benefits of adapting to them, will be part of a manager's job from now on.

This brings us back to a theme that emerged in our discussion above: *plan adaptation actions cooperatively*. It will be important to deal with adaptation challenges within a *whole-of-organization* framework, rather than within traditional silos of autonomy.

Fortunately, overall, this is good news for a proactive organization. The best news is that a number of adaptation actions are feasible to undertake in the near term. Because adaptation challenges may be unavoidable, it is heartening to know that things can be done now that will reduce vulnerability and help ensure continuity of operations.

Three Tools for Planning

Management tools are a regular part of an organization's day-to-day operations. They're engrained in how many managers are trained—most Master of Business Administration programs teach a variety of useful methods and tools that can be applied across dozens of challenges. So there's already a foundation in place for facing the new challenges that climate change will present, with many of the tools already in use and ready for application. In particular, though, three analytical tools are essential for decision making: cost-benefit analysis, risk assessment, and life-cycle analysis.

Cost-Benefit Analysis

Many opportunities are available to reduce GHG emissions, and sifting through them will require managers to consider costs and benefits. Specifically, managers need to think about how much it will cost to implement the technology and how it will further the organization's goals of reducing GHG emissions. Traditional cost-benefit analysis, then, will be an essential tool in this process. It'll be particularly useful when management clearly dictates that particular investments must yield positive results financially.

No doubt it'll be challenging to secure reliable data for calculations, but that's true of all cost-benefit analysis. Also, managers will be confronted with how to best represent uncertainties. Applying sensitivity analyses should help deal with this. In short, the challenges posed by cost-benefit analysis of mitigation/adaptation opportunities are no more daunting than those of most other forms of investment.

In other instances, managers may be employing a somewhat different but closely related form of analysis—namely, cost-effectiveness. This will be more appropriate if the objective is to secure a given amount of GHG reduction or a certain level of protection for key assets. The job will be to distinguish among alternatives the least costly methods of achieving a particular physical goal or the strategy that will yield the greatest effect for a given amount spent.

As analysis of cost benefit or effectiveness begins, one issue will be determining the discount factor to use. An organization faces a given cost of capital, and this cost normally sets the standard for evaluating any given investment project. However, climate change involves the welfare of future generations, and some economists have argued that a lower social discount rate is appropriate.[1]

Here's the point at which managers need to clearly understand what is motivating the organization's efforts to mitigate or adapt to climate change. Mitigation may be required, for example, by a presidential executive order or it may be voluntary. If mandatory, it would seem that it is being done for societal purposes, and a social discount rate makes sense. If, on the other hand, it is voluntary, how management frames the objective—if for the good of society, for example—might mean that again a social discount rate makes sense. If, however, the effort is motivated by the organization's private interests, then its own cost of capital would be the right standard.

The cost-benefit analysis will be a challenge in another way: the interaction between efforts to mitigate and adapt. Just because an organization invests energy and resources into mitigation does not mean it remains independent of the same investments in adaptation (and vice versa).

Of course, these interactions can be positive. For example, shifting from oil products to renewable fuels may not only lower GHG emissions in the near term but may also make an energy supply less vulnerable to future weather-related disruptions. This also means that assets that use energy will also be less vulnerable in the future. In other words, the organization will have increased its adaptation capabilities even as it has increased mitigation.

Sometimes, though, mitigation actions might also make it harder to adapt. If, for example, an organization chooses to rely more heavily on biofuels for energy supply, it may be able to substantially reduce emissions. At the same time, though, this energy supply is more vulnerable to the impact that weather can have on fuel feedstock crops. That would force the organization to commit more resources toward hedging against areas of vulnerability. This implies that a cost-benefit analysis of a biofuels option should include not only the mitigation gains but also factor in the cost of the additional adaptation hedge.

Risk Assessment

A second important tool to use in developing and implementing plans to deal with climate change is *risk assessment*—especially for adaptation plans. Climate change will introduce a collection of risks probably not experienced before, such as sea level rise. It will also change the risk profile of the challenges normally encountered. For example, the risk of snow load on the roofs of an organization's facilities may differ from what it has been traditionally.

Several steps are associated with risk assessment. Briefly, they involve developing a probability of (or probability distribution around) discrete climate-related events that are likely to affect an organization's assets, assessing the physical impacts of such events if they occur, and determining the vulnerability of the assets to those impacts.

The first step requires an understanding of the physical characteristics of an event and what it would do to a particular geographic area. The second step involves figuring out the damage this event would do to the assets an organization possesses in that area. The third and final step is an assessment of the likelihood of some particular event, along with the uncertainty surrounding that likelihood. This isn't an easy process, but proper assessment of risk requires the effort.

Risks don't all run in one direction, though. For example, a warming trend would reduce health problems related to hypothermia. Adaptation strategies would seek to exploit such positive trends, too.

A confluence of events, including melting ice, thawing permafrost, and increased storm surge has led some areas of Alaska to experience devastating shoreline erosion.

Of course, risk profiles and changes to them will vary from location to location. As discussed in several chapters, Alaska already suffers from the impacts of greatly increased temperatures over the past few decades, whereas other regions show little or no significant change. The threat of increased damage from storm surge is only important to an organization if that organization's facilities are near the coast (although long-term changes in coastlines themselves are expected as sea level rises). A risk analysis should recognize the different aspects of the challenges within the context of the working environment.

Adaptation challenges and the subsequent risks will change with time. In general, climate change is expected to become more pronounced. As a result, the impacts will grow more challenging as time passes.

Sea level rise, for example, may be the most predictable of these time-varying challenges. Even so, it is a challenge to estimate sea level year to year. It is even more challenging to estimate how the frequency and intensity of storms will change and harder still to stipulate how changes in numbers and types of insect infestations will affect hardwood supplies. Even the most likely information on these eventualities will change over time. Risk assessments—when associated with climate change and its implications for adaptation strategies—will be subject to varying levels of uncertainty. This makes continual updates a requirement.

Life-Cycle Analysis

The third recommended tool is life-cycle analysis (LCA), and it may be the one with which managers are the least familiar. LCA is of particular use in analyzing the GHG consequences of acquiring, using, and disposing of a particular product. This book examines LCA in depth when discussing the supply chain; that's where it is essential in helping clarify where there are GHG emissions—certainly when it comes to making and distributing a particular product, but also those associated with that product's consumption and disposal.

LCA involves a full understanding of material and energy inputs to products. Many times, this analytic process can help identify more efficient uses of resources and lead to significant cost savings. In this context, however, LCA can be viewed as helping to identify opportunities to reduce GHG emissions or to adapt to climate change.

What comes to mind? For one, by analyzing inputs and use of a product, and the associated GHG, LCA provides an opportunity to discover what plays the biggest role in an organization's emissions profile. This in turn may point toward the best ways to achieve reductions.

For example, transporting a product an organization uses may be especially GHG intensive. However, combining shipments, ordering less frequently, or substituting one form of transport for another may reduce the impacts of transport. Alternatively, persuading a supplier to alter the product's material makeup to curb its associated GHG may be possible. Still another possibility is identifying recycling opportunities that previously had escaped notice.

LCA compares the environmental performance of products and services.

However, LCA is not limited to mitigation applications. This approach also can help with adaptation through the identification of vulnerabilities in supply chains. This is done through a detailed analysis of sources and sinks in the production process. LCA can then lead to better strategies to hedge against the risks of climate change–related impacts to the supply chain.

As with the two other tools, the mitigation and adaptation opportunities identified by LCA are dynamic, not just because supply chains are constantly changing, but because climate change itself will create new mitigation opportunities and adaptation challenges. LCA will also reveal where, within context, opportunities and threats could complicate investment strategies and other actions.

Interactions

One final theme demands discussion—it has been lurking in the background, but now needs to be emphasized. In working to mitigate and adapt to climate change, managers are likely to be interacting with a host of other stakeholders, and if managers are aware of these interactions and take advantage of them, their effectiveness can be substantially enhanced—perhaps many times over.

Working Toward the Same Goals

The relationship between mitigation and adaptation means that the more stakeholders collectively mitigate GHG now, the less adaption will be needed in the future. This seems unremarkable, a "*no, duh*" line of thinking. But it's reasonable to argue that what one person does to mitigate GHG, even someone in a large organization with important functional responsibilities, can't be much of a factor in changing the trajectory of the earth's destiny with respect to climate change. Indeed, if no one organization or country acting alone can make much difference, how can one individual contribute much?

Here, it's important to recognize a basic truth: while it's true that no one person acting alone can solve the entire problem, many individuals are working to mitigate GHG emissions, as well as many organizations and many nations. This is not an exclusive, lonely task.

Take encouragement from the fact that millions of people around the world take seriously the threats involved and are doing their part to reduce them. In this sense, even someone working alone in an organization would still be working in parallel with many other counterparts who are trying to achieve the same ends. Likewise, everyone can take encouragement from another's efforts. This is one kind of interaction that benefits all involved: the reinforcement that comes from knowing that others are working toward the same end.

Activities in One Functional Area Will Affect Others

Another concept discussed briefly in the opening chapter is that every mitigation or adaptation step will affect others. For example, the projects an infrastructure manager will undertake to protect transport facilities (such as roadbeds, bridges, tunnels, and ports) will affect the options available to a vehicle manager. In turn, those in charge of an organization's supply chain will view these actions with keen interest, collecting and organizing data on suppliers, which in turn will bring information and communication technologies managers into the loop as they seek to provide the necessary architecture for meeting this goal. It's all connected, and never done in a vacuum.

Many other such interactions are conceivable. Someone who manages lands near the shore will benefit from infrastructure projects that protect coastal areas from flooding and erosion. Steps taken by land managers to control crop pests and insect infestations may, in turn, affect public health. Efforts by public health officials to reduce adverse health effects associated with a warming climate figure to enhance workforce productivity. Again, it's all connected.

This is synergy—and when it arises between functions, it demands some common planning. For example, the "Information" chapter recommends that an information manager be part of an organization's adaptation planning in order to know the priorities to establish, what others in the organization expect from the information department in the event of a catastrophic weather event, and how best to make use of social media to respond. The "Security" chapter recommends that national security stakeholders work with other functional specialists (certainly in the private-sector contracting community) to maximize the United States' ability to respond to the ramifications of climate change events abroad. Common planning of these sorts is likely to enhance an organization's ability to respond to climate change, making every contribution more extensive and effective.

Interacting through Information Flows

"Knowledge is power." Sharing it, then, is revolutionary. This is another way stakeholders likely will interact as they seek to mitigate GHG and adapt to climate change—by trading information with others who are doing the same things. By extending and expanding a common knowledge base, access is gained to lessons learned, new ideas and innovations, and a baseline for discussing emerging challenges. Every functional manager will find unique ways to mitigate GHG and, by interacting with others, managers will likely hear of things previously unheard of while offering ideas of their own. Similarly, it may be possible that people elsewhere who manage similar functions—and some who manage other functions as well—will be thinking about adaptation actions differently, with ideas that may improve all viewpoints.

The "Information" chapter talks about using social media to make use of the "crowd," meaning others with knowledge and perspective that may prove useful in accomplishing the things an organization wants to accomplish. Crowdsourcing is a kind of group interaction concept, a little like a social circle except that with a purpose. A question or challenge is being addressed.

Not all information received via crowdsourcing is useful, and it will take some skill and effort to separate the wheat from the chaff, but listening to others who may have found ways to attack the very challenges managers face can be enlightening.

Summing Up

Stakeholder interests, and the efforts of functional managers responsible for climate mitigation and adaptation, are likely to pay off even more if managers take advantage of interaction opportunities:

- Recognizing what other managers in the organization are doing to mitigate and adapt
- Discussing what everyone is doing
- Coordinating planning efforts where the payoff is evident
- Exchanging ideas with others who have similar responsibilities on how best to mitigate and adapt
- Making use of the "crowd"—the collective intelligence of many others—to further knowledge and help address particular challenges.

Understanding what others are saying will take time, and trying to discern whether what they offer is useful will often be frustrating. Figuring out who can be trusted to come up with good ideas, who will admit it if they can't, and who is making progress in mitigating or adapting will help. So, dealing with climate change is no different than working together to take on other work-related challenges.

But because taking concrete steps to deal with climate change is still a new activity, one likely to go on for a long time, a good deal remains to be learned. Interaction is a way to enhance knowledge, and by doing so, boost effectiveness. This interaction will allow stakeholders to build knowledge, improve performance, and make the organization a leader in the effort to deal with climate change.

These are exciting times. The challenges are great, which means the solutions must be greater. But the collective commitment to attacking the climate change issue is indisputable—and functional managers now get their shot at making an impact. Good luck!

---

[1] See for example Richard Newell and William Pizer, "Discounting the Benefits of Climate Change Mitigation—How Much do Uncertain Rates Increase Valuations?" Pew Center on Global Change, December 2001.

Biographies

Jeremey Alcorn

Jeremey M. Alcorn led development of chapter 8 (National Security). He has an extensive background in environmental security, sustainability planning, and greenhouse gas accounting. Since January 2010, he has been a research consultant on the staff of LMI's Energy and Environment group. Before that, he was an environmental engineer at Concurrent Technologies Corporation, where he conducted the Army's first bottom-up greenhouse gas inventory at several installations. Mr. Alcorn also spent 6 years at Science Applications International Corporation, providing agency sustainability support to the National Aeronautics and Space Administration, Air Force, and Environmental Protection Agency. He has an MS in environmental science from George Mason University.

Julian Bentley

Julian A. Bentley led development of chapter 6 (Vehicles). He is a member of LMI's Energy and Environment group, where he specializes in the environmental, political, and economical implications of biofuels and electric vehicle technologies. Before LMI, Mr. Bentley worked in environmental regulatory development, strategy consulting, and energy marketing. As a chemical engineer for Radian Corporation, he provided technical and management support for the Environmental Protection Agency's development of the Clean Water Act regulation for the transportation equipment cleaning industry. Mr. Bentley has an MBA from the University of North Carolina–Chapel Hill.

Virginia Bostock

Virginia A. Bostock has professional experience and expertise in the areas of environmental policy, sustainability, greenhouse gas management, energy management, and Leadership in Energy and Environmental Design assessments. She is a research consultant in LMI's Energy and Environment group, where she supports corporate efforts to formulate federal government–wide guidance on greenhouse gas accounting and management and helps agencies to implement it. Her work in climate change includes coauthoring the *Greenhouse Gas Protocol for the U.S. Public Sector: Interpreting the Corporate Standard for U.S. Public Sector Organizations*. She has a master's degree in environmental management from Duke University's Nicholas School of the Environment.

Michael Canes

Michael E. Canes, PhD, is a valued member of LMI's Energy and Environment group, where he focuses on the economics of energy and climate policy. He is an internationally known economist, having nearly 30 years of experience in energy and environmental economics. Before LMI, he was the vice president and chief economist at the American Petroleum Institute, where he guided efforts to deal with legislative and regulatory issues such as environmental requirements and fuels-related mandates. He has published numerous studies on climate policy and developed a method for estimating policy choice impacts on greenhouse gas emissions. Dr. Canes is an officer in the United States Association for Energy Economics. He has a PhD in economics from the University of California–Los Angeles.

Matt Daigle

Matthew J. Daigle joined LMI's Communications group after an extensive career in media relations and journalism. He was a general assignment reporter for KNDU-TV, an NBC affiliate in Washington State, where he provided acclaimed coverage of local, state, and national political news, as well as the Department of Energy's Hanford Site. He also served as press secretary and official spokesperson for the Washington, DC, office of Congressman Greg Walden of Oregon. At LMI, Mr. Daigle is supporting the company's public and corporate communications goals as well as promoting visibility of LMI's expert personnel. He holds an MS in journalism from Northwestern University.

Stu Funk, *Contributor*

Stuart D. Funk has served for more than 35 years in government and private industry in staff and senior management positions, focusing on program management, energy and logistics planning and execution, climate change planning and management, strategic planning (including organizational studies and alignment), facility planning and recapitalization, weapon-system acquisition, and resource analysis. He has expertise in the bulk petroleum and energy communities garnered from eight assignments over nearly 26 years. He leads the energy and climate change practice at LMI, with past and present clients in the Departments of Defense, Energy, Commerce, and State, General Services Administration, and Architect of the Capitol. He has performed high-level analyses and briefed senior leaders on the results. Mr. Funk serves on the Energy Committee of the National Defense Industrial Association. He has a bachelor's degree in aerospace engineering from the Naval Academy and master's degrees in petroleum management from the University of Kansas and national security and strategic studies from the Naval War College.

Rachael Jonassen

Rachael G. Jonassen, PhD, is an authority on climate change science. Before joining LMI's Energy and Environment group, she served as program director at the National Science Foundation, where she led the organization's carbon cycle research efforts for the U.S. Global Change Research Program. As a member of the interagency working group for the carbon cycle, she helped manage the first State of the Carbon Cycle Report, organized the first all-investigators meeting on the carbon cycle, and promoted international cooperation on bilateral climate change agreements. Dr. Jonassen was a lead author of *A Federal Leader's Guide to Climate Change: Policy, Adaptation, and Mitigation* (2009) and the *Strategic Plan for the U.S. Climate Change Science Program* (2003). She has a PhD in geology from The Pennsylvania State University.

Francis J. Reilly, Jr.

Francis J. Reilly Jr. led development of chapter 4 (Land). He joined LMI's Energy and Environment group from the Reilly Group, his own company, which provided environmental consulting to government, industrial, and academic clients. He was the executive director of the Watersheds and Wetlands working group and performed a variety of services, including curriculum and program development, risk assessment, and permitting support. At LMI, Mr. Reilly has become a subject matter expert in land-use planning and ecological risk assessment. He has an MS in biology is from East Carolina University.

John Selman, *Book Project Director*

John R. Selman, program director of LMI's Energy and Environment group, has more than 20 years of experience supporting federal energy, environment, and facilities programs. Before joining LMI, he led a team of 90 environmental consulting professionals as a principal at Booz Allen Hamilton. Mr. Selman has also worked at the Department of Energy, Project Performance Corporation, and US Senate. He holds an MS in environmental engineering from Johns Hopkins University and an MPA from Syracuse University.

Rich Skulte

Richard A. Skulte led development of chapter 5 (Structure). He is a program manager in the Infrastructure and Engineering Management group and has more than 20 years of experience in program and project management, facilities engineering, asset management, and business process improvement. He oversees multiple projects, including infrastructure condition assessments, real property master planning, business transformation, and recapitalization program support. Before his work at LMI, Mr. Skulte was a management consultant with PricewaterhouseCoopers, where he advised federal agencies in process improvements. He was also a strategy consultant in Internet intelligence for Cyveillance, where he advised Fortune 50 companies in brand management, partner networking, and competitive intelligence. Mr. Skulte began his career with Johnson Controls, Inc., where he focused on design and implementation of building automation, HVAC, and facilities management systems. He has an MBA from Georgetown University.

Taylor Wilkerson

Taylor H. Wilkerson led development of chapter 7 (Supply Chain). He is a member of LMI's Supply Chain Management group, where he provides clients with comprehensive supply chain analysis, systems and performance assessments, and training. Much of his recent work has focused on sustainable supply chain operation, and he helped create LMI's GAIA Supply Chain Sustainability Maturity Model to help corporate and public managers reduce environmental impact throughout supply chain operations. Before LMI, Mr. Wilkerson was an environmental consultant at SECOR International, Inc., where he focused on industry-level environmental compliance. He has an MBA from the University of Maryland.

John Yasalonis

John W. Yasalonis led development of chapter 2 (Health). He has been a research consultant and program manager in LMI's Energy and Environment group for the past 15 years. He has an extensive background in occupational and environmental health management after nearly 25 years developing, managing, and directing programs at all levels for the Army. Mr. Yasalonis also served as the surgeon general's industrial hygiene and medical supply consultant, developing policies, goals, and objectives unique to military occupational and environmental exposures. He has an MS in environmental science from Drexel University.

Reviewers

We are grateful to the following people, who provided expert review of this guide.

Navid Ahdieh
Project Leader, Market Transformation Center
National Renewable Energy Laboratory,
U.S. Department of Energy

Dana Arnold
Director, Program Analysis Division
Federal Acquisition Service Office of Acquisition Management, General Services Administration

Ramon Barquin, Ph.D.
President & Chief Executive Officer
Barquin International

Lawrence Betts, M.D., Ph.D., CIH, FACOEM, Fellow of the AIHA
Professor, Family & Community Medicine, and Clinical Professor, Physiological Sciences
Eastern Virginia Medical School

Michael Biddick
President & Chief Technology Officer
Fusion PPT, LLC

Ann H. Brennan
Director, Market Transformation Center
National Renewable Energy Laboratory,
U.S. Department of Energy

Brett Brunk, Ph.D., PMP
Chief Enterprise Architect
General Services Administration

Joshua Busby, Ph.D.
Asst. Professor and Crook Scholar at the Robert S. Strauss Center for International Security and Law
Lyndon B. Johnson School of Public Affairs,
University of Texas at Austin

David A. Butler, Ph.D.
Director, Medical Follow-up Agency
The Institute of Medicine of the National Academy of Sciences

David D. Close
Virginia Cooperative Extension State Master Gardener Coordinator
Virginia Polytechnic Institute and State University

Robert Crosslin, Ph.D.
Founder & President
Technology Economics International, Inc.

Cynthia Cummis
Senior Associate
World Resources Institute

Ryan Daley
Federal Fleet Program Task Leader
National Renewable Energy Laboratory,
U.S. Department of Energy

Steven J. Davis, J.D., Ph.D.
Senior Research Associate
Carnegie Institution for Science

Robert DeFraites, M.D., MPH
Past Director, Department of Preventive Medicine and Biometrics
Uniformed Services University of the Health Sciences

Frank L. Eichorn, DM, PMP, CCP
President
Eichorn Consulting

Frederick (Rick) Erdtmann, M.D., MPH
Director, Board on the Health of Select Populations
The Institute of Medicine of the National Academies

William Flanagan, Ph.D.
Ecoassessment Leader
GE Global Research

Larisa Ford, Ph.D., MPA
Co-Acting Southern Border Coordinator
U.S. Fish & Wildlife Service,
U.S. Department of the Interior

John A. Goolsby, Ph.D.
Research Entomologist, Biological Control and Integrated Pest Management of Pests and Weeds
Agricultural Research Service,
U.S. Department of Agriculture

Bill Goran
Director, Center for the Advancement of Sustainability Innovations
Engineer Research and Development Center
Construction Engineering Research Laboratory,
U.S. Army Corps of Engineers

Leila J. Hamdan, Ph.D.
Research Microbial Ecologist
United States Naval Research Laboratory

Lisa Harrington
President, The LHarrington Group, LLC
Adjunct Professor, Supply Chain Management,
Robert H. Smith School of Business,
University of Maryland

Tim Harvey
Chief, Park Facility Management Division
National Park Service

Jennifer Hazen, Ph.D.
Research Fellow
Lyndon B. Johnson School of Public Affairs,
University of Texas at Austin

Michael Helwig, Ph.D.
Senior Engineer
National Renewable Energy Laboratory,
U.S. Department of Energy

Don Hewitt, CISSP
Principal Consultant
Proconsul, Inc.

Marcus Jones
Associate Director, Office of Science, Office of Safety, Security, and Infrastructure
U.S. Department of Energy

Jessica Katz, LEED A.P.
Engineer
National Renewable Energy Laboratory,
U.S. Department of Energy

Brian F. Keane
President
SmartPower

Marc Kodack, Ph.D., PMP, RPA
Program/Project Manager
Office of the Assistant Secretary of the Army for Energy and Sustainability, U.S. Department of Defense

Robert ter Kuile
Senior Director, Environmental Sustainability,
Global Public Policy
PepsiCo, Inc.

James Leatherwood, SES
Director, Environmental Management Division
National Aeronautics and Space Administration

Becca Madsen, MEM
Biodiversity Program Manager
Ecosystem Marketplace

Esther McClure (CDR, USN, Ret.)
Strategy Action Officer
Office of the Secretary of Defense,
U.S. Department of Defense

E. Lile Murphree, Jr., Ph.D., P.E.
Professor of Engineering Management
and Systems Engineering
The George Washington University

Peter Murray, CIRM
Project Leader, Biofuels Integrated Supply Chain;
Corporate Program Leader, Sustainable Supply
Chain
DuPont

Christine Parthemore
Fellow
Center for a New American Security

Carlos Peña, Jr., P.E.
Principal Engineer
U.S. International Boundary and Water Commission

Kevin Percival
Branch Chief, Facility Planning
National Park Service

Verena Radulovic
Director, Partnerships and Supply Chain,
Climate Leaders Program
United States Environmental Protection Agency

Bob Ramba
Purchasing and Supply Management Specialist
United States Postal Service

Mark Reichhardt
Lead for Fleet Management and Greenhouse Gas
Management
Federal Energy Management Program,
U.S. Department of Energy

Robert S. Ryczak, Ph.D.
Preventive Medicine Planner
Office of the Surgeon General–Proponency Office
for Preventive Medicine, U.S. Army

Amanda Sahl
Energy Technology Program Specialist
Federal Energy Management Program, U.S. Department of Energy

LtCol Paul Schimpf
Strategist for Energy & Environment
Office of the Secretary of Defense,
U.S. Department of Defense

David Spitzley
Product Sustainability Manager
Kimberly-Clark Corporation

Barb Stewart
Fire Communication Specialist, Northeast
and National Capital Regions
National Park Service

Jim Villani
Senior Software Engineer
ASSETT, Inc.

Linda Wennerberg, Ph.D.
Environmental Assurance for NASA Systems
National Aeronautics and Space Administration

Nicole Willis, PMP
Senior Enterprise Architect
U.S. Department of Homeland Security

Scott Wright, Ph.D., P.E., PMP
Assistant Professor of Project Management
University of Wisconsin–Platteville

Anonymous Reviewer

Credits

Chapter 1, Action

Temperature anomalies
Adapted from U.S. Environmental Protection Agency, *Climate Change Indicators in the United States*, April 2010, EPA 430-R-10-107.

Venn diagram
Created by LMI.

Weather station records
NOAA Environmental Visualization Laboratory, *Heat Defines the Country in July*, July 2011, http://www.nnvl.noaa.gov/MediaDetail.php?MediaID=795&MediaTypeID=1. Note: Not all July records have been received by the National Climatic Data Center.

Relation of chapters
Created by LMI.

Mitigation/adaptation clock
Created by LMI.

Figure 1-1. *Future Increases in Temperature Depend upon Choices Made Today*
Adapted from "Summary for Policymakers" in *Climate Change 2007: The Physical Science Basis. Working Group I Contribution to the Fourth Assessment Report of the Intergovernmental Panel on Climate Change.* Figure SPM-5, Cambridge University Press, Cambridge, United Kingdom, and New York, HY, USA, www.ipcc.ch/pdf/assessment-report/ar4/wg1/ar4-wgl-spm.pdf. Used by permission of the IPCC.

Iceberg
©z576/Fotolia #14775652.

Figure 1-2. *Schematic Notion of the Growth of the Mitigation and Adaptation Challenged Over Time*
Created by LMI.

Table 1-1. *Characteristics of the Principal Kyoto Greenhouse Gases*
Adapted from T.J. Blasing and K. Smith, *Recent Greenhouse Gas Concentrations*, Carbon Dioxide Information Analysis Center (CDIAC), July 2006, cdiac.ornl.gov/pns/current_ ghg.html.

Table 1-2. *Major Agency GHG Mitigation Goals (Percentage Reduction) for 2020 under EO 13514*
Created by LMI.

Rainy day
©Demydenko Mykhailo/Fotolia #3189918.

Airport
©Sandor Jackal/Fotolia.com.

Table 1-3. *Principal Sources of GHG Emissions We Discuss, Sources Are Grouped within "Scopes," Which Are Controlled by Rules and Regulations in Different Ways*
Created by LMI.

Table 1-4. *Criteria for Mitigation Recommendations and Definitions of Our Assessment Levels*
Created by LMI.

Table 1-5. *Description of New Adaptation Potential Criteria applied Only to Adaptation Recommendations.*
Created by LMI.

Performance

Chapter 2, Health

Hazard-related deaths in the U.S.
Adapted from Borden, K.A. and S.L. Cutter, 2008: Spatial Patterns of Natural Hazards Mortality in the United States. *International Journal of Health Geographics*, 7:64, doi:10.1186/1476-072X-7-64.

Figure 2-1. *Projected Change in U.S. Population Age Structure to 2050*
Data from U.S. Census Bureau, *Census 2000 Summary File 1 and 2010 Census Summary File 1*, http://www.census.gov/prod/cen2000/doc/sf1.pdf, http://www.census.gov/compendia/statab/2011/tables/11s0008.pdf.

Mud cracks

Ozone concentrations vs. temperature
Data from National Assessment Synthesis Team (NAST), 2001: *Climate Change Impacts on the United States: The Potential Consequences of Climate Variability and Change*. Cambridge University Press, Cambridge, UK, and New York, 612 pp, http://www.usgcrp.gov/usgcrp/Library/nationalassessment/.

Ragweed pollen concentrations vs. CO_2 concentrations
Data from, Ziska, L.H. and F.A. Caulfield, 2000: Rising CO_2 and pollen production of common ragweed *(Ambrosia artemisiifolia L.)*, a known allergy-inducing species: implications for public health. *Australian Journal of Plant Physiology*, 27(10), 893-898.

Red flour beetle
U.S. Department of Agriculture Agricultural Research Service, *Tribolium genetics*, Photograph by Peggy Greb, December 17, 2009, www.ars.usda.gov/npa/cgahr/spiru/ tribolium. Courtesy of U.S. Department of Agriculture.

Mosquito and tick
Centers for Disease Control and Prevention, Division of Vector-Borne Diseases, *Promoting Health and Quality of Life by Preventing and Controlling Vector-Borne Diseases*, 2011, http://www.cdc.gov/ ncezid/dvbd/pdf/dvbd-pamphlet-2011.pdf.

Obesity trends among adults
Adapted from Centers for Disease Control and Prevention, *Obesity Trends Among U.S. Adults Between 1985 and 2010*, http://www.cdc.gov/obesity/downloads/obesity_trends_2010.ppt.

Cell phone
Adapted from © rangizzz/Fotolia # 27725245 and mangostock/Fotolia # 22488421.

Figure 2-2. *Significant Energy Use Burden from Healthcare Operations*
Data from U.S. Energy Information Administration, *2008 Commercial Buildings Energy Consumption Survey*, Table 2.9, http://www.eia.gov/totalenergy/data/ annual/ txt/ptb0209.html.

Figure 2-3. *Healthcare Energy Use per Square Foot is Significant*
Data from U.S. Energy Information Administration, *Commercial Buildings Energy Consumption 2003*, Table E3A, http://www.eia.gov/totalenergy/data/annual/ txt/ptb0211.html.

Public health support diagram
Created by LMI.

Table 2-1. *Reprocessing Medical Equipment Eliminates Waste*
Data from Lars Thording, Senior Director, Marketing & Public Affairs, Stryker Sustainability Solutions, personal communications with the author, October 19, 2011.

HEATLINE call center
Philadelphia Corporation for Aging, PCACares.org, *PCA HEATLINE – 215-765-9040- ACTIVATED MAY 31*, May 31, 2011, http://www.pcacares.org/News_List.aspx?newsID=437A16G1L5&orgID=230Q26D15H53.

Figure 2-4. *Since 1999, West Nile Virus Has Spread Across the United States*
Adapted from Centers for Disease Control and Prevention, *Final 2009 West Nile Virus Activity in the United States*, http://www.cdc.gov/ncidod/dvbid/westnile/Mapsactivity/ surv&control09Maps.htm.

Omaha wastewater treatment plant
City of Omaha, Public Works Department. Courtesy of the City of Omaha.

FEMA shelter
Federal Emergency Management Agency Photo Library, *Photograph by Andrea Booher taken on 09/02/2005 in Texas*, http://www.fema.gov/photolibrary/photo_details.do?id= 14506.

Chapter 3, Information

Figure 3-1. *New York City's Challenge of Integrating Energy Data*
Created by LMI.

Table 3-1. *Projected Increases in World and ICT GHG Emissions, 2007-2020*
Data from Greenpeace International, *Make IT Green, Cloud Computing and its Contribution to Climate Change*, 2010, http://www.greenpeace.org/usa/Global/usa/report/2010/3/make-it-green-cloud-computing.pdf.

Google search
Google Search Engine, 2011, http://www.google.com/.

Telecommuting quote
Martha N. Johnson, Administrator, U.S. General Services Administration, speaking at Telework Exchange Fall Town Hall Meeting in Washington, DC, 7 October, 2010. http://www.teleworkresearchnetwork.com/ telecommuting-gsa-gets-it/5120.

Video teleconference
Tandberg, *Telepresence*, 2008, http://www.tandberg.com/press-room/video-conferencing-telepresence-photo-gallery.jsp. Courtesy of TANDBERG Corporation.

Telecommuting and CO_2 Emissions
TIAX LLC, The Energy and Greenhouse Gas Emissions Impacts of Telecommuting and e-Commerce, Final Report by TIAX LLC to the Consumer Electronics Association (CEA), July 2007, http://www.ce.org/Energy_ and_Greenhouse_Gas_Emissions_Impact_CEA_July_ 2007.pdf. Used by permission of the Consumer Electronics Association.

Figure 3-2. *Enterprise Architecture Looks across the Organization to Achieve Efficiencies*
Created by LMI, *EA Practice Handbook*, 2009.

Cloud Computing
Adapted from Wikipedia, *Cloud Computing*, Sam Johnston, March 2, 2009, http://en.wikipedia.org/wiki/File:Cloud_computing.svg.

Punch cards
Wikipedia, *Punched Card Program Deck*, Arnold Reinhold, October 3, 2006, http://en.wikipedia.org/wiki/File:Punched_card_program_deck.agr.jpg.

Google data center
John Markoff and Saul Hansell, *Hiding in Plain Sight, Google Seeks More Power*, June 14, 2006, Photograph by Melanie Conner, *New York Times*, http://www.nytimes.com/2006/06/14/technology/14search.html?scp=2&sq=hiding%20in%20plain%20site&st=cse. Used by permission of Melanie Conner.

An innovative system at Union Hospital in Terra Haute, IN links data from surgery scheduling to the building automation system and determines which of the two HVAC system modes should be used: surgery or setback.
Union Hospital Marketing and Public Relations, 2011, http://www.myunionhospital.org/unionhospital-terre-haute/surgery/surgery_davinci.php. Used by permission of Union Hospital Marketing and Public Relations.

PlaNYC
NYC.gov, *Biography*, Michael R. Bloomberg, 2012, http://www.nyc.gov/portal/site/nycgov/menuitem.e985cf5219821bc3f7393cd401c789a0/.

Dorm energy use at Oberlin College
Oberlin College, Building Dashboard, *Residence Hall Electricity Comparisons*, screen shot, 2011, http://www.oberlin.edu/dormenergy/. Used by Permission of Oberlin College.

Volt digital display on smart phones
Flickr®, *Felix Volt*, Felix Kramer, January 14, 2011, http://www.flickr.com/photos/56727147@N00/5360899917/in/set-72157625838522104/.

Figure 3-3. *The Internet of Things Allows Objects to Communicate and Respond Without Human Intervention*
Adapted from Connected Environmental Ltd, Pachube, *Welcome to Pachube – Data Infrastructure Internet of Things*, February 24, 2008, http://community.pachube.com/about. Used by permission of Pachube.

Figure 3-4. *Trends in Natural Disasters and Victims*
EM-DAT: The OFDA/CRED International Disaster Database – www.emdat.be, Université Catholique de Louvain, Brussels (Belgium), November 7, 2011, http://www.emdat.be/result-country-profile.

Kennedy Space Center
©Joseph Chiapputo/Fotolia #4828009.

Social media usage
flickr®, *United States*, Photograph by Eric Fischer, http://www.flickr.com/photos/walkingsf. Courtesy of Eric Fischer.

Crowdsourcing
Adapted from Wikipedia, Wikimedia, *The Crowdsourcing Process in Eight Steps*, Daren C. Brabham, http://upload.wikimedia.org/wikipedia/commons/a/ae/Crowdsourcing_process2.jpg. Courtesy of Daren C. Brabham.

Chapter 4, Land

Figure 4-1. *North American Carbon Storage Compartments*
Adapted from King, A.W.; Dilling, L.; Zimmerman, G.P.; Fairman, D.M.; Houghton, R.A.; Marland, G.; Rose, A.Z.; Wilbanks, T.J., eds. *The First State of the Carbon Cycle Report (SOCCR): the North American Carbon Budget and Implications for the Global Carbon Cycle*, 2007, a Report by the US Climate Change Science Program and the Subcommittee on Global Change Research. Asheville, NC: National Oceanic and Atmospheric Administration, National Climatic Data Center.

Figure 4-2. *Proportion of GHG Emissions from Various Human Activities*
Adapted from IPCC. 2006 IPCC Guidelines for National Greenhouse Gas Inventories, Vol. 4, *Agriculture, Forestry and Other Land Use*. Hayama, Japan: IGES, 2006.

Table 4-1. *Mitigation Activities that Can Be Accomplished at Various Levels of Government*
Created by LMI.

Highway cloverleaf
©OrdinaryLight/Fotolia # 6145985.

Green roof cross section
Adapted from Safeguard Europe Ltd, *Flat Green Roofs*, 2011, http://www.safeguardeurope.com/applications/green_roofs_flat.php. Used by permission of Safeguard Ltd.

Prescribed burn
©Jim Parkin /Fotolia # 131816.

Table 4-2. *Additional Number of People Flooded as a Result of Storm Surges Due to Sea-level Rise at 4°C, Using the IPCC A1B Scenario*
Adapted from MetOffice, Additional Number of People Flooded by Region, Year, and Level of Rise, May 2011, http://www.metoffice.gov.uk/climate-change/guide/impacts/high-end/sea-level. © British Crown Copyright 2011, the Met Office. Used by permission of the Met Office.

Switchgrass
Samuel Roberts Noble Foundation, Switchgrass Image, © 2010. Used by permission of Samuel Roberts Noble Foundation.

Damaged trees
U.S. Fish and Wildlife Service, *Wyoming: 'Perfect Storm' Fuels Mountain Pine Beetle Epidemic*, Jennifer Strickland, May 18, 2011, http://www.fws.gov/news/blog/index.cfm/ 2011/5/18/Wyoming-Perfect-Storm-Fuels-Mountain-Pine-Beetle-Epidemic. Used by permission of USFWS.

Table 4-3. *A Change of Only a Few Degrees Can Cause Significant Changes in Agriculture*
Adapted from IPCC, *Contribution of Working Group II to the Fourth Assessment Report of the Intergovernmental Panel on Climate Change*, 2007 M.L. Parry, O.F. Canziani, J.P. Palutikof, P.J. van der Linden and C.E. Hanson (eds) Cambridge University Press, Cambridge, United Kingdom and New York, NY, USA. Used by permission of the IPCC.

Andean potatoes
U.S. Department of Agriculture, Agricultural Research Service, Image Number K9152-1, Photograph by Scott Bauer, 2006, http://www.ars.usda.gov/is/graphics/photos/dec00/k9152-1.htm. Courtesy of U.S. Department of Agriculture.

Schooner Bayou Control Structure
Wikipedia, *Schooner Bayou Control Structure*, 1994, http://en.wikipedia.org/wiki/Schooner_Bayou_Control_Structure.

Chapter 5, Structure

Nature of Infrastructure
K.E. Purdy, *Flood 2008 Louisa County, Start Building Levee*, Photograph by Ken Purdy, June 10, 2008, http://www.kenpurdy.com/FloodWeb/Flood%20Web-Pages/Image0.html. Used by permission of Ken Purdy.

Table 5-1. *Infrastructure Sub-Components are Broad and Varied*
Created by LMI.

Figure 5-1. *The Major Infrastructure Components by Percentage of Total Value*
Data from J. Heintz, R. Pollin, H. Garrett-Peltier, *How Infrastructure Investments Support the U.S. Economy*, January 2009, Political economy Research Institute.

Table 5-2. *Estimated 5-Year Infrastructure Investment Needs in Billions of Dollars*
Adapted from American Society of Civil Engineers, *2009 Report Card for America's Infrastructure*, March 25, 2009, http://www.infrastructurereportcard.org/sites/default/files/ RC2009_full_report.pdf. Used by permission of the ASCE.

Mitigation
Created by LMI.

Green building definition
U.S. Environmental Protection Agency, *Definition of Green Building*, 2010, http://www.epa.gov/greenbuilding/pubs/about.htm.

Carpool sign
©trekandshoot/Fotolia # 33671148.

Low emission building (rendering)
Smith Group Architects, *Addition to Defense Intelligence Analysis Center*, 2011, http://www.wbdg.org/tools/leed_atfp.php. Images courtesy of SmithGroupJJR.

Road crew
© Dwight Smith /Fotolia # 4933734.

Figure 5-2. *Incentivized Mitigation Must be Applied across All Phases of the Infrastructure Life Cycle*
Created by LMI.

Smart tablets
©berc/Fotolia # 34165921.

Risk Equation
Created by LMI.

Figure 5-3. *Three-Dimensional Representation of Risk Analysis*
Created by LMI.

Table 5-3. *Adaptation Actions Appropriate for Various Stakeholders*
Created by LMI.

Flooded school
©mark yuill /Fotolia # 578918.

Storm surge barrier
©Marcel Mooij /Fotolia # 4645748.

White roof
U.S. Department of Energy Federal Energy Management Plan, *Cool Roof Resources for Federal Agencies*, April 11, 2011, http://www1.eere.energy.gov/femp/features/cool_ roof_resources.html. Courtesy of the U.S. Department of Energy.

Chapter 6, Vehicles

Table 6-1. U.S. Tailpipe GHG Emissions from Transportation Activities (Tg CO_2 eq)
US Environmental Protection Agency, *Inventory of US Greenhouse Gas Emissions and Sinks: 1990–2009*, EPA 430-R-11-005, April 2011.

Figure 6-1. *Components of Direct and Indirect Life-Cycle GHG Emissions*
Adapted from S. M. Lenski, G.A. Keoleian, and K.M. Bolon, "The Impact of 'Cash for Clunkers' on Greenhouse Gas Emissions: A Life Cycle Perspective," *Environmental Research Letters*, 5 (October-December 2010) 044003.

Figure 6-2. *GHG Emissions from Fuel and Vehicle Cycles Relative to Tailpipe Emissions*
Adapted from U.S. Environmental Protection Agency, *Greenhouse Gas Emissions from the US Transportation Sector*, 1990–2003, EPA 420 R 06 003, March 2006. Used by permission of the EPA.

Table 6-2. *GHG Emissions from fuel combustion*
Adapted from US Environmental Protection Agency, *Inventory of US Greenhouse Gas Emissions and Sinks: 1990–2008*, April 2010. Used by permission of the EPA.

Emissions equation
Created by LMI.

Light-duty vehicle fuel economy standards
Adapted from White House Obama Administration, *Driving Efficiency: Cutting Costs for Families at the Pump and Slashing Dependence on Oil*, 2011, http://www.whitehouse.gov/sites/default/files/fuel_ economy_report.pdf .

TSE Site Locator
Shorepower Technologies, 2011, *Truck Stop Electrification*, http://www.shorepower.com/tse.html. Used by permission of Shorepower Technologies.

Idaho National Lab buses
Department of Energy Idaho National Laboratory, *INL Bus Driver First Female to Reach Important Milestone*, John Howze, 2011, https://inlportal.inl.gov/portal/server.pt?open= 514&objID=1269&mode=2&featurestory= DA_575221. Used by permission of the DOE Idaho National Laboratory.

Figure 6-3. *Comparison Life-cycle GHG Emissions for Alternative Fuels Relative to Conventional Fuels*
Adapted from US Environmental Protection Agency, *Greenhouse Gas Impacts of Expanded Renewable and Alternative Fuels Use*, EPA420-F-07-035, April 2007. Used by permission of the EPA.

Electric vehicle recharging sign
©Stephen Finn/Fotolia # 4763314.

Recharging electric car
©Tom-Hanisch.de/Fotolia # 30104301.

Humvee refueling
U.S. Army, *Net Zero is an Army Sustainability Goal*, Photograph by Maj. Jeff Parker, July 29, 2011, http://www.army.mil/media/214754. Photo courtesy of U.S. Army.

Table 6-3. *Estimates of Life-cycle GHG Emissions by Transportation Mode*
Adapted from U.S. Department of Transportation Center for Climate Change and Environmental Forecasting, *Transportation's Role in Reducing U.S. Greenhouse Gas Emissions, Volume 1and Volume 2: Synthesis Report*, April 2010, http://ntl.bts.gov/lib/32000/32700/32779/DOT_Climate_Change_Report_-_April_2010_-_Volume_1_and_2.pdf.

Flooded roadway
©The Josh/Fotolia # 10480942.

NYC airports map
Adapted from from J.L. Weiss, J.T. Overpeck, and B. Strauss. 2011. Implications of Recent Sea Level Rise Science for Low-elevation Areas in Coastal Cities of the Conterminous U.S.A. *Climatic Change* 105: 635-645. Used by permission of the University of Arizona.

Permafrost-damaged road
http://www.istockphoto.com/stock-photo-9346042-permafrost-damaged-road-near-yellowknife.php?st=67036c6.

Table 6-4. *Example Climate Change Impacts on Fleet Operations*
Adapted from National Research Council Transportation Research Board, *Potential Impacts of Climate Change on U.S. Transportation*, A Report by the Committee on Climate Change and U.S. Transportation, Transportation Research Board, and Division on Earth and Life Sciences, 2008, http://onlinepubs.trb.org/onlinepubs/sr/sr290.pdf.

Flooded gas station
FEMA, *Flooded Gas Station in Missouri*, Photograph by Jocelyn Augustino, March 23, 2008, http://www.fema.gov/photolibrary/photo_details.do?id=34539.Used by permission of FEMA.

Chapter 7, Supply

Steps of the supply chain
Created by LMI.

Parts of an iPhone
Ifixit.com, *iPhone 4 Verizon Teardown*, 2011, http://www.ifixit.com/Teardown/iPhone-4-Verizon-Teardown/4693/1. Used by permission of Ifixit.com.

Textile supply chain
Adapted from Systain Consulting GmbH, Blog Action Day: *What's the Climate Impact of Your Wardrobe*, October 15, 2009, http://taliweinberg.wordpress.com/2009/10/15/whats-the-climate-impact-of-your-wardrobe/. Used by permission of Systain Consulting GmbH.

Resilience equation
Created by LMI.

Figure 7-1. *Supply Chain Mitigation Process*
Created by LMI.

Figure 7-2. *Cradle-to-Grave Elements of a Product Life Cycle*
Adapted from *Life Cycle Management - A Business Guide to Sustainability*, 2007, United Nations Environment Programme, http://www.unep.org. Used by permission of the UNEP.

Figure 7-3. *Two Tiers of a Simple Supply Chain*
Created by LMI.

Figure 7-4. *Patagonia's "Footprint Chronicles" for their Nano-Puff Pullover*
Patagonia, Inc., *The Footprint Chronicles, Nano Puff® Pullover*, 2011, http://www.patagonia.com/us/footprint/index.jsp. Used by Permission of Patagonia, Inc.

Table 7-1. *Four Widely Used Standards Relevant to Supply Chain GHG Emissions*
Created by LMI.

Table 7-2. *Common Supply Chain GHG Reduction Actions*
Created by LMI.

Figure 7-5. *Greater Partner Engagement Can Lead to Greater GHG Emission Reduction*
Created by LMI.

Figure 7-6. *Supply Chain Risk Management Approach*
Created by LMI.

Figure 7-7. *Overview of Climate Change Supply Chain Risks*
Created by LMI.

Table 7-3. *Climate Change Risks*
Created by LMI.

Figure 7-8. *Risk Prioritization Scheme*
Created by LMI.

Table 7-4. *Example Treatment Actions*
Created by LMI.

Sierra Nevada barley field
Courtesy of Sierra Nevada Brewing Co.

Egyptian cotton
Associated Press, An Egyptian family works in its cotton field in Fayoum, some 60 miles, 100 kms, south of Cairo, April, 29, 2003, Photograph by Mohamas al-Shehety. Used by permission of The Associated Press.

Chapter 8, Security

Figure 8-1. *U.S. Petroleum Consumption, Production, and Import Trends (1949-2010)*
Adapted from U.S. Energy Information Administration, *Monthly Energy Review* (May 2011), preliminary data, and *Annual Energy Review 2009*, Table 5.1 (August 2010). Used by permission of the EIA.

Figure 8-2. *Energy Security and Climate Mitigation Potential of Technologies*
World Resources Institute, *Weighing U.S. Energy Options: The WRI Bubble Chart*, Jeffrey Logan and John Venezia, July 2007, http://www.wri.org/publication/us-energy-options. Used by permission of WRI.

Figure 8-3. *Risk Matrix*
Adapted from NASA, *Guidelines for Risk Management*, S3001, 03/25/2009, 15, http://www.nasa.gov/centers/ivv/pdf/209213main_S3001.pdf.

X-48B aircraft
NASA, *Fact Sheet: X-48B Blended Wing-Body*, April 2009. http://www.nasa.gov/centers/dryden/news/FactSheets/FS-090-DFRC.html.

Forward operating base
U.S. Department of Defense, *Reporter's Notebook: Forward Operating Base Gardez 'Feels' Like Afghanistan*, Fred W. Baker III, February 20, 2009, http://www.defense.gov/news/newsarticle.aspx?id=53019. Courtesy of the U.S. Department of Defense.

Flight simulator
Wikimedia Commons, *Flight Simulator*, 2010, Photograph by Rama, http://en.wikipedia.org/wiki/File:Flight_simulator-IMG_5639.jpg.

Green Hornet
U.S. Navy, F/A-18 Super Hornet, http://www.navy.mil/navydata/aircraft/fa18/fa18ewep.jpg

Irrigation system
©Brenda Carson/Fotolia # 16222568.

Table 8-1. *Comparison of Water Consumption by Energy Types*
Adapted from *Linking Water, Energy & Climate Change: A proposed water and energy policy initiative for the UN Climate Change Conference*, COP15, in Copenhagen 2009, DHI, Draft Concept Note, January 2008, http://www.semide.net/media_server/files/Y/I/water-energy-climatechance_nexus.pdf.

Figure 8-4. *Melting Sea Ice within the Arctic Circle Breeds Disputes about National Borders*
Adapted from "Joint Development of Arctic Ocean Oil and Gas Resources and the United Nations Convention on the Law of the Sea," A presentation given by John Abrahamson, Australian National University, at the *International Boundaries Research Unit Conference on The State of Sovereignty*, Durham University, April 2009, http://www.dur.ac.uk/resources/ ibru/conferences/sos/john_abrahamson_powerpoint.pdf. Used by permission of the BBC.

Water stress
UNEP/GRID-Arendal Maps and Graphics Library, *Increased Global Water Stress*, Philippe Rekacewicz, 2009, http://maps.grida.no/go/graphic/increased-global-water-stress. Used by permission of the UNEP/GRID-Arendal.

Desperate Somalis
UN, *Famine in Somalia*, Stuart Price (UN Photo), 2010, http://www.un.org/ecosocdev/geninfo/afrec/newrels/famine-somalia.html.

Figure 8-5. *Change in Land Cover at Air Force Dare County Bombing Range (DCBR)*
Adapted from U.S. Department of Defense, *Sea Level Rise Risk Assessment for DoD Coastal Installations*, Project # 08-410, 2009, Robert Mickler, Alion Science and Technology, http://www. dodworkshops.org/files/ClimateChange/Fact_Sheet_Sea_Level_Rise_ Risk_Assessment_ for_DoD_Coastal_Installations.pdf. Used by permission of Alion Science and Technology.

Unified Quest war game
U.S. Military, *'Unified Quest' Focuses on Future Persistent Conflict*, Chris Gardner, http://www.army.mil/article/9017/unified-quest-focuses-on-future-persistent-conflict.

USCG Cutter Healy
U.S. Army, Research, Development, and Engineering Command, 2011, *Researchers on Arctic Field Campaign*, Photograph by Md. Marie C. Darling, http://www.army.mil/media/221651/.

UK Met Office web site
Met Office, *Impacts of 'high-end' climate change*, Kristy Lewis, September 14, 2011, http://www.metoffice.gov.uk/climate-change/guide/impacts/high-end.

Chapter 9, Initiative

Table 9-1. *Summary of Mitigation Recommendations from Each Chapter*
Created by LMI.

Table 9-2. *Summary of Adaptation Recommendations from Each Chapter*
Created by LMI.

Table 9-3. *Total Numbers of Stars per Recommendation*
Created by LMI.

Table 9-4. *Summary of Ratings for Mitigation Recommendations Using Four Criteria Important to Managers*
Created by LMI.

Table 9-5. *Summary of Ratings for Adaptations Recommendations Using Four Criteria Important to Managers*
Created by LMI.

Falling house
U.S. Geological Service, *Erosion Doubles Along Part of Alaska's Arctic Coast: Cultural and Historical Sites Lost*, Photograph by Benjamin Jones, 2009, http://www.usgs.gov/newsroom/images/2009_02_17/P1010067.JPG. Courtesy of USGS.

Life cycle analysis
Green Options, *Life Cycle Assessments*, http://www.greenoptions.com/a/life-cycle-assessments.